Basic Electromagnetism
and Materials

Basic Electromagnetism
and Materials

André Moliton

Université de Limoges
Limoges, France

 Springer

André Moliton
Laboratory Xlim-MINACOM
Faculte des Sciences et Techniques
123 Avenue Albert Thomas
87060 Limoges Cedex
France
andre.moliton@unilim.fr

Library of Congress Control Number: 2005939183

ISBN-10: 0-387-30284-0
ISBN-13: 978-0387-30284-3

9 8 7 6 5 4 3 2 1

springer.com

Foreword

Electromagnetism has been studied and applied to science and technology for the past century. It is mature field with a well developed theory and a myriad of applications. Understanding of electromagnetism at a deep level is important for core understanding of physics and engineering fields and is an asset in related fields of chemistry and biology. The passing of the knowledge of electromagnetism and the skills in application of electromagnetic theory is usually done in university classrooms.

Despite the well developed literature and existing texts there is a need for a volume that can introduce electromagnetism to students in the early mid portion of their university training, typically in their second or third year. This requires bringing together many pieces of mathematics and physics knowledge and having the students understand how to integrate this information and apply the information and concepts to problems.

André Moliton has written a very clear account of electromagnetic radiation generation, and its propagation in free space and various dielectric and conducting media of limited and also infinite dimensions. Absorption and reflection of radiation is also described. The book begins by reminding the reader in an approachable way of the mathematics necessary to understand and apply electromagnetic theory. Thus chapter one refreshes the reader's knowledge of operators and gradients in an understandable and concise manner. The figures, summaries of important formulas, and schematic illustrations throughout the text are very useful aids to the reader. Coulomb's law describing the force between two charges separated by a distance r and the concept of electric field produced at a distance from a charge are introduced. The scalar potential, Gauss's theorem, and Poisson's equation are introduced using figures that provide clarity to the concepts. The application of these concepts for a number of different geometries, dimensionalities and conditions is particularly useful in cementing the reader's understanding. Similarly in chapter one, Ohm's law, Drude model, and drift velocity of charges are clearly introduced. Ohm's law and its limits at high frequency are described. The author's comments provide a useful perspective for the reader.

The introduction of magnetostatics and the relationships between current flow, magnetic fields, and vector potential, and Ampère's theorem are also introduced in chapter one. Again the figures, summaries of important formulas, and author's comments are particularly helpful. The questions and detailed answers are useful for retaining and deepening understanding of the knowledge gained.

Chapter two provides a useful and practical introduction to dielectrics. The roles of dipolar charges, discontinuities, space charges, and free charges as well as homo- and hetero-charges are illustrated in Figure 2.2. The applications of the formulas introduced in chapter one and the corresponding problems and solutions complete this chapter. Similarly magnetic properties of materials are introduced in chapter three. Dielectric and magnetic materials are introduced in Figure 3. The properties of magnetic dielectric materials are described in Figure 4 together with a number of useful figures, summaries, problems, and solutions.

Maxwell's equations together with oscillating electromagnetic fields propagating in materials of limited dimensions are introduced and described in chapters five through seven. Propagation of oscillating electromagnetic waves in plasmas, and dielectric, magnetic, and metallic materials is described in chapter eight. The problems and solutions at the end of each chapter will be particularly helpful.

The generation of electromagnetic radiation by dipole antennae is described in chapter nine, with emphasis on electric dipole emission. Absorption and emission of radiation from materials follow in chapters ten and eleven. Chapter twelve concludes with propagation of electromagnetic radiation in confined dimensions such as coaxial cables and rectangular waveguides.

In sum, I recommend this book for those interested in the field of electromagnetic radiation and its interaction with matter. The presentation of mathematical derivations combined with comments, figures, descriptions, problems, and solutions results in a refreshing approach to a difficult subject. Both students and researchers will find this book useful and enlightening.

Arthur J. Epstein
Distinguished University Professor
The Ohio State University
Columbus, Ohio
October 2006

Preface

This volume deals with the course work and problems that are common to basic electromagnetism teaching at the second- and third-year university level. The subjects covered will be of use to students who will go on to study the physical sciences, including materials science, chemistry, electronics and applied electronics, automated technologies, and engineering.

Throughout the book full use has been made of constructive exercises and problems, designed to reassure the student of the reliability of the results. Above all, we have tried to demystify the physical origins of electromagnetism such as polarization charges and displacement and Amperian currents ("equivalent" to magnetization).

In concrete terms, the volume starts with a chapter recalling the basics of electromagnetism in a vacuum, so as to give all students the same high level at the start of the course. The formalism of the operators used in vectorial analysis is immediately broached and applied so as to help all students be well familiarized with this tool.

The definitions and basic theories of electrostatics and magnetostatics then are established. Gauss's and Ampère's theories permit the calculation—by a simple route—of the electric and magnetic fields in a material. The calculations for charges due to polarization and Amperian currents caused by magnetization are detailed, with attention paid to their physical origins, and the polarization and magnetization intensity vectors, respectively.

A chapter is dedicated to the description of dielectric and magnetic media such as insulators, electrets, piezoelectrets, ferroelectrics, diamagnets, paramagnets, ferromagnets, antiferromagnets, and ferrimagnets.

Oscillating environments are then described. As is the tradition, slowly oscillating systems—which approximate to quasistationary states—are distinguished from higher-frequency systems. The physical origin of displacement currents are detailed and the Maxwell equations for media are established. The general properties of electromagnetic waves are presented following a study of their propagation in a vacuum. Particular attention has been paid to two different types of notation—used by dielectricians and opticians—to describe what in effect is the same wave. The properties of waves propagating in infinitely large materials are then described along with the description of a general method allowing determination of dynamic polarization in a material that disperses and absorbs the waves. The Poynting vector and its use in determining the energy of an electromagnetic wave is then detailed, followed by the behavior of waves in the more widely encountered materials such as dielectrics, plasmas, metals, and uncharged magnetic materials.

The following two chapters are dedicated to the analysis of electromagnetic field sources. An initial development of the equations used to describe dipole radiation in a vacuum is made. The interaction of radiation with electrons in a material is detailed in terms of the processes of diffusion, notably Rayleigh diffusion, and absorption. From this the diffusion of radiation by charged particles is used to explain the different colors of the sky at midday and sunset. Rutherford diffusion along with the various origins of radiation also are presented. The theory for absorption is derived using a semiquantic approximation based on the quantification of a material but not the applied electromagnetic field. This part, which is rather outside a normal first-degree course, can be left out on a first reading by undergraduate students (even given its importance in materials science). It ends with an introduction to spectroscopy based on absorption phenomena of electromagnetic waves, which also is presented in a more classic format in a forthcoming volume entitled *Applied Electromagnetism and Materials*.

The last two chapters look at the propagation of waves in media of limited dimensions. The study of reflection and refraction of waves at interfaces between materials is dominated by the optical point of view. Fresnel's relations are established in detail along with classic applications such as frustrated total internal reflection and the Malus law. Reflection by an absorbing medium, in particular metallic reflections, is treated along with studies of reflection at magnetic layers and in antiradar structures. Guided propagation is introduced with an example of a coaxial structure; then, along with total reflection, both metallic wave guides and optical guides are studied. The use of limiting conditions allows the equations of propagation of electromagnetic waves to be elaborated. Modal solutions are presented for a symmetrical guide. The problem of signal attenuation, i.e., signal losses, is related finally to a material's infrared and optical spectral characteristics.

I would like to offer my special thanks to the translator of this text, Dr. Roger C. Hiorns. Dr. Hiorns is following post-doctorial studies into the synthesis of polymers for electroluminescent and photovoltaic applications at the Laboratoire de Physico-Chimie des Polymères (Université de Pau et des Pays de l'Adour, France).

Contents

Chapter 1

Introduction to the Fundamental Equations of Electrostatics and Magnetostatics in vacuums and Conductors

1.1. Vectorial Analysis

1.1.1. Operators

1.1.1.1. Gradients

A gradient is the vectorial magnitude of a scalar. For example, the scalar (Φ) which at point P has coordinates x,y,z, takes on a value Φ(x,y,z). By definition, at the point P (see Figure 1.1), the gradient of Φ is given by a vectorial magnitude, as in:

$$\vec{G}(P) = \overrightarrow{grad}_P \ \Phi(x,y,z), \text{which is such that} \ \overrightarrow{grad}_P \ \Phi \ = \ \vec{i}\frac{\partial\Phi}{\partial x} + \vec{j}\frac{\partial\Phi}{\partial y} + \vec{k}\frac{\partial\Phi}{\partial z}$$

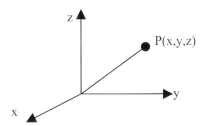

Figure 1.1 *Coordinates of the point P.*

1.1.1.2. Divergence

A divergence is an operator that is a scalar magnitude of a vector, for example, $\vec{A}$, and indicates its tendency. Here $A_x(x,y,z)$, $A_y(x,y,z)$ and $A_z(x,y,z)$ are the components of this vector. By definition, the divergence at the point P of $\vec{A}$ is given by: $\overline{\text{div}_p \vec{A}} = \dfrac{\partial A_x}{\partial x} + \dfrac{\partial A_y}{\partial y} + \dfrac{\partial A_z}{\partial z}$.

1.1.1.3. Rotational

A rotational operator is a vectorial magnitude of a vector. By denoting the components of $\vec{A}$ by A_x, A_y, and A_z, the rotational for $\vec{A}$ for P is by definition:

$$\overrightarrow{\text{rot}_p}\,\vec{A} = \vec{i}\left(\frac{\partial A_z}{\partial y} - \frac{\partial A_y}{\partial z}\right) + \vec{j}\left(\frac{\partial A_x}{\partial z} - \frac{\partial A_z}{\partial x}\right) + \vec{k}\left(\frac{\partial A_y}{\partial x} - \frac{\partial A_x}{\partial y}\right)$$

$$= \begin{pmatrix} \vec{i} & \vec{j} & \vec{k} \\ \frac{\partial}{\partial x} & \frac{\partial}{\partial y} & \frac{\partial}{\partial z} \\ Ax & Ay & Az \end{pmatrix}$$

1.1.1.4. Laplacian

The Laplacian, a differential operator, is a scalar magnitude of a scalar, and by definition the Laplacian of the scalar $\Phi(x,y,z)$ is given by the scalar for a point P as

$$\Delta\Phi = \frac{\partial^2\Phi}{\partial x^2} + \frac{\partial^2\Phi}{\partial y^2} + \frac{\partial^2\Phi}{\partial z^2} \ .$$

1.1.1.5. Laplacian vector

The Laplacian vector is a vector magnitude operating on a vector. By definition the Laplacian vector of $\vec{A}$ with components A_x, A_y, and A_z is given by:

$$\overrightarrow{\Delta\vec{A}} = \vec{i}\,\Delta A_x + \vec{j}\,\Delta A_y + \vec{k}\,\Delta A_z .$$

1.1.1.6. Comment

These definitions can be used for all types of coordinates, whether cylindrical or spherical. For example, the cylindrical coordinates such that $\overrightarrow{OP} = r\,\vec{e}_r + z\,\vec{e}_z$, and with p being defined by the projection of a point P onto a flat surface Oxy so that $r = Op$, $z = pP$, and $\theta = \left(\overrightarrow{Ox}, \overrightarrow{e_r}\right)$ as detailed in Figure 2.1, the components of the gradient vector are:

for $\overrightarrow{e_r}$: $\dfrac{\partial \Phi}{\partial r}$,

for $\overrightarrow{e_\theta}$: $\dfrac{1}{r}\dfrac{\partial \Phi}{\partial \theta}$,

and for $\overrightarrow{e_z}$: $\dfrac{\partial \Phi}{\partial z}$.

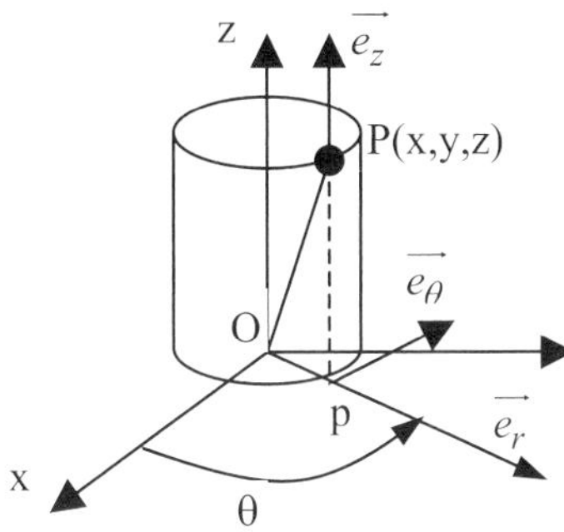

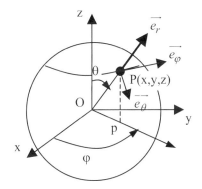

Figure 1.2. *Cylindrical coordinates.*

Using the unit vectors $\overrightarrow{e_r}$, $\overrightarrow{e_\theta}$ and $\overrightarrow{e_\varphi}$ defined in Figure 1.3, the similar spherical coordinates are given by (with $r = OP$):

for $\overrightarrow{e_r}$: $\dfrac{\partial \Phi}{\partial r}$,

for $\overrightarrow{e_\theta}$: $\dfrac{1}{r}\dfrac{\partial \Phi}{\partial \theta}$,

and for $\overrightarrow{e_\varphi}$: $\dfrac{1}{r\,\sin\theta}\dfrac{\partial \Phi}{\partial \varphi}$.

Figure 1.3. *Spherical coordinates.*

Given the coordinates r, θ, and φ in Figure 1.3, the total volume of a sphere of radius R can be defined using the limits of r [0, R], of θ [0, π], and of φ [0, 2π]. If θ were to be defined by $\theta = \left(\overrightarrow{Op}, \overrightarrow{OP}\right)$, then its limits would be [-π/2, π/2], in which case the gradient component for $\overrightarrow{e_\varphi}$ is $\dfrac{1}{r\,\cos\theta}\dfrac{\partial \Phi}{\partial \varphi}$.

1.1.2. *Important formulae*

The gradient, divergence, and rotational operators all bring into action differentials so that if λ is a constant, then

$$\frac{\partial(\lambda V)}{\partial x} = \lambda \frac{\partial V}{\partial x} \ ,$$

and if u and v are functions of x, then

$$\frac{\partial(uv)}{\partial x} = u\frac{\partial v}{\partial x} + v\frac{\partial u}{\partial x}$$

$$\overrightarrow{grad}\ (\Phi\Psi) = \Phi\ \overrightarrow{grad}\ \Psi + \Psi\ \overrightarrow{grad}\ \Phi$$

$$div\ (a\overrightarrow{A}) = a\ div\overrightarrow{A} + (\overrightarrow{grad}\ a).\overrightarrow{A}$$

$$div\ (\overrightarrow{A} \times \overrightarrow{B}) = (\overrightarrow{rot}A).B - \overrightarrow{A}.(\overrightarrow{rot}\ B)$$

$$\overrightarrow{rot}(a\ \overrightarrow{A}) = a\overrightarrow{rot}\overrightarrow{A} + \overrightarrow{grad}\ a \times \overrightarrow{A}\ .$$

The schematization below can help in memorizing these important formulae:

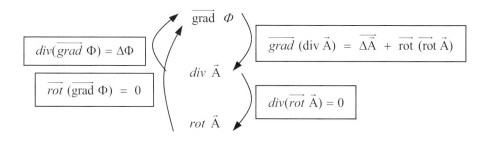

$$div(\overrightarrow{grad}\ \Phi) = \Delta\Phi$$

$$\overrightarrow{rot}\ (\overrightarrow{grad}\ \Phi) = 0$$

$$\overrightarrow{grad}\ (div\ \overrightarrow{A}) = \Delta\overrightarrow{A} + \overrightarrow{rot}\ (\overrightarrow{rot}\ \overrightarrow{A})$$

$$div(\overrightarrow{rot}\ \overrightarrow{A}) = 0$$

1.1.3.   *Vectorial integrations*

1.1.3.1. *Vector circulation or "curl"*

By definition, the circulation of a vector denoted $\vec{F}$ around an open curve (C) is given by (see Figure 1.4a): $T = \int_{(C)} \vec{F}.\overrightarrow{dl} = \int_{M}^{P} \vec{F}.\overrightarrow{dl}$.

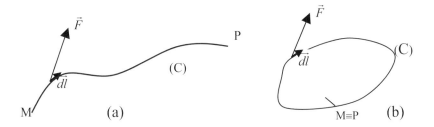

Figure 1.4. $\vec{F}$ *circulating around (a) an open curve and (b) a closed curve.*

For a closed curve, as shown in Figure 1.4b, the same circulation is given by $T = \oint_{(C)} \vec{F}.\vec{dl}$.

If $\vec{F}$ is a gradient, i.e., $\vec{F} = \overrightarrow{grad}\ \Phi$ for example, then

$$T = \int_M^P \overrightarrow{grad}\ \Phi.\ \vec{dl} = \Phi(P) - \Phi(M)\ .$$

A simple verification of this result can be performed in one dimension (1D). For a given axis (Ox) that defines the single dimension, the unit vector can be denoted as $\vec{i}$ so that:

$$\overrightarrow{grad}\ \Phi\ =\ \vec{i}\frac{\partial \Phi}{\partial x}\ \text{and}\ \vec{dl}\ =\ \vec{i}\ dx$$

and hence

$$T = \int_M^P \overrightarrow{grad}\ \Phi.\ \vec{dl} = \int_M^P \vec{i}\frac{\partial \Phi}{\partial x}.\vec{i}dx\ =\ \int_M^P d\Phi\ =\ \Phi(P) - \Phi(M).$$

In three dimensions (3D), the definition for a scalar product directly yields $\overrightarrow{grad}\ \Phi.\ \vec{dl}\ =\ d\Phi$, which is in effect the same result.

From this can be derived directly that the curl of a gradient around a closed curve is zero, as $M \equiv P$ (see Figure 1.4b), so that $\Phi(P) = \Phi(M)$, from which $T = 0$.

1.1.3.2. The flux of the vector $\vec{A}$

The flux of $\vec{A}$ through an open surface (S) that is not limited by a certain volume is given by $\phi = \iint_S \vec{A}.\overrightarrow{dS}$, where $\overrightarrow{dS} = dS\,\vec{n}$ with $\vec{n}$ being the normal external to an element of the surface described by dS.

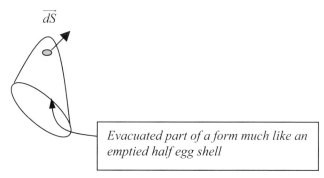

Figure 1.5a. *Flux through an open surface.*

The flux of $\vec{A}$ through a closed surface (S), delimited by a certain volume, is given by $\phi = \oiint_S \vec{A}.\overrightarrow{dS}$ where $\overrightarrow{dS} = dS\,\vec{n}$, and $\vec{n}$ is the normal external to an element of the surface denoted dS.

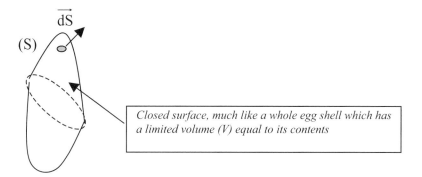

Figure 1.5b. *Flux through a closed surface.*

1.1.3.3. Integrated transformations

In most courses concerning vectorial analysis, the following are encountered:

1. Ostrogradsky's theory $\iiint\limits_V \mathrm{div}_M \, \vec{A} \, d\tau = \oiint\limits_{S\ \text{fermée}} \vec{A}_{(m)}.\overrightarrow{dS}$ (Figure 1.6a)

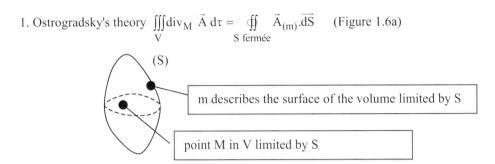

Figure 1.6a. *Points m and M relating to the surface and volume integrals.*

2. Stokes' theorem $\oint\limits_{C\ \text{closed}} \vec{A}_{(m)}.\overrightarrow{dl} = \iint\limits_S \overrightarrow{\mathrm{rot}}_M \vec{A}.\overrightarrow{dS}$ (Figure 1.6b)

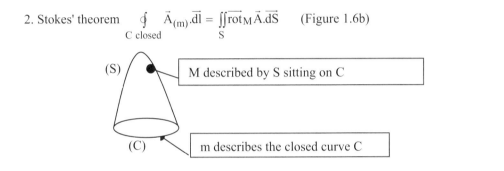

Figure 1.6b. *Points m and M relating to the curvilinear and surface integrations.*

3. Gradient formula give by $\iiint\limits_V \overrightarrow{\mathrm{grad}} \, \Phi \, d\tau = \oiint\limits_S \Phi \, \overrightarrow{dS}$; and

4. The rotational formula for the rotational, used in magnetism when determining the expression for Ampèrian currents, as in $\iiint\limits_V \overrightarrow{\mathrm{rot}} \, \vec{A} \, d\tau = \oiint\limits_S \overrightarrow{dS} \times \vec{A}$.

1.1.4. *Terminology*

1. $\vec{A}$ is *said* to be derived from scalar potential if there is a scalar (Φ) such that $\vec{A} = -\overrightarrow{\mathrm{grad}}\Phi$. By consequence, if $\vec{A} = -\overrightarrow{\mathrm{grad}}\Phi$, then:

• as $\oint \overrightarrow{\mathrm{grad}} \, \Phi \, \overrightarrow{dl} = 0$, then $\oint \vec{A}.\overrightarrow{dl} = 0$ as an integral property of a vector derived from a scalar potential; and

- as $\oint_{C\,closed} \vec{A}_{(m)}.\vec{dl} = \iint_S \overrightarrow{rot}_M\vec{A}.\vec{dS}$ (Stokes' theorem), then for the same conditions,

$\overrightarrow{rot}\,\vec{A} = 0$ in a localized property of vectors derived from a scalar potentials. This can also be understood by considering that if $\vec{A}$ is derived from a scalar potential, its rotational is zero. This property can be directly obtained from the general formula, $\overrightarrow{rot}(\overrightarrow{grad}\,\Phi) = 0$.

2. The vector $\vec{B}$ is *said* to be derived from a potential vector if there is a vector ($\vec{A}$) such that it is possible to state $\vec{B} = \overrightarrow{rot}\vec{A}$. As a consequence, if $\vec{B} = \overrightarrow{rot}\vec{A}$ then:

- Ostrogradsky's theorem makes it possible to state that $\oiint\vec{B}.\vec{dS} = \iiint div\,\vec{B}\,d\tau = \iiint div(\overrightarrow{rot}\,\vec{A})\,d\tau = 0$. In addition to which, if the vector $\vec{B}$ is derived from a vector potential, then the flux is conserved (as $\oiint\vec{B}.\vec{dS} = O$ is verified). This is an inherent property of vectors derived from vector potentials.

- generally speaking, from $div(\overrightarrow{rot}\,\vec{A}) = 0$, then in this case, $div\,\vec{B} = 0$, showing the localized property of vectors derived from a vector potential.

1.2. Electrostatics and Vacuums

1.2.1. Coulomb's law

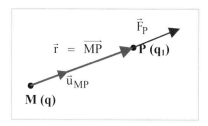

Figure 1.7. *Coulomb's law.*

This is an empirical law that describes that when two charges interact they exert on each other a force. For a charge (q_1) situated at a point (P) and another charge (q) situated at another point (M), and by making $\vec{r} = \overrightarrow{MP}$ as shown in Figure 1.7, the force exerted on q_1 by the presence of q is given with Coulomb's law as:

$$\vec{F}_P = \frac{1}{4\pi\varepsilon_0}\frac{q\,q_1}{r^2}\frac{\vec{r}}{r} = \frac{1}{4\pi\varepsilon_0}\frac{q\,q_1}{r^2}\vec{u}_{MP}$$

where $\vec{u} = \vec{u}_{MP}$ is the unit vector in the direction MP, which is directed from q toward q_1 and ε_0 is the dielectric permittivity given by

$$\varepsilon_0 = \frac{1}{36\,\pi\,10^9} \text{ SI} \approx 8.854 \times 10^{-12} \text{ F m}^{-1},$$

or in other terms, $\dfrac{1}{4\pi\varepsilon_0} = 9 \times 10^9$ SI .

1.2.2. The electric field: local properties and its integral

1.2.2.1. Form of the electric field

The force $\vec{F}_P$ can be written as $\vec{F}_P = q_1 \vec{E}$, where

$$\vec{E} = \frac{q}{4\pi\varepsilon_0} \frac{\vec{r}}{r^3}$$

is the electric field generated at P by the charge q at M.

Additionally, on recalling that $\overrightarrow{grad}_P \dfrac{1}{r} = -\dfrac{\vec{r}}{r^3}$, it is possible to equate $\vec{E}$ in the

form: $\vec{E} = -\dfrac{q}{4\pi\varepsilon_0} \overrightarrow{grad}_P \dfrac{1}{r} = -\overrightarrow{grad}_P \dfrac{q}{4\pi\varepsilon_0 r} = -\overrightarrow{grad}_P V$ where

$$V = \frac{q}{4\pi\varepsilon_0 r}$$

appears as the scalar potential from which is derived the electric field $\vec{E}$ (see also the terminology introduced in Section 1.1.4).

1.2.2.1. Local and integral properties of an electric field

As the electric field is derived from a scalar potential, it is possible to state that:

♦ $\displaystyle\oint_{\text{Closed}} \vec{E}.\overrightarrow{dl} = 0$ which is an integral property of $\vec{E}$. If the curve C is not

closed, then $\displaystyle\int_A^B \vec{E}.\overrightarrow{dl} = -\int_A^B \overrightarrow{grad}\ V.\overrightarrow{dl} = V_A - V_B$; and

♦ $\overrightarrow{rot}\ \vec{E} = 0$ is a local property of $\vec{E}$.

1.2.3. **Gauss's theorem**

1.2.3.1. The solid angle

In simple terms, a plane angle (θ) is defined in two dimensions by $\theta = \ell / R$, as shown in Figure 1.8a.

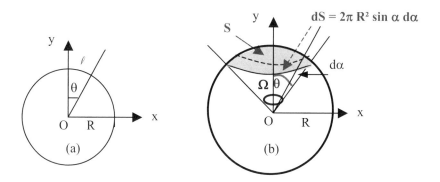

Figure 1.8. *(a) Plane and (b) solid angles.*

Similarly, in three dimensions, a solid angle (Ω) can be represented as an angle generated by a plane angle (θ) when it rotates in space around an axis Oy. By definition, and as shown in Figure 1.8b, $\Omega = \dfrac{S}{R^2}$, where S represents the surface at the interception of a half angle cone at the vertex θ on a sphere of radius R.

With $S = \int_0^\theta 2\pi R^2 \sin\alpha \, d\alpha = 2\pi R^2 (1 - \cos\theta)$, the expression for the half angle cone at the vertex θ, which defines the solid angle, is given by $\Omega = 2\pi (1 - \cos\theta)$.

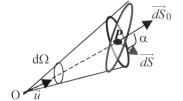

Figure 1.9. *Solid angle where there is a surface dS.*

In more general terms, we can look for the solid angle (dΩ) through which from a point O can be seen the surface element $\overrightarrow{dS}$ such that at a point P is r = OP, as in Figure 1.9. If dS_0 represents at P a straight section of the cone, and given the preceding definition, we should arrive at

$d\Omega = \dfrac{dS_0}{r^2}$. From this it is possible to state that $d\Omega = \dfrac{dS \cdot \cos\alpha}{r^2} = \dfrac{\vec{u}.\overrightarrow{dS}}{r^2}$,

where $\vec{u}$ denotes the unit vector in the direction OP. The solid angle through which all space can be seen is 4π steradians, as

$$\Omega_{space} = \int\limits_{all\ space} d\Omega = \int \dfrac{\vec{u}.\overrightarrow{dS}}{r^2} = \int \dfrac{dS_0}{r^2} = \int\limits_{\theta=0}^{\theta=\pi} 2\pi\sin\theta\ d\theta = 4\pi .$$

1.2.3.2. Flux created by an electric field generated by a charge outside of a given closed surface

In general terms, an electric field generated by a charge (q) at a point (M) on another point (P) such that $\vec{r} = \overrightarrow{MP} = r\,\vec{u}$ is given by $\vec{E} = \dfrac{1}{4\pi\varepsilon_0}\dfrac{q}{r^2}\vec{u}$

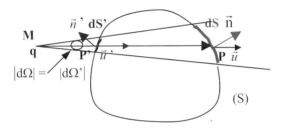

Figure 1.10a *Calculation of the flux generated by a charge outside of a surface.*

In order to calculate the electric field generated by such a charge through a closed surface (S) which itself contains no charge, it is possible to associate opposite elements of S as seen from the point M, as shown in Figure 1.10a. In other words, the elements of the form $\overrightarrow{dS}$ and $\overrightarrow{dS'}$ are associated. The corresponding flux elements are:

- through $\overrightarrow{dS}$, $d\Phi = \vec{E}.\overrightarrow{dS} = \dfrac{q}{4\pi\varepsilon_0}\dfrac{\vec{u}.\overrightarrow{dS}}{r^2} = \dfrac{q}{4\pi\varepsilon_0}d\Omega$; and

- through $\overrightarrow{dS'}$, $d\Phi' = \vec{E'}.\overrightarrow{dS'} = \dfrac{q}{4\pi\varepsilon_0}\dfrac{\vec{u'}.\overrightarrow{dS'}}{r'^2} = \dfrac{q}{4\pi\varepsilon_0}d\Omega'$

$$= -\dfrac{q}{4\pi\varepsilon_0}d\Omega .$$

In effect, the negative sign is introduced because the angle $\theta' = (\vec{u}', \vec{n}')$ is greater than $\pi/2$; in other terms $\cos \theta' < 0$. Thus $d\Omega'$ is negative as $|d\Omega| = |d\Omega'|$, while $d\Phi' = -d\Phi$.

The total flux for the two elements therefore is given by $d\Phi_T = d\Phi + d\Phi' = 0$, and the resultant flux through the whole of S therefore is also zero.

1.2.3.3. Electric field flux generated by a charge inside a given closed surface

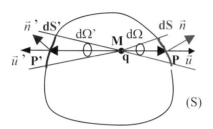

Figure 1.10b *Calculation of the flux generated by a charge inside a closed surface.*

The fundamental expressions for flux remain

$$d\Phi = \vec{E}.\overrightarrow{dS} = \frac{q}{4\pi\varepsilon_0} \frac{\vec{u}.\overrightarrow{dS}}{r^2} = \frac{q}{4\pi\varepsilon_0} d\Omega \text{ and}$$

$$d\Phi' = \vec{E}'.\overrightarrow{dS'} = \frac{q}{4\pi\varepsilon_0} \frac{\vec{u}'.\overrightarrow{dS'}}{r'^2} = \frac{q}{4\pi\varepsilon_0} d\Omega',$$

but here, as shown in Figure 1.10b, $\theta' < \pi/2$. The two flux elements no longer cancel each other out and the total flux can be simply given by

$$\Phi = \int\limits_{\text{all space}} d\Phi = \frac{q}{4\pi\varepsilon_0} \int\limits_{\text{all space}} d\Omega = \frac{q}{\varepsilon_0}.$$

1.2.3.4. The integrated form of Gauss's general theorem

1.2.3.4.1 Classic or "standard" form
For a distribution of charges, only those that contribute to a flux from the inside of a surface are considered. Each internal charge (q_i) will contribute $\dfrac{q_i}{\varepsilon_0}$ to the resultant flux (Φ_T), which can be described by:

$$\Phi_T = \oiint_S \vec{E}.\,\overrightarrow{dS} = \sum_i \frac{q_i}{\varepsilon_0}$$

where q_i represents the charges inside the closed surface (S).

    If the system of charges is distributed in a continuous manner, with a charge volume density denoted by ρ, then

$$\Phi_T = \oiint_S \vec{E}.\,\overrightarrow{dS} = \frac{1}{\varepsilon_0}\iiint_V \rho\, d\tau = \frac{Q}{\varepsilon_0}$$

where Q represents the resultant charge inside a volume (V) delimited by S for a uniform spread of charges.

1.2.3.4.2. The merit of the integrated form of Gauss's theorem

The integrated form is of particular use in determining the electric fields inside symmetrical systems. For example, if $\left|\vec{E}\right|$ = constant on a surface, the flux through the surface is simple to calculate, as $\Phi = ES$, and it can be expressed as a direct function of E.

1.2.3.4.3. The particular case of a surface charge distribution

If the charges under examination are superficial ones at S, the solid angle through which a charge given by $dq = \sigma\, dS$ can see the surface dS is equal to 2π, due to its view being through a half space of the surface dS, as shown in Figure 1.11. From this, therefore, $d\Phi = \dfrac{\sigma\, dS}{4\pi\varepsilon_0}2\pi$, so that $\Phi = \oiint_S \vec{E}.\,\overrightarrow{dS} = \dfrac{1}{2\varepsilon_0}\int\sigma\, dS = \dfrac{Q_s}{2\varepsilon_0}$.

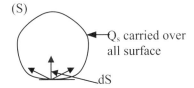

(S)

Q_s carried over all surface

dS

Figure 1.11. *Solid angle observed by charges at a surface.*

1.2.3.5. Local form of Gauss's theorem

This can be obtained via Ostrogradsky's theorem, which makes it possible to state that $\Phi_T = \oiint_S \vec{E}.\,\overrightarrow{dS} = \iiint_V \text{div}\,\vec{E}\, d\tau$. As above detailed, $\Phi_T = \dfrac{1}{\varepsilon_0}\iiint_V \rho\, d\tau$, and therefore:

$$\boxed{\text{div}\vec{E} = \frac{\rho}{\varepsilon_0}}.$$

It is worth noting that this equation concerns only the point P around which the divergence is calculated, so symmetry is no longer part of the problem.

1.2.4. The Laplace and Poisson equations

1.2.4.1. Laplace's equation

1.2.4.1.1. Mathematical form

In a vacuum in which there are no electrical charges distributed so that $\rho = 0$, for a point P, Gauss's local theorem makes it possible to state that $\text{div}_p\vec{E} = 0$. As $\vec{E} = -\overrightarrow{\text{grad}}_p V$, it is possible to derive that

$$\boxed{\Delta_p\ V = 0},$$

which is the Laplace equation for a vacuum.

A similar formula can be obtained for an electric field. In order to do this, the notable equation, $\overrightarrow{\Delta\vec{E}} = \overrightarrow{\text{grad}} (\text{div } \vec{E}) - \overrightarrow{\text{rot}} (\overrightarrow{\text{rot}} \vec{E})$, is used. As in a vacuum, $\text{div}_p\vec{E} = 0$, and in electrostatics, $\overrightarrow{\text{rot}} \vec{E} = 0$, we immediately find

$$\boxed{\overrightarrow{\Delta\vec{E}} = 0}.$$

1.2.4.1.2. The consequence: flux conservation throughout the length of a "tubular" electric field

Before going any further, it can be noted that the lines of an electric field are defined at each point P in space by a curve that is tangential to the electric field vector at that point. The equation for the field lines is thus given by $\overrightarrow{dl} \times \vec{E} = 0$ (see Figure 1.12). As $\vec{E}$ is collinear to $\overrightarrow{\text{grad}}_p V$ and the gradient vector is normal to the equipotentials (as between 2 points denoted A and B there is an equipotential given by $\int_A^B \overrightarrow{\text{grad}} V.\overrightarrow{ds} = V(B) - V(A) = 0$ so that $\overrightarrow{\text{grad}}V \perp \overrightarrow{ds}$), $\vec{E}$ is perpendicular to the equipotentials and directed toward decreasing potentials.

A collection of field lines acting on a closed contour constitute a field tube, as described in Figure 1.12. The contour denoted C1 goes onto become C2.

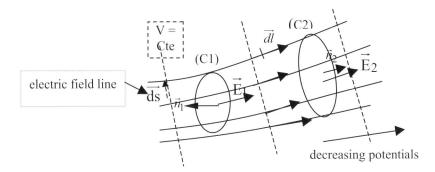

Figure 1.12. *A tube of electric field.*

The electric field flux through the closed surface formed by the field tube (lateral surface) and the two surfaces denoted as S_1 (delimited by C1) and S_2 (delimited by C2) is of the form:

$$\Phi = \iint_{S\ lateral} \vec{E}.\vec{dS} + \iint_{S_1}\vec{E}.\vec{dS_1} + \iint_{S_2}\vec{E}.\vec{dS_2} = 0 + \Phi s_1 + \Phi s_2 ,$$

where Φs_1 and Φs_2 denote the fields exiting from the surfaces S_1 and S_2, respectively.

Figure 1.10 shows that $\Phi s_1 < 0$ [as $\pi/2 < (\vec{n}_1, \vec{E}_1) < \pi$] while $\Phi s2 > 0$. Given that the flux traversing S_1 is given by $\Phi r_1 = \iint_{S_1}\vec{E}.(-\vec{n}_1)dS = - \iint_{S_1}\vec{E}.\vec{dS_1} = - \Phi s_1$, it is possible to directly derive $\Phi = \Phi_{s2} - \Phi_{r1}$.

From the integrated form of Gauss's theorem and given that here that $\rho = 0$, it can be deduced that $\boxed{\Phi = 0}$, from which it can be stated in more general terms for a vacuum that $\boxed{\Phi r = \Phi s}$. In effect, the flux entering by one side (Φ_r) of the field tube is the same as that leaving by the other (Φ_s).

1.2.4.2. Poisson's equation

In the presence of a volume charge distribution ($\rho \neq 0$), carrying forward $\vec{E} = -\overrightarrow{grad}_p V$ into the local Gauss theorem ($div_p\vec{E} = \dfrac{\rho}{\varepsilon_0}$) directly yields

$$\boxed{\Delta_P V + \dfrac{\rho}{\varepsilon_0} = 0} .$$

1.3. The Current Density Vector: Conducting Media and Electric Currents

1.3.1. The current density vector

1.3.1.1. Current through a section

Figure 1.13a shows a conducting wire with an elementary section given by $\overrightarrow{dS}$; ρ and $\vec{v}$ denote the density of mobile charges contained therein and their average velocity, respectively.

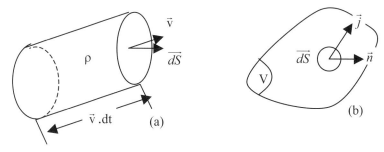

Figure 1.13. *Current flowing through (a) a section and (b) through a closed surface.*

During an interval of time (dt) between the time (t) and t + dt, the quantity of mobile charge (dq_m) that traverses the surface dS initially can be found in the volume given by $d\tau = \overrightarrow{dS}.\vec{v} \, dt$. Therefore, $dq_m = \rho \, d\tau = \rho\vec{v} \, dt.\overrightarrow{dS}$.

By definition, the current density vector $(\vec{j})$ is $\vec{j} = \rho\vec{v}$.

This relationship indicates that the quantity of charge traversing the surface $\overrightarrow{dS}$ in the time dt = 1 second is $(dq_m)_{dt=1} = \vec{j}.\overrightarrow{dS}$. The elementary intensity is thus

$$dI = \vec{j}.\overrightarrow{dS} = \frac{dq_m}{dt}$$, and $I = \iint_S \vec{j}.\overrightarrow{dS}$ therefore represents the quantity of charge that

traverses S per unit time and is the intensity of electric current across the S.

This last equation shows that the intensity appears as a flux of $\vec{j}$ through S.

1.3.1.2. *Comment*

The density ρ that is used above corresponds to the algebraic volume mobile charge density (ρ_m) and is different from the total volume density (ρ_T), which is generally zero in a conductor. Thus, $\rho_T = \rho_m + \rho_f$, where ρ_m is typically the (mobile) electron volume density and ρ_f is the volume density of ions sitting at fixed nodes in a lattice. As these ions only vibrate around their equilibrium positions   due to thermal energy (phonons), their average velocity (v_f) is such that $v_f = 0$, and the corresponding current (j_f) is $j_f = 0$.

1.3.1.3. Current traversing a closed surface

For a volume (V) delimited by a closed surface ($\bar{S}$), and yet orientated toward the exterior as shown in Figure 1.13(b), the total charge (Q) of V is not necessarily constant and can *a priori* change with the flux traversing $\bar{S}$.

Therefore, if dQ_m represents the quantity of mobile charges traversing $\bar{S}$ toward the exterior during the period t to t + dt, then

$$I = \frac{dQ_m}{dt} = \oiint_S \vec{j}.\overrightarrow{dS} = \oiint_S \vec{j}.n\overrightarrow{dS}.$$

If dQ_m exits the volume V, the conservation of charge implies that the total charge in V varies as $dQ = - dQ_m$. Thus $-\dfrac{dQ}{dt} = \dfrac{dQ_m}{dt} = I$, which can be rewritten as $I + \dfrac{dQ}{dt} = 0$, where dQ is the variation in internal charge during the given period of time. This equation means that there is no accumulation of charge at certain points in a circuit.

1.3.2. Equation for conservation of local charge

Thus $I = \oiint_S \vec{j}.\overrightarrow{dS} = \iiint \text{div } \vec{j} \, d\tau$

$$= -\frac{dQ}{dt} = -\frac{d}{dt} \iiint \rho \, d\tau = -\iiint \frac{d\rho}{dt} d\tau \qquad \Biggr\} \quad \Rightarrow$$

with $Q = \iiint \rho \, d\tau$

$$\iiint \left(\text{div } \vec{j} + \frac{\partial \rho}{\partial t} \right) d\tau = 0 \text{, so that} \quad \boxed{ \text{div } \vec{j} + \frac{\partial \rho}{\partial t} = 0 }$$

This formula is called the *continuity equation* and also represents a conservation of charge.

1.3.3. Stationary regimes

The title of this section indicates time-independent regimes, where the distribution of charge and current is time independent and $\dfrac{\partial \rho}{\partial t} = 0$. The condition $\dfrac{\partial \rho}{\partial t} = 0$ implies that $\rho = \text{constant}$, so that the charge contained in V is renewed (and maintained) by the passage of current (continuously produced by a generator) in a circuit that maintains a constant charge.

1.3.3.1. Current lines and tubes

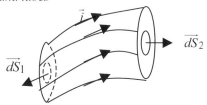

Figure 1.14. Current lines and tube.

By definition, a current line is at a tangent at all points to the vector $\vec{j}$ shown in Figure 1.14. For its part, a current tube is the surface generated by lines of current applied to a closed contour.

1.3.3.2. Flux due to $\vec{j}$ under stationary regimes

As a consequence of the stationary regime, the flux provided by $\vec{j}$ through the lateral faces of a current tube is zero (Figure 1.14), and $\Phi_{\text{Slateral}} = 0$.

In addition, the continuity equation can be reduced to $\operatorname{div}\vec{j} = 0$ and Ostrogradsky's theorem permits $\Phi = \oiint \vec{j}.\vec{n}dS = 0$. Thus, we find $\Phi = \Phi_{\text{Slateral}} + \Phi_{S1} + \Phi_{S2} = 0$ with $\Phi_{S1} = \iint\limits_{S1} \vec{j}.\overrightarrow{dS_1}$ as the flux leaving S_1, and likewise for S_2. From this it is determined that $-\Phi_{S1} = \Phi_{S2}$ and the flux entering S_1 is equal to $\Phi_{r1} = \iint\limits_{S1} \vec{j}.\left(-\overrightarrow{dS_1}\right) = -\Phi_{S1}$; therefore it can be concluded that the flux entering a section of a current tube is equal to that leaving by another section. In other terms, under a stationary regime, the current intensity is the same throughout all sections of the current tube.

1.3.3.3. Properties of $\vec{j}$ under a stationary state

At the interface between two media, there is conservation of the normal component (j_N) of $\vec{j}$. In effect, by denoting $\vec{j_1}$ and $\vec{j_2}$ as the vector $\vec{j}$ in media 1 and 2, respectively, we have:

$$\oiint \vec{j}.\vec{n}dS = 0 = \iint\limits_{S1} \vec{j_1}.\vec{n}_2 dS + \iint\limits_{S2} \vec{j_2}.\vec{n}_1 dS + \iint\limits_{\text{Slateral}} \vec{j}.\overrightarrow{dS} \text{ , where}$$

$$\iint\limits_{\text{Slateral}} \vec{j}.\overrightarrow{dS} \approx \varepsilon \approx 0 \ .$$

[If the problem is considered simply in terms of the opposing sides of the interface (Figure 1.15), then the extensions $A_1 \rightarrow A_2$ and $S_{lateral} \rightarrow \epsilon$ are made].

We now arrive at $\iint_{S1} \vec{j}_1 . \vec{n}_2 dS + \iint_{S2} \vec{j}_2 . \vec{n}_1 dS = 0$. By making $\vec{n} = \vec{n}_1 = -\vec{n}_2$, and by noting that for the two parts on either side of the interface, that $S_1 = S_2 = S$, it can be written that:

$$\iint_S \vec{j}_2 . \vec{n} \ dS - \iint_S \vec{j}_1 . \vec{n} \ dS = \iint_S (\vec{j}_2 . \vec{n} \ - \ \vec{j}_1 . \vec{n}) dS = 0 \ , \text{ so that}$$

$$\boxed{j_{1n} = j_{2n}}$$

where the components of $\vec{j}_1$ and $\vec{j}_2$ at the normal $\vec{n}$ are continued at the interface between the two materials.

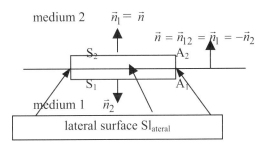

Figure 1.15. *Vector $\vec{j}$ at an interface.*

If one of the materials is an insulators, take for example here medium 1, then $\vec{j}_1 = 0$ and j_{1n} is also zero. The continuity equation given above details that in the second medium the value for j_{2n} is also zero. From this can be deduced that in a conductor in the neighborhood of an insulator, j_n is zero. The current density vector ($\vec{j}$) can therefore only be at a tangent to the dividing surface. As a consequence, inside the conductor, $E_T = \dfrac{j_T}{\sigma} \neq 0$, where σ is the conductivity to be recalled in the definition given in Section 1.3.4 below. As there is continuity in the tangential component (E_T) at the interface between the conductor and the insulator, the external field no longer is normal to the conductor as otherwise would be found at equilibrium, i.e., when there is no current.

1.3.3.4. Comments

Under a quasistationary regime, ρ varies with time but sufficiently little so that it can be assumed that $\dfrac{\partial \rho}{\partial t} \ll \text{div}\,\vec{j}$. Hence, once again $\text{div}\,\vec{j} = 0$ (as found for sinusoidal currents at low frequencies).

Under a rapidly varying regime, ρ varies rapidly with time, so that $\dfrac{\partial \rho}{\partial t}$ is no longer negligible with respect to $\text{div}\,\vec{j}$. Therefore, $\vec{j}$ is no longer a conservative flux but accords to $\text{div}\,\vec{j} + \dfrac{\partial \rho}{\partial t} = 0$, and the current lines do not fold in on themselves. We find therefore that $\iint_{S_1} \vec{j}.\vec{dS_1} \neq \text{constant}$, and after one cycle $\vec{j}$ differs from that traversing S_1.

1.3.4. *Ohm's law and its limits*

1.3.4.1. The model (Drude)

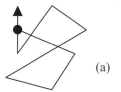
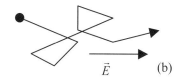

(a)

$\vec{E}$ (b)

Figure 1.16. *Trajectory of a carrier (a) in the absence and (b) in the presence*

of an electric field.

In the absence of an applied field, free charges collide and the average (Brownian) displacement is zero, so that $\langle \vec{v} \rangle = 0$, and as there is no current, $\vec{j} = 0$. It is worth noting that the instantaneous velocity is non-zero and that it corresponds to the thermal velocity (v_{th}), that is very high.

In the presence of an electric field, the trajectories are deformed and an average derivation occurs; thus charges are transported and a current appears.

1.3.4.2. Determination of conductivity (σ)

In the presence of an electric field ($\vec{E}$), the applied dynamic fundamental formula for a charge (q), which is typically an electron so that $q = - e$, is given by

$$m\dfrac{d\vec{v}}{dt} = q\vec{E}.$$

The differential equation, $d\vec{v} = \dfrac{q\vec{E}}{m} dt$, can be integrated from between the initial instant when $t_0 = 0$ and a time t. By assuming that $\vec{E}$ is uniform and therefore constant between these two times, then $\vec{v}(t) = \vec{v}(0) + \dfrac{q\vec{E}}{m}(t - 0)$ where $\vec{v}(0)$ is the initial velocity of the carrier and $\vec{v}(t)$ is the carrier velocity at the later time t. Between two successive collisions, the average value of the velocity is given by: $<\vec{v}> = <\vec{v}(0)> + \dfrac{q\vec{E}}{m} <t>$ where $<t>$, denoted below more simply as τ, is the average time between two successive collisions. Given that the impacts are random, the initial velocity of an electron is zero, although this average can be over a high number of electrons and unless of course if the collisions are orientated, which can result in a high average velocity under a strong field.

So, with $<t> = \tau$ (called the relaxation time) and $<\vec{v}(0)> = 0$, we finally have:

$$<\vec{v}> = \dfrac{q\vec{E}}{m}\tau = \mu \vec{E} = \vec{v}_d$$

by convention, this velocity is denoted $\vec{v}_d$ and is termed "drift velocity"

with

$$\mu = \dfrac{q\tau}{m}$$ as the charge mobility expressed in $cm^2\ V^{-1}\ sec^{-1}$.

In addition, by introducing the value of velocity into the formula for current density, $\vec{j} = \rho\vec{v} = nq\vec{v}$, where n is the carrier density, we obtain:

$$\vec{j} = \dfrac{nq^2\tau}{m}\vec{E} = \sigma\vec{E}$$, where $\sigma = \dfrac{nq^2\tau}{m}$ is the conductivity in units $\Omega^{-1}\ m^{-1}$.

1.3.4.3. Order of scale

For copper, the conductivity $\sigma \approx 6 \times 10^7\ \Omega^{-1}m^{-1}$ ($6 \times 10^5\ \Omega^{-1}cm^{-1}$) and the carrier density $n = \dfrac{\rho N}{M_a}$ is approximately 8.5×10^{28} electrons m^{-3} where N is Avogadro's number and M_a is the atomic mass. From this can be determined that $\tau = \dfrac{m\sigma}{nq^2} \approx 2 \times 10^{-14}\ s$. Given a value of $j \approx 10\ A/mm^2$, then $v_d = j/\rho \approx 0.3\ mm/s$ and the mobility $\mu = q\tau/m \approx 35\ cm^2/V.s$.

1.3.4.4. Limits to Ohm's law

For an intense field, the velocities $<v_d> \approx \dfrac{qE}{m}\langle\tau\rangle$ are relatively high and alter the character of impacts. It is no longer possible to state therefore that $<v_0> = 0$. For fields that are extremely intense, the impacts are orientated in the sense of the field, so that the trajectory is practically parallel to the field and the velocities approach the thermal velocity (v_{th}) as indicated in Figure 1.17.

In addition, if $\vec{E}$ is very intense, then ionization phenomena can occur, resulting in a non-Ohmic avalanche.

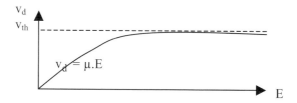

Figure 1.17. *Variation of velocity with electric field intensity.*

If $\vec{E}$ varies too rapidly, then the integration of the differential equation in Section 1.3.4.2 between two instants $t_0 = 0$ (start of impact) and $t = \tau$ (statistical end of impact, equal to the average time between successive impacts) assumes that $\vec{E}$ is constant during the interval τ. The frequency of $\vec{E}$ therefore must be below $v = 1/\tau \approx 5 \times 10^{13}$ Hz, a frequency that corresponds to that of an electromagnetic wave with wavelength given by $\lambda = c\tau \approx 6\ \mu m$ (infrared). As a consequence, Ohm's law is valid in metals for electrotechnical and radioelectrical frequencies but not optical frequencies where $(v \approx 10^{15}$ Hz$)$.

1.3.4.5. Macroscopic form of Ohm's law

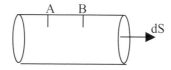

Figure 1.18. *Cross section of a current tube.*

Given a conductor with a cross section (dS) perpendicular to the current lines. The intensity of the current traversing dS is given by $dI = j\ dS = \sigma\ E\ dS$, so that

$E = \dfrac{1}{\sigma}\dfrac{dI}{dS}$. Between two points A and B along a current line, we thus have:

$\displaystyle\int_A^B dV = \int_A^B - E.dl = -\int_A^B \dfrac{1}{\sigma}\dfrac{dI}{dS}dl$. By making $r = \int_A^B \dfrac{1}{\sigma}\dfrac{dl}{dS}$, we find that (with dI

being a constant between A and B under a stationary state): $V_A - V_B = dI \displaystyle\int_A^B \dfrac{1}{\sigma}\dfrac{dl}{dS}$,

so that $dI = \dfrac{V_A - V_B}{r}$.

Making $R = \dfrac{1}{\iint \dfrac{1}{r}}$, it is possible to state that $I = \displaystyle\iint_S dI = \iint_S \dfrac{V_A - V_B}{r} = \dfrac{V_A - V_B}{R}$.

Therefore $V_A - V_B = RI$.

1.3.5. Relaxation of a conductor

On introducing the relation $\vec{j} = \sigma\vec{E}$ into the general equation of charge

conservation, $\operatorname{div}\vec{j} + \dfrac{\partial\rho}{\partial t} = 0$, we find $\sigma\operatorname{div}\vec{E} + \dfrac{\partial\rho}{\partial t} = 0$.

Using the local form of Gauss's theorem gives $\dfrac{\partial\rho}{\partial t} + \dfrac{\sigma}{\varepsilon_0}\rho = 0$.

By making $\tau = \dfrac{\varepsilon_0}{\sigma}$, we obtain $\rho = \rho_0\,e^{-t/\tau}$.

In the volume charge density of a conductor, there are both interventions due to free electron charges and charges associated with ions. Under a field effect, the electronic charges and the ions are separated and a localized charge can appear, given by the volume density ρ, which thus includes electron and ion charges ($\rho \equiv \rho_T$).

Nevertheless, the integration of the differential equation shows that for periods of the order of several τ, ρ tends toward zero. Physically, this result can be understood if there is a localized excess of charge appearing in the conductor, returning electric forces act between opposite charges and if the material is sufficiently conducting, a return to electrical neutrality will occur quickly. With $\sigma \approx 10^7\ \Omega^{-1}\ m^{-1}$, we have $\tau \approx 10^{-18}$ s . However, given the use of Ohm's law here, the result is unacceptable for periods below 10^{-14} s, and in practical terms, inside a homogeneous conductor, the total volume charge density can be assumed to be zero across all Hertzian frequency domains; that is to say from the stationary state to the infrared.

1.3.6. *Comment: definition of surface current (current sheet)*

Figure 1.19. *(a) volume and (b) surface current densities.*

It has been noted above that the elementary intensity (dI) of current (with $I = \dfrac{dQ_m}{dt}$)
traversing a section (dS) of a volume (dV), as in Figure 1.19, is given by
$dI = \dfrac{d^2 Q_m}{dt} = \dfrac{\rho \vec{v}.dt.\overrightarrow{dS}}{dt} = \rho \vec{v}.\overrightarrow{dS} = \vec{j}.\overrightarrow{dS}$. If we flatten the cylinder, as in Figure
1.19b, dI is an intensity that no longer crosses dS but dl where dl is perpendicular to
the current lines ($\vec{j_s}$). Crosses $\vec{j_s}$ is therefore defined by $\vec{j_s} = \sigma_s \vec{v}$ where σ_s
represents the surface density of free charge. Thus, $dI = \vec{j_s}.\vec{n}\, dl$, so that $\vec{j_s} = \dfrac{dI}{dl}\vec{n}$
where $\vec{n}$ is normal to dl, and dI is the intensity traversing the segment dl
perpendicular to the current lines. Considering that the line AB perpendicular at all
points to $\vec{j_s}$, the current across the flat sheet is defined by $I = \int_A^B \vec{j_s}.\vec{n}\, dl = \int_A^B j_s.dl$. As
j_s is uniform, we find that $|j_s| = \dfrac{I}{AB}$, which has units $A\, m^{-1}$.

1.4. Magnetostatics

By definition, magnetostatics is the study of magnetic fields due to steady current
distributions, or in other terms, a spread of volume, surface, or wire currents
independent of time. The circuits through which the current circulates continuously
also are assumed to be fixed.

   In fact, magnetostatics is not a study of statics in the strictest sense, as the
electrons, or holes, are mobile, but what is steady (or stationary and independent of
time) is the spread of charges. As a consequence, we always find $\rho = $ constant with
respect to time, and therefore $\dfrac{\partial \rho}{\partial t} = 0$.

   The equation for the conservation of charge therefore can be reduced in
magnetostatics to $div \vec{j} = 0$. According to Ostrogradsky's theorem, it can be

deduced that $\oiint \vec{j}.\overline{dS} = \iiint \text{div } \vec{j} \, d\tau = 0$, indicating that the incoming flux is equal to the outgoing flux, and that the intensity is constant across sections of the current tube.

1.4.1. Magnetic field formed by a current

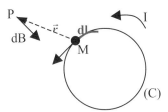

Figure 1.20. *Magnetic field produced by a circuit (C).*

Figure 1.20 shows a circuit (C) through which runs a current (I). The charges moving at a velocity ($\vec{v}$) through the wire interact with other charges (q). The action of I on q at a point P is given by Laplace's equation in the form $\vec{F} = q\vec{v} \times \vec{B}$. The vector $\vec{B}$ thus generated by the circuit is called the *magnetic field* (or even the *magnetic induction*). Historically, it is the vector $\vec{H}$ that is given by $\vec{B} = \mu_0 \vec{H}$, which was first called the magnetic field vector, even though it is simpler to say the "$\vec{B}$ vector" or the "$\vec{H}$ vector" to denote them. $\mu_0 = 4 \pi \, 10^{-7}$ MKS is the magnetic permeability of a vacuum, and $\vec{B}$ is expressed in Tesla or in Weber m^{-2}.

For the various types of current, whether in a wire, a surface, or a volume, the expression for the element $d\vec{B}$ at a point P situated at $\overline{MP} = \vec{r}$ with respect to the point M which defines an element of the circuit, whether it be dl, dS, or dτ, is given by the Biot-Savart empirical law:

-for a current in a wire

$$\overline{dB} = \frac{\mu_0}{4\pi} I \, \overline{dl} \times \frac{\vec{r}}{r^3} \text{ (elementary field produced by } \overline{dl}\text{), and } \vec{B} = \int_C \overline{dB};$$

- for a surface current $\overline{dB} = \frac{\mu_0}{4\pi} (\vec{j}_s dS) \times \frac{\vec{r}}{r^3}$, and $\vec{B} = \int_S \overline{dB}$; and

- for a volume current $\overline{dB} = \frac{\mu_0}{4\pi} (\vec{j} \, d\tau) \times \frac{\vec{r}}{r^3}$, and $\vec{B} = \int_V \overline{dB}$.

1.4.2. *Vector potential*

For a wire circuit, we have:

$$\vec{B} = \frac{\mu_0 I}{4\pi} \int_C \vec{dl} \times \frac{\vec{r}}{r^3} = \frac{\mu_0 I}{4\pi} \int_C \overrightarrow{grad}_P \left(\frac{1}{r}\right) \times \vec{dl} \ .$$

As $\overrightarrow{rot}_P \left(\dfrac{\vec{dl}}{r}\right) = \dfrac{1}{r} \overrightarrow{rot}_P \ \vec{dl} + \overrightarrow{grad}_P \left(\dfrac{1}{r}\right) \times \vec{dl} = \overrightarrow{grad}_P \left(\dfrac{1}{r}\right) \times \vec{dl}$, it is possible to

state: $\underbrace{\qquad} = 0$ as $\vec{dl}$ is fixed independently of P

$$\vec{B} = \frac{\mu_0 I}{4\pi} \int_C \overrightarrow{rot}_P \left(\frac{\vec{dl}}{r}\right) .$$

By making $\boxed{\vec{A} = \dfrac{\mu_0 I}{4\pi} \int_C \dfrac{\vec{dl}}{r}}$, we can write that $\boxed{\vec{B} = \overrightarrow{rot}_P \ \vec{A}}$ The vector $\vec{B}$ is thus

derived from the vector potential $\vec{A}$, as given above.

For a surface circuit, we obtain $\boxed{\vec{A} = \dfrac{\mu_0}{4\pi} \iint_S \dfrac{\vec{j_s} \ dS}{r}}$

For a volume surface $\boxed{\vec{A} = \dfrac{\mu_0}{4\pi} \iiint_V \dfrac{\vec{j} \ d\tau}{r}} .$

1.4.3.   *Local properties and the integral of* $\vec{B}$

Local property: as $\vec{B} = \overrightarrow{rot}_P \ \vec{A}$, we have $div_P \ \vec{B} = div_P \ \overrightarrow{rot}_P \ \vec{A} = 0$, so that in general terms:

$$\boxed{div \ \vec{B} = 0} .$$

Integral property: as $\oiint \vec{B}. \ \vec{dS} = \iiint div \ \vec{B} \ d\tau = 0$, we have $\boxed{\oiint \vec{B}. \ \vec{dS} = 0}$.

1.4.4.   *Poisson's equation for the vector potential*

Here we will use the same reasoning as that for electrostatic formulas. While this does not constitute a very rigorous approach, it does at least limit the need for long-winded procedures.

First, compare the form of the electrostatic potential, $V = \dfrac{1}{4\pi\varepsilon_0} \iiint \dfrac{\rho}{r} d\tau$, with

that of the Ox component of the potential vector $\vec{A}$, as in $A_x = \dfrac{\mu_0}{4\pi} \iiint \dfrac{j_x d\tau}{r} .$

The forms of A_x (along with A_y and A_z too) and of V are the same, with the exception of μ_0, which corresponds to $1/\varepsilon_0$ and j_x to ρ. We can write then for A_x a Poisson's equation in the form $\Delta A_x + \mu_0 \, j_x = 0$. The three equations for the three components then can be condensed into a single equation, as in

$$\boxed{\Delta \vec{A} + \mu_0 \vec{j} = 0}\;.$$

1.4.5.   Properties of the vector potential $\vec{A}$

1.4.5.1. In magnetostatics, $div_P\ \vec{A} = 0$

In magnetostatics, we have $div_M \, \vec{j} = 0$, where M is a point on a circuit through which $\vec{j}$ moves. Recalling that at the surface separating a conductor and an insulator, by conservation of j_N (which is zero in an insulator), $\vec{j}$ can only be at a tangent to the interface.

In addition, with $\vec{A} = \dfrac{\mu_0}{4\pi} \iiint \dfrac{\vec{j}\ d\tau}{r}$, and by calculating

$div_P\ \vec{A} = \dfrac{\mu_0}{4\pi} \int_V div_P \left(\dfrac{\vec{j}}{r} \right) d\tau$, and using the fact that

$div_P \left(\dfrac{\vec{j}}{r} \right) = \dfrac{1}{r}\ div_P \vec{j} + \underbrace{\vec{j}.\overrightarrow{grad}_P \left(\dfrac{1}{r} \right)} = \vec{j}.\overrightarrow{grad}_P \left(\dfrac{1}{r} \right)$,   and therefore:

> $= 0$ as $\vec{j}$ depends only on the coordinates of M points on the circuit and not those of P (see Figure 1.20).

$$div_P \vec{A} = \dfrac{\mu_0}{4\pi} \int_V \vec{j}.\ \overrightarrow{grad}_P \left(\dfrac{1}{r} \right) d\tau = -\dfrac{\mu_0}{4\pi} \int_V \vec{j}.\ \overrightarrow{grad}_M \left(\dfrac{1}{r} \right) d\tau\;.$$

On again using the identity (but only for a point M):

$div_M \left(\dfrac{\vec{j}}{r} \right) = \dfrac{1}{r}\ div_M \vec{j} + \vec{j}.\overrightarrow{grad}_M \left(\dfrac{1}{r} \right)$, and the fact that (magnetostatic)

$div_M \vec{j} = 0$,   gives $div_M \left(\dfrac{\vec{j}}{r} \right) = \vec{j}.\overrightarrow{grad}_M \left(\dfrac{1}{r} \right)$.

By moving this result into the expression for $div_P\ \vec{A}$, finally:

$$div_P\ \vec{A} = -\dfrac{\mu_0}{4\pi} \int_V div_M \left(\dfrac{\vec{j}}{r} \right) d\tau = -\dfrac{\mu_0}{4\pi} \oiint_S \dfrac{\vec{j}}{r}\overrightarrow{dS}\;.$$

Given that the vector $\vec{j}$ is at a tangent to the surface of the conductor it is normal to $\overrightarrow{dS}$, which means that

$$\boxed{\operatorname{div}_P \vec{A} = 0}.$$

1.4.5.2. Indeterminability of the vector potential

The vector potential $\vec{A}$ is defined simply by the relationship $\vec{B} = \overrightarrow{\operatorname{rot}}_P \vec{A}$, which only defines the derivatives of $\vec{A}$ so that in reality we can find an infinite number of $\vec{A}'$ vectors each different from $\vec{A}$ and fitting into $\vec{B} = \overrightarrow{\operatorname{rot}}_P \vec{A}'$.

Therefore $\overrightarrow{\operatorname{rot}}_P (\vec{A}' - \vec{A}) = 0$, which shows that $(\vec{A}' - \vec{A})$ is derived from a scalar potential. It is possible therefore to take $(\vec{A}' - \vec{A})$ in the form $(\vec{A}' - \vec{A}) = \overrightarrow{\operatorname{grad}}_P \varphi$, so that:

$$\vec{A}' = \vec{A} + \overrightarrow{\operatorname{grad}}_P \varphi.$$

As $\operatorname{div}_P \vec{A}' = 0$ (just as $\operatorname{div}_P \vec{A} = 0$), then $\operatorname{div}_P \overrightarrow{\operatorname{grad}}_P \varphi = \Delta_P \varphi = 0$ must be true where the φ potentials are such that $\Delta\varphi = 0$ and are called the Newtonian potentials. Finally, the $\vec{A}'$ vectors are such that:

$$\boxed{\vec{A}' = \vec{A} + \overrightarrow{\operatorname{grad}}_P \varphi, \text{ with } \Delta_p\varphi = 0}.$$

1.4.6. The Maxwell-Ampere relation

1.4.6.1. Localized forms and the integral of Ampere's theorem

By using the following relationships

$$\begin{cases} \Delta_P\vec{A} = \overrightarrow{\operatorname{grad}}_P \operatorname{div}_P \vec{A} - \overrightarrow{\operatorname{rot}}_P \overrightarrow{\operatorname{rot}}_P \vec{A} = - \overrightarrow{\operatorname{rot}}_P \vec{B} \\ \qquad\qquad (\text{as } \operatorname{div}_P \vec{A} = 0 \text{ and } \vec{B} = \overrightarrow{\operatorname{rot}}_P \vec{A}) \\ \Delta\vec{A} = - \mu_0\vec{j} \end{cases}$$

it is possible to deduce that in a vacuum $\overrightarrow{\operatorname{rot}}_P \vec{B} = \mu_0\vec{j}$, so that in turn:

• the local form of Ampère's theorem is give by $\boxed{\overrightarrow{\operatorname{rot}}_P \vec{H} = \vec{j}}$; and

• the integral form of Ampère's theorem is

$$I = \iint_S \vec{j}.\overrightarrow{dS} = \iint_S \overrightarrow{\operatorname{rot}}_P \vec{H}.\, \overrightarrow{dS} = \boxed{\oint_C \vec{H}.\, \overrightarrow{ds} = I},$$

where $\overrightarrow{ds}$ is an element in the curve (C) as shown in Figure 1.21.

1.4.6.2. Simple example of the use of Ampère's theorem

Consider a wire circuit perpendicular to the plane of Figure 1.21 through which runs I. Here we shall calculate the $\vec{H}$ vector at a point P anywhere in space.

The sense and direction of $\vec{H}$ are given by Ampère's right-hand rule (which results from the vectorial product $\vec{dl} \times \vec{r}$ introduced into the Biot-Savart law with $\vec{r} = \overrightarrow{MP}$; here $\vec{dl}$ is perpendicular to the plane of Figure 1.21 and runs across the plane as the intensity (I) given by the symbol $\otimes$).

In order to calculate H, the integral form of Ampere's theorem may be used, given also that H is a constant at a distance r from M. In the plane of Figure 1.21, the length of the circle C, which has radius r = MP, we find that H = constant. In addition, $\vec{H}$ is a tangent to all points on the circle and $I = \oint_C \vec{H}.\overrightarrow{ds} = \oint_C H.ds = H\oint_C ds = 2\pi\, r\, H$, from which $H_{(P)} = \dfrac{I}{2\pi r}$.

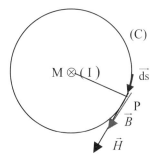

Figure 1.21. $\vec{H}$ *field due to C through which moves I.*

In terms of vectors, we can write that $\vec{H} = \dfrac{I}{2\pi r}\vec{u}_\theta$, where $\vec{u}_\theta = \vec{u}_I \times \vec{u}_{MP}$ (where $\vec{u}_I$ is the unit vector carried by the conducting wire in the same sense as the intensity I, and $\vec{u}_{MP}$ is the unit vector of $\vec{r} = \overrightarrow{MP}$).

1.4.6.3. Physical significance of rotation

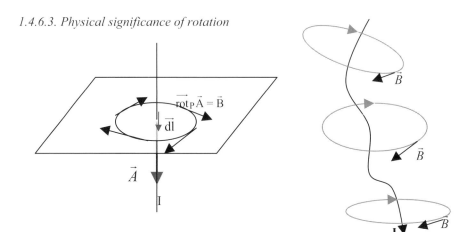

Figure 1.22. *Rotational sense of* $\vec{B}$ *lines for (a) a rectilinear current and (b) a twisting current.*

In the simple example treated above in Section 1.4.6.2, the potential vector $\vec{A} = \dfrac{\mu_0 I}{4\pi} \displaystyle\int_C \dfrac{\overline{dl}}{r}$ is carried by the conducting wire ($\vec{A}\,//\,\overline{dl}$). As $\vec{B}\,//\,\vec{H}$ and $\vec{B} = \overrightarrow{rot}_P\,\vec{A}$, we can see in Figures 1.21 and 1.22a that the vector $\overrightarrow{rot}_P\,\vec{A}$ turns around the vector $\vec{A}$.

For its part, the vector $\vec{B}$ (or $\vec{H}$) exhibits a twisting character. The rotational sense of the $\vec{B}$ (or $\vec{H}$) lines with respect to I is given by Stokes' integral, as shown in Figure 1.22b.

1.5 Problems

1.5.1. Calculations. A vector given by $\vec{r} = \overrightarrow{MP}$ has components:

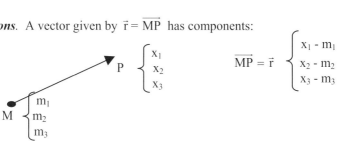

$$\overrightarrow{MP} = \vec{r} \begin{cases} x_1 - m_1 \\ x_2 - m_2 \\ x_3 - m_3 \end{cases}$$

This vector is such that:

$$\vec{r}^2 = (x_1 - m_1)^2 + (x_2 - m_2)^2 + (x_3 - m_3)^2$$

$$= u(x_1, x_2, x_3) \text{ if the calculation for the operator is for point P}$$
$$= u(m_1, m_2, m_3) \text{ if the calculation for the operator is for point M}$$
$$\Bigg\} \Rightarrow r = u^{1/2}$$

Verify the following results:

$$\overrightarrow{grad}_P r = -\overrightarrow{grad}_M r = \frac{\vec{r}}{r} \ ,$$

$$\overrightarrow{grad}_P \frac{1}{r} = -\overrightarrow{grad}_M \frac{1}{r} = -\frac{\vec{r}}{r^3} \ ,$$

$$div_P \vec{r} = -div_M \vec{r} = 3 \quad \text{(3D space)} ,$$

$$\overrightarrow{rot}_{M \text{ or } P} \vec{r} = 0 \ , \ \Delta\left(\frac{1}{r}\right) = 0 , \text{ and}$$

$$div\left(\frac{\vec{r}}{r^3}\right) = 0 ; \quad \text{what can be said about the flux of the vector } \frac{\vec{r}}{r^3} \ ?$$

Answers
1.

$$\overrightarrow{grad}_P (r) \begin{cases} \dfrac{\partial r}{\partial x_1} \\[2mm] \dfrac{\partial r}{\partial x_2} \\[2mm] \dfrac{\partial r}{\partial x_3} \end{cases}$$

$$\frac{\partial r}{\partial x_1} = \frac{\partial u^{1/2}(x_1, x_2, x_3)}{\partial x_1} = \frac{1}{2} u^{-1/2} \, 2(x_1 - m_1) = \frac{(x_1 - m_1)}{r}$$

Similarly, $\dfrac{\partial r}{\partial x_2} = \dfrac{(x_2 - m_2)}{r}$ and $\dfrac{\partial r}{\partial x_3} = \dfrac{(x_3 - m_3)}{r}.$

$$\Rightarrow \overrightarrow{grad}_P r = \frac{\vec{r}}{r}$$

2.

$$\overrightarrow{grad}_P(\frac{1}{r})\begin{cases}\dfrac{\partial}{\partial x_1}\left(\dfrac{1}{r}\right)\\[2mm]\dfrac{\partial}{\partial x_2}\left(\dfrac{1}{r}\right)\\[2mm]\dfrac{\partial r}{\partial x_3}\left(\dfrac{1}{r}\right)\end{cases}$$

We have $\dfrac{\partial}{\partial x_1}\left(\dfrac{1}{r}\right)=\dfrac{\partial u^{-1/2}(x_1,x_2,x_3)}{\partial x_1}=-\dfrac{1}{2}u^{-3/2}u'_{x_1}$

with $u'_{x_1}=2(x_1-m_1)$ and $u^{-3/2}=r^{-3}$ $\Rightarrow$

$\dfrac{\partial}{\partial x_1}\left(\dfrac{1}{r}\right)=-\dfrac{1}{2}r^{-3}\,2(x_1-m_1)=-\dfrac{x_1-m_1}{r^3}$

Finally

$$\overrightarrow{grad}_P(\frac{1}{r})\begin{cases}\dfrac{\partial}{\partial x_1}\left(\dfrac{1}{r}\right)=-\dfrac{x_1-m_1}{r^3}\\[2mm]\dfrac{\partial}{\partial x_2}\left(\dfrac{1}{r}\right)=-\dfrac{x_2-m_2}{r^3}\\[2mm]\dfrac{\partial r}{\partial x_3}\left(\dfrac{1}{r}\right)=-\dfrac{x_3-m_3}{r^3}\end{cases}\qquad\boxed{\Rightarrow\ \overrightarrow{grad}_P(\frac{1}{r})=-\dfrac{\vec{r}}{r^3}}$$

And also

$$\overrightarrow{grad}_M(\frac{1}{r})\begin{cases}\dfrac{\partial}{\partial m_1}\left(\dfrac{1}{r}\right)=-\dfrac{1}{2}u^{-3/2}u'_{m_1}=\dfrac{x_1-m_1}{r^3}\ \text{as}\ u'_{m_1}=2(x_1-m_1)(-1)\\[2mm]\dfrac{\partial}{\partial m_2}\left(\dfrac{1}{r}\right)=\dfrac{x_2-m_2}{r^3}\\[2mm]\dfrac{\partial r}{\partial m_3}\left(\dfrac{1}{r}\right)=\dfrac{x_3-m_3}{r^3}\end{cases}\qquad\boxed{\Rightarrow\ \overrightarrow{grad}_M(\frac{1}{r})=\dfrac{\vec{r}}{r^3}}$$

3. $div_P\vec{r}=\dfrac{\partial}{\partial x_1}(x_1-m_1)+\dfrac{\partial}{\partial x_2}(x_2-m_2)+\dfrac{\partial}{\partial x_3}(x_3-m_3)=1+1+1=3$

$div_M\vec{r}=\dfrac{\partial}{\partial m_1}(x_1-m_1)+\dfrac{\partial}{\partial m_2}(x_2-m_2)+\dfrac{\partial}{\partial m_3}(x_3-m_3)=-1-1-1=-3$

4. $\overrightarrow{rot}_{M\ or\ P}\ \vec{r}=0$ as, for example, $\left[\overrightarrow{rot}_P\vec{r}\right]_{x1}=\dfrac{\partial}{\partial x_2}(x_3-m_3)-\dfrac{\partial}{\partial x_3}(x_2-m_2)=0$.

5. $\Delta_P\left(\dfrac{1}{r}\right) = \dfrac{\partial^2}{\partial x_1^2}\left(\dfrac{1}{r}\right) + \dfrac{\partial^2}{\partial x_2^2}\left(\dfrac{1}{r}\right) + \dfrac{\partial^2}{\partial x_3^2}\left(\dfrac{1}{r}\right)$ so that following a rather long

calculation, we find:

$$\dfrac{\partial^2}{\partial x_1^2}\left(\dfrac{1}{r}\right) = -\dfrac{1}{r^3} + \dfrac{3(x_1 - m_1)^2}{r^5} \implies$$

$$\Delta_P\left(\dfrac{1}{r}\right) = -\dfrac{3}{r^3} + \dfrac{3}{r^5}\left[(x_1 - m_1)^2 + (x_2 - m_2)^2 + (x_3 - m_3)^2\right] = -\dfrac{3}{r^3} + \dfrac{3\,r^2}{r^5} = 0\,.$$

As $\overrightarrow{\text{grad}}_P(\dfrac{1}{r}) = -\dfrac{\vec{r}}{r^3}$, then $\text{div}_P\,\dfrac{\vec{r}}{r^3} = -\,\text{div}_P\,\overrightarrow{\text{grad}}_P(\dfrac{1}{r}) = -\Delta_P\left(\dfrac{1}{r}\right) = 0$ meaning that

$\dfrac{\vec{r}}{r^3}$ is a stationary flux.

1.5.2. Field and potential generated inside and outside of a charged sphere

For a sphere with a center denoted by O and of radius r, uniformly charged in its volume (charge volume density denoted by ρ):

1. Show that the field at a point (P) outside the sphere ($OP = r > R$) is in the form

$$\vec{E}_{ext} = \dfrac{\rho.R^3}{3\varepsilon_0 r^2}\,\vec{u}_{OP}\,.$$

2. Show that for a point P inside the sphere ($OP < R$) we have $\vec{E}_{int} = \dfrac{\rho r}{3\varepsilon_0}\,\vec{u}_{OP}\,.$

3. Plot the curve $\vec{E} = f(r)\,.$

4. Show that when $r > R$, the equation for the electric potential is $V_{ext} = \dfrac{\rho.R^3}{3\varepsilon_0 r}\,.$

Similarly, show that when $r < R$, we have $V_{int} = \dfrac{\rho R^2}{2\varepsilon_0}\left[1 - \dfrac{r^2}{3R^2}\right]$ and

plot $V = f(r)\,.$

5. Show how the calculation for the potential can also be made using Poisson's equation.

Answers

1. Given that the charge is spread symmetrically, the resulting electric field also will have the same symmetry, as at any point P in a given sphere with radius r = OP (point P_e or P_i shown in Figure 1) and in space there is the same charge distribution. Thus, for a sphere with a radius denoted by r, the modulus of the field (E) is independent of P on the sphere; that is to say the angles θ and φ given to define P in terms of spherical coordinates (Figure 2). The E is therefore only dependent on r, so that it is possible to write in terms of modulus that $E = E(r)$. In terms of vectors, and again due to symmetry, $\vec{E}$ is parallel to $\vec{e}_r$, a unit vector of the radial direction under consideration, which corresponds to the normal ($\vec{n}$) outside of the sphere (for $\rho > 0$).

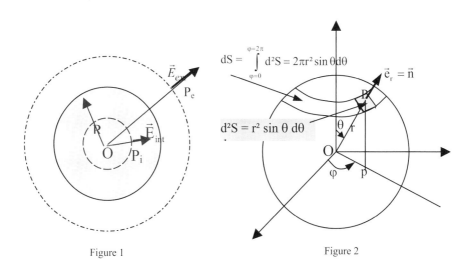

Figure 1   Figure 2

Gaussian surface. For the calculation of the field $\vec{E}$ at P, which is at a distance from O given by $r = OP$, the sphere with radius r around the center O is called the Gaussian surface, for which E is independent of P or in other terms, E is a constant for a given value of r.

The electric field flux (Φ) through a Gaussian surface is represented by a sphere with a radius given by r, and thus is

$$\Phi = \oiint_{\theta,\varphi} \vec{E}.\overrightarrow{d^2S} = \oiint_{\theta,\varphi} E.d^2S$$

as $\vec{E}$ and $\overrightarrow{d^2S}$ are with respect to the external normal, in that $\vec{n} = \vec{e}_r$.

With $d^2S = r^2 \sin\theta \, d\theta \, d\varphi$, we have:

$$\Phi = r^2 \, E \oiint_{\theta,\varphi} \sin\theta \, d\theta \, d\varphi = r^2 \, E \int_{\theta=0}^{\theta=\pi} \sin\theta \, d\theta \int_{\varphi=0}^{2\pi} d\varphi = 4 \pi \, r^2 \, E$$

The calculation of the field denoted E_{ext} at P_e, which is such that $OP_e = r > R$ may be performed using Gauss's theorem, which gives here

$$\Phi = 4\pi r^2 E_{ext} = \sum_i \frac{Q_{int}}{\varepsilon_0} = \frac{\displaystyle\int_{r=0}^{r=R} \int_{\theta=0}^{\theta=\pi} \int_{\varphi=0}^{\varphi=2\pi} \rho d\tau}{\varepsilon_0} \quad ,$$

so that with $\rho = Cte$ and $d\tau = r^2 \sin\theta \, dr \, d\theta \, d\varphi$, then

$$\Phi = 4\pi r^2 E_{ext} = \frac{4}{3}\pi R^3 \frac{\rho}{\varepsilon_0} \quad . \text{ From this can be deduced that } \vec{E}_{ext} = \frac{R^3}{3r^2\varepsilon_0}\rho \, \vec{e}_r .$$

With $Q = \rho \dfrac{4}{3}\pi R^3$, we also can write that

$$\vec{E}_{ext} = \frac{Q}{4\pi\varepsilon_0 r^2} \vec{e}_r$$

and the electric field corresponds to that of a charge (Q) placed at O at a distance $r = OP_e$ from P_e. In other words, the electric field generated at P_e can be considered due to a point charge at O.

2. For a point denoted P_i inside a charged sphere, which is such that $OP_i = r < R$, Gauss's theorem gives:

$$\Phi = 4\pi r^2 E_{int} = \sum_i \frac{Q_{int}}{\varepsilon_0} = \frac{\displaystyle\int_{r=0}^{r=OP_i} \int_{\theta=0}^{\theta=\pi} \int_{\varphi=0}^{\varphi=2\pi} \rho d\tau}{\varepsilon_0} = \frac{4}{3}\pi \, OP_i^3 \frac{\rho}{\varepsilon_0} = \frac{4}{3}\pi r^3 \frac{\rho}{\varepsilon_0} ,$$

from which can be determined that

$$\vec{E}_{int} = \frac{\rho \, r}{3\varepsilon_0} \vec{e}_r .$$

3. The plot of $E = f(r)$ is given in Figure 3.

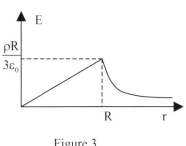

Figure 3

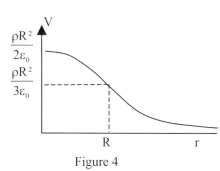

Figure 4

4. With the electric field being radial, it is possible to state that: $dV = -\int \vec{E}.\vec{dr} = -\int E.dr$, and that outside the sphere, we have:

$V_{ext} = -\int E_{ext} dr = -\dfrac{\rho R^3}{3\varepsilon_0} \int \dfrac{dr}{r^2} = \dfrac{\rho R^3}{3\varepsilon_0 r} + C$. The integral constant (C) can be

determined using the limiting condition $V_{ext}\,(r \rightarrow \infty) \rightarrow 0$, from which $C = 0$ and

$$V_{ext} = \dfrac{\rho R^3}{3\varepsilon_0 r}.$$

Inside, $V_{int} = -\int E_{int} dr = -\dfrac{\rho}{3\varepsilon_0} \int r\,dr = -\dfrac{\rho r^2}{6\varepsilon_0} + D$. The constant denoted D is

determined using the continuous potential condition for $r = R$, so that

$V_{ext}\,(R) = V_{int}\,(R)$ and $\dfrac{\rho R^3}{3\varepsilon_0 R} = -\dfrac{\rho R^2}{6\varepsilon_0} + D$, from which $D = \dfrac{\rho R^2}{2\varepsilon_0}$. Finally,

$$V_{int} = \dfrac{\rho R^2}{2\varepsilon_0}\left[1 - \dfrac{r^2}{3R^2}\right].$$

The plot of $V = g(r)$ is shown in Figure 4.

5. Poisson's equation, written in the form $\Delta V + \dfrac{\rho}{\varepsilon_0} = 0$, is written in spherical

coordinates for a function, which is independent of the variables θ and φ; thus

$\dfrac{1}{r}\dfrac{\partial^2(rV)}{\partial r^2} = -\dfrac{\rho}{\varepsilon_0}$, so that:

- when $r \geq R$, with $\rho = 0$: $-\dfrac{1}{r}\dfrac{\partial^2 (rV_{ext})}{\partial r^2} = 0 \Rightarrow rV_{ext} = Ar + B$, so that

$V_{ext} = A_1 + \dfrac{A_2}{r}$; the condition $V_{ext}(r \rightarrow \infty) \rightarrow 0$ gives $A_1 = 0$, from which

$V_{ext} = \dfrac{A_2}{r}$;

- when $r \leq R$, with "$\rho = \rho \neq 0$", $-\dfrac{1}{r}\dfrac{\partial^2(rV_{int})}{\partial r^2} = -\dfrac{\rho}{\varepsilon_0} \Rightarrow \dfrac{\partial^2(rV_{int})}{\partial r^2} = -\dfrac{\rho r}{\varepsilon_0}$ so that:

$\dfrac{\partial(rV_{nt})}{\partial r^2} = -\dfrac{\rho r^2}{2\varepsilon_0} + B_1$. The result is that:

$rV_{int} = -\dfrac{\rho r^3}{6\varepsilon_0} + B_1 r + B_2 \Rightarrow V_{int} = -\dfrac{\rho r^2}{6\varepsilon_0} + B_1 + \dfrac{B_2}{r}$.

In physical terms, the former potential retains a finite value and cannot diverge from $r = 0$, from which $B_2 = 0$.

The determination of the two constants denoted A_2 and B_1 necessitates two equations with two unknowns, which can be contained by writing the continuity, for $r = R$, in terms of the potential (V), as in $(V_{ext} = V_{int})_{r=R}$, and in terms of the electric field (E), as in $E = - dV/dr$.

From these can be determined that

$$A_2 = \dfrac{\rho R^3}{6\varepsilon_0} \text{ and } B_1 = \dfrac{\rho R^2}{2\varepsilon_0},$$

from which the same forms as those of V_{ext} and V_{int}.

Chapter 2

Electrostatics of Dielectric Materials

2.1.   Introduction: Dielectrics and Their Polarization

2.1.1.   Definition of a dielectric and the nature of the charges

A dielectric is effectively an insulator, meaning that *a priori* it does not have free
and mobile charges. Nevertheless, different types of charges can be found, as shown
in Figure 2.1. They are:

1. Dipolar charges, which are attached to a molecule that makes up the dielectric;
examples are the dipoles attached to HCl molecules where the center of negative
electronic charge is displaced towards the Cl atom, while the H has an excess
positive charge, thus giving rise to the dipole H^+Cl^-. These charges are inseparable,
tied to the bonded H and Cl, and are called *bound charges*.

2. Charges due to discontinuities such as interfaces between aggregates. Where the
solid dielectrics exhibit defaults, charges can accumulate giving rise to particular
electronic phenomena, such as the Maxwell-Wagner-Sillars effect at low
frequencies.

3. Homocharges, which have the same sign as the electrodes to which they are
adjacent.

4. Heterocharges, which have the opposite sign to the electrode at which they are
near.

5. Space charges, which are charges localised within a region of space.

6. Free charges, in principle, are little or not present in dielectrics. They can appear,
however, when there is a breakdown caused the application of an electric field and a
sudden loss in the ability of the material to insulate. When the current is relatively
weak it is called a leak current and it can be due to a wide range of causes, such as
impurities in the dielectric and so forth.

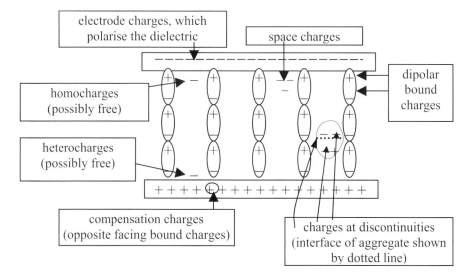

Figure 2.1. *Various types of charges in a dielectric.*

2.1.2.   Characteristics of dipoles

2.1.2.1. Electric potential produced by an electric dipole

If two charges bound to an electric dipole, detailed in Figure 2.2a, are placed at A (-q) and at B (+q) so that $\overrightarrow{AB} = \overrightarrow{dl}$ (in the sense going from the negative to the positive charge), the dipolar moment is by definition

$$\boxed{\overrightarrow{d\mu} = q\overrightarrow{dl}}$$

To calculate the potential (dV) for a point (P) produced by the dipole that has a center at M, the following equation can be directly obtained:

$$dV = \frac{1}{4\pi\varepsilon_0}\left(\frac{q}{r_B} - \frac{q}{r_A}\right) = \frac{q\,(r_A - r_B)}{4\pi\varepsilon_0 r_A r_B} \approx \frac{q\,dl\cos\theta}{4\pi\varepsilon_0 r^2}.$$

Accordingly:

$$\boxed{dV = \frac{1}{4\pi\varepsilon_0}\frac{\vec{r}}{r^3}\overrightarrow{d\mu}}.$$

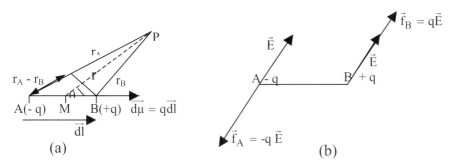

Figure 2.2. *(a) Electric dipole; and (b) force couple.*

2.1.2.2. *Energy of an electric dipole in a uniform electric field ($\vec{E}$)*

It is worth recalling here that an electric charge placed at a point N at an electric potential (V_N) takes on a potential energy (E_p) $E_p = W(q) = q\, V_N$. The energy associated with a dipole placed as detailed in Figure 2.2, with –q at A and +q at B, is therefore $E_p = -q\, V(A) + q\, V(B) = q\, (V_B - V_A)$. In addition, it is possible to write

that: $V_B - V_A = \int_B^A - dV = \int_B^A \vec{E}.\overline{dl}$, so that with $\vec{E}$ being uniform (over a space $\overline{dl}$),

$$V_B - V_A = \vec{E}\int_B^A \overline{dl} = \vec{E}\left[\overline{OA} - \overline{OB}\right] = -\vec{E}.\overline{AB} .$$

Then with $\overline{d\mu} = q\overline{dl} = q\,\overline{AB}$ we obtain:

$E_p = q\,(V_B - V_A) = -q\overline{AB}.\vec{E} = -\overline{d\mu}.\vec{E}$. For a dipole with a moment $\vec{\mu} = q\,\vec{l}$,
then

$$\boxed{E_P = -\vec{\mu}.\vec{E}}$$

2.1.2.3. *Electric couple*

This section refers to Figure 2.2(b). The moment of the couple of forces ($\vec{f}_A, \vec{f}_B = -\vec{f}_A$) is by definition: $\vec{C} = \overline{AB} \times \vec{f}_B = \overline{AB} \times q\vec{E} = q\,\overline{AB} \times \vec{E}$, so that with $\vec{\mu} = q\,\overline{AB}$,

$$\boxed{\vec{C} = \vec{\mu} \times \vec{E}}$$

2.1.3. Dielectric in a condenser

2.1.3.1. Study of a condenser with a vacuum between the armatures

Figure 2.3 shows a condenser that has armatures carrying the charges +Q and –Q: The potential gradient is therefore only between the armatures that have a known surface area (S), outside of which the electric field can be considered zero. Gauss's theorem, applied across the surface Σ, gives rise to

$$\begin{cases} \Phi = \oiint_{\Sigma} \vec{E}.\overrightarrow{dS} = \iint_S \vec{E}.\overrightarrow{dS} = E\,S \text{ (assuming that } \vec{E} \text{ and } \overrightarrow{dS} \text{ are collinear)} \\ = \dfrac{Q}{\varepsilon_0} = \dfrac{\sigma\,S}{\varepsilon_0} \qquad \text{where } \sigma \text{ is the superficial charge density at the electrodes,} \end{cases}$$

from which can be determined : $\boxed{E = \dfrac{\sigma}{\varepsilon_0}}$.

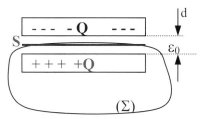

Figure 2.3. *Vacuum-based condenser.*

2.1.3.2. The presence of a depolarizing field in the dielectric

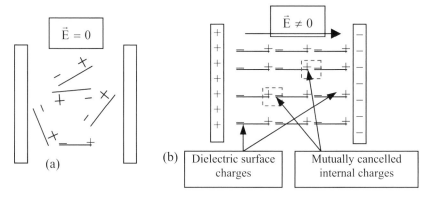

Figure 2.4. *Dipole orientation: (a) without field and (b) with an applied field.*

If a dielectric is inserted into an unpolarized condenser (zero field between the armatures), as schematized in Figure 2.4a, then the dipoles associated with the molecules of the dielectric are randomly distributed. Statistically, the opposite charges cancel each other out and the material is electrically neutral overall.

If the dielectric is inserted into an electric field, as shown in Figure 2.4b, then the positive charges are pulled in on direction and the negative charges in the other. The dipoles $d\vec{\mu}$ are aligned in parallel in such a way that their potential energy (E_p) is minimized, as in $E_P = -d\vec{\mu}.\vec{E}$. As the bonded charges cannot move, the result is a chain of dipoles in the volume; however, at the surface of the dielectric, charges opposite to that of the armatures appear.

If an applied external field ($\vec{E}_{ext}$) is in the sense indicated in Figure 2.4c, given the presence of surface dielectric charges, then an opposing field appears called the depolarizing field ($\vec{E}_d$). The result is an effective field, which can more precisely be termed an external effective field, and is such that:

$$\vec{E}_a = \vec{E}_{ext} + \vec{E}_d .$$

As $\vec{E}_d$ is antiparallel to $\vec{E}_{ext}$, the field $\vec{E}_a$ is less than the external applied field $\vec{E}_{ext}$.

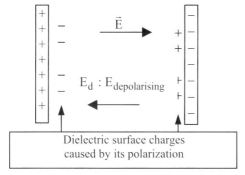

Figure 2.4(c). *The action of a depolarizing field.*

2.1.4. *The polarization vector*

In a dielectric sits a small element with volume $d\tau$, length dl, and cross section dS, such that it is parallel to the surface of the armatures, as shown in Figure 2.5. If the volume of this element is equivalent to that of a dipole, with respect to the armatures, then it is possible to consider that the elementary charges (q) of the dipoles are such that $q = \sigma_p dS$, where σ_p represents the superficial charge density compared with the armature and therefore the superficial polarization charge density.

The moment of dipoles compared to the armatures can be written as $d\vec{\mu} = q\,\vec{dl} = \sigma_P\,dS\,\vec{dl}$, and as $d\mu = \sigma_P\,d\tau$, we can state that $\sigma_P = \dfrac{d\mu}{d\tau}$.

The magnitude of the dielectric polarization is used to define the polarization vector ($\vec{P}$), which is such that $\left|\vec{P}\right| = \sigma_P = \dfrac{d\mu}{d\tau}$, and has as apparent modulus the dipolar moment of the unit volume. In terms of vectors

$$\boxed{\vec{P} = \dfrac{\overrightarrow{d\mu}}{d\tau}}.$$

As $d\vec{\mu}\,//\,\vec{E}_{ext}$ or $\vec{E}_a$, we also have $\vec{P}\,//\,\vec{E}_{ext}$ or $\vec{E}_a$. Accordingly, when we speak of the electric field ($\vec{E}$) without any further precision, it is the effective field $\vec{E}_a$ which is the subject of discussion.

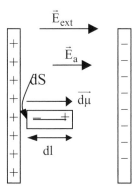

Figure 2.5. *Dipolar polarization and orientation.*

2.2.   Polarization Equivalent Charges

2.2.1.   *Calculation for charges equivalent to the polarization*

2.2.1.1. *The potential generated at P by a finite volume of polarized dielectric*
The above figures indicate that a polarized dielectric can be represented by a vacuum, of permitivity ε_0, in which is placed dipoles orientated by an applied field. So, to calculate the potential generated at a point (P) by a volume (V) of a dielectric, as shown in Figure 2.6, it suffices to calculate the potential *generated in a vacuum* by the orientated dipoles ($d\vec{\mu}$) occupying a fraction ($d\tau$) of V and then to integrate over the total volume V.

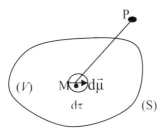

Figure 2.6. *Calculation of charges equivalent to a polarization.*

Thus, the element of volume $d\tau$ taken from around a point M forms at P a potential

$$dV = \frac{1}{4\pi\varepsilon_0}\frac{\vec{r}}{r^3}\overrightarrow{d\mu} \text{ (result obtained in Section 2.1.2.1) which also can be written as:}$$

$$dV = \frac{1}{4\pi\varepsilon_0}\overrightarrow{grad}_M\left(\frac{1}{r}\right)\overrightarrow{d\mu} = \frac{1}{4\pi\varepsilon_0}\overrightarrow{grad}_M\left(\frac{1}{r}\right)\vec{P}d\tau .$$

The potential V(P) formed at P by the total volume V is therefore:

$$V(P) = \frac{1}{4\pi\varepsilon_0}\iiint_V\overrightarrow{grad}_M\left(\frac{1}{r}\right).\vec{P}d\tau .$$

Now using the notable equation $\text{div}\,(a \cdot \vec{A}) = a\,\text{div}\,\vec{A} + \vec{A}\,.\,\overrightarrow{grad}\,a$, for which here $a = 1/r$ and $\vec{A} = \vec{P}$, we arrive at:

$$\overrightarrow{grad}_M\left(\frac{1}{r}\right).\vec{P} = \text{div}_M\frac{\vec{P}}{r} - \frac{1}{r}\,\text{div}_M\vec{P} .$$

Substituting this into V(P), we have

$$V(P) = \frac{1}{4\pi\varepsilon_0}\left[\iiint_V\text{div}_M\frac{\vec{P}}{r}d\tau - \iiint_V\frac{1}{r}\text{div}_M\vec{P}\,d\tau\right] , \text{ and by using the Ostrogradsky}$$

equation on the first triple integral, we obtain:

$$V(P) = \frac{1}{4\pi\varepsilon_0}\left[\oiint_S\frac{\vec{P}.\overrightarrow{dS}}{r} - \iiint_V\frac{1}{r}\text{div}_M\vec{P}\,d\tau\right] .$$

If we make $\boxed{\sigma_P = \vec{P}.\vec{n} = P_N \text{ and } \rho_P = -\text{div}_M\vec{P}}$, we finally arrive at:

$$V(P) = \frac{1}{4\pi\varepsilon_0} \left[\oiint_S \frac{\sigma_P}{r} dS + \iiint_V \frac{\rho_P}{r} d\tau \right].$$

2.2.1.2. Polarization equivalent charges

We can conclude by stating that a volume V of polarized dielectric is equal to the sum in V of a volume distribution of charge density $\rho_P = -\text{div}_M \vec{P}$ along with a surface charge density $\sigma_P = \vec{P}.\vec{n} = P_N$ over the total surface (S) of the dielectric, with all charges distributed in a vacuum.

The volume (ρ_P) and surface (σ_P) densities thus appear as charges *equivalent to the polarization, with the condition that they are in a vacuum!* There is nothing imaginary about them and are in complete contrast to the term "imaginary charges" which has been used to describe them. In other words, a polarized material which takes up a space (*E*) can be represented as taking up the equivalent vacuum space (*E*) in which are distributed volume and surface charge densities ρ_P and σ_P, respectively, which represent the polarization effect of the dipoles making up the material, evidently from an electrical point of view.

If additional (real) charges with a volume density ρ_ℓ and surface density σ_ℓ are added to the dielectric, the potential at a point P and that generated by the polarized dielectric can be termed in the generality:

$$V(P) = \frac{1}{4\pi\varepsilon_0} \left[\oiint_S \frac{\sigma_\ell + \sigma_P}{r} dS + \iiint_V \frac{\rho_\ell + \rho_P}{r} d\tau \right].$$

2.2.2.   Physical characteristics of polarization and polarization charge distribution

Up to now, the equivalence of polarization and polarization charge distributions has been essentially mathematical because the determination of ρ_P and σ_P requires particularly long and abstract calculations.

2.2.2.1. Preliminary comments: charge displacement and the corresponding charge polarizations

Comment 1:   *How moving a charge q_i over $\vec{\delta}_i$ is the same as applying (superimposing) a dipole moment $\delta\vec{\mu}_i = q_i \vec{\delta}_i$*

To verify the proposition, a schematic verification can be used to show how the two transformations are equivalent in that going from the same initial state (q_i at O) both arrive at the same final state (q_i at O', which is such that $\overrightarrow{OO'} = \vec{\delta}_i$).

<u>First transformation</u> displacement of q_i by $\vec{\delta}_i$:

O (q_i) $\vec{\delta}_i$ O'
initial state: q_i at O

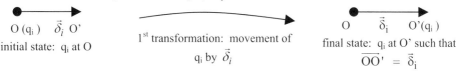

1st transformation: movement of
q_i by $\vec{\delta}_i$

O $\vec{\delta}_i$ O'(q_i)
final state: q_i at O' such that
$\overline{OO'} = \vec{\delta}_i$

<u>Second transformation</u> superposition on starting system of a dipole of moment $\delta\vec{\mu}_i = q_i \vec{\delta}_i$:

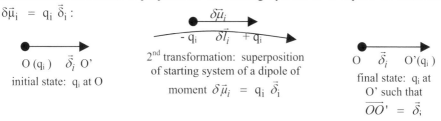

O (q_i) $\vec{\delta}_i$ O'
initial state: q_i at O

$\delta\vec{\mu}_i$
- q_i $\delta\vec{l}_i$ + q_i

2nd transformation: superposition
of starting system of a dipole of
moment $\delta\vec{\mu}_i = q_i \vec{\delta}_i$

O $\vec{\delta}_i$ O'(q_i)
final state: q_i at
O' such that
$\overline{OO'} = \vec{\delta}_i$

Comment 2. Calculation of the polarization associated with a charge displacement
To arrive at a solution to this problem, it suffices to calculate the dipole moment per unit volume which is apparent following the displacement of q_i through $\vec{\delta}_i$. If the number of q_i per unit volume is n_i, the polarization that would come about following the displacement by $\vec{\delta}_i$ of only the q_i charges would be $n_i q_i \vec{\delta}_i$, which is the dipolar moment per unit volume due to the charge displacement.

If in the dielectric there are several types of charge (q_i) such as q_1, q_2, q_v and so on, then the dipolar moment per unit volume that would appear following a movement of all the charges, i.e. q_1 being moved by $\vec{\delta}_1$, q_v by $\vec{\delta}_v$ and so on, is $\sum_i n_i q_i \vec{\delta}_i = \vec{P}$, which is the system polarization and is the dipole moment per unit volume.

2.2.2.2. The corollary of charge movement: Number of charges entering and leaving the dielectric volume following polarization and the volume charge density equal to the polarization

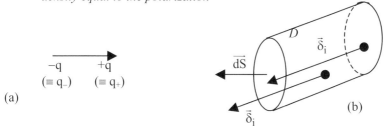

−q   +q
($\equiv q_-$)   ($\equiv q_+$)

(a)

$\overline{dS}$ $\vec{\delta}_i$ D

$\vec{\delta}_i$

(b)

Figure 2.7. *(a) Characteristics of the electric charges and (b) their displacement.*

A neutral zone (D) in a dielectric contains various type q_i charges, where in reality each charge ($q_- = -q$, and $q_+ = + q$, with identical densities n_- and n_+) meets at each dipole, as shown in Figure 2.7, with n_- and n_+ so that D is definitely neutral.

In the absence of an electric field and also therefore a polarization, q_i are not moved ($\vec{\delta}_i = 0$) and the polarization $\vec{P} = \sum_i n_i \, q_i \, \vec{\delta}_i$ is zero.

Under the effect of an applied electric field, q_i are moved (by $\vec{\delta}_i \neq 0$), and the polarization vector becomes $\vec{P} = \sum_i n_i \, q_i \, \vec{\delta}_i = (\sum_{i=-,\,+} n_i \, q_i \, \vec{\delta}_i)$ as the summation is performed over all charges $q_- (\equiv -q)$ and $q_+ (\equiv + q)$.

The algebraic flux for q_i traversing a part ($\overrightarrow{dS}$) of the surface following the algebraic displacements ($\vec{\delta}_i$) indicated in Figure 2.7(b) is equal to the number of carriers contained in a cylinder with base $\overrightarrow{dS}$ and length $\vec{\delta}_i$, and therefore a volume $= \vec{\delta}_i . \overrightarrow{dS}$. The number of carriers therefore is equal to $dN_i = n_i \cdot \vec{\delta}_i \cdot \overrightarrow{dS}$, and the corresponding amount of charge is $dQ_i = q_i \cdot n_i \cdot \vec{\delta}_i \cdot \overrightarrow{dS}$.

With respect to the various charges q_i, which include both q_- and q_+, the total algebraic charge traversing the surface element $\overrightarrow{dS}$ thus is $dQ = \sum_{i=-,\,+} n_i \, q_i \, \vec{\delta}_i \cdot \overrightarrow{dS} = \vec{P} \cdot \overrightarrow{dS}$. The algebraic charge Q crossing the closed surface S therefore is $Q = \oiint dQ = \oiint_S \vec{P}.\overrightarrow{dS}$ and the zone D, which initially was neutral, now contains after the polarization a total charge (Q_P) opposite to Q. In order to calculate the number of charges gained by entering into D, the reentrant surface $-\overrightarrow{dS}$ can be used.

The final result is that $Q_P = - \oiint_S \vec{P}.\overrightarrow{dS}$, so that $Q_P = - \iiint_D \text{div} \, \vec{P} \, d\tau$.

We thus have shown that on a macroscopic scale, polarization is equal to a volume density of charge $\rho_P = - \text{div} \, \vec{P}$, which is indeed the same result as that obtained from the calculation for a potential generated by a polarized dielectric.

2.2.2.3. Surface charges
Following a polarization there is an accumulation of charges at a surface, as detailed in Section 2.1.4 and described in Figure 2.8.

On calculating $\vec{P} \cdot \vec{n} = \dfrac{d\vec{\mu}}{d\tau} \cdot \vec{n}$ with $d\vec{\mu} = q \cdot d\vec{l} = \sigma_P \, dS \, d\vec{l}$, we find $\rho_P = \vec{P} \cdot \vec{n}$. Once again, this time for surface polarization charges, the result is equivalent to that for the charge generated by a polarized dielectric.

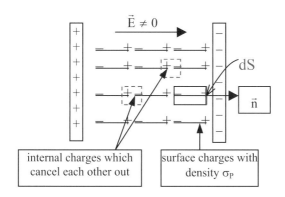

Figure 2.8. *Surface polarization charges.*

2.2.2.4. Conclusion

The physical evidence for polarization charges shows that the densities σ_P and ρ_P are not associated with imaginary charges or simple mathematical equivalent, but are the result of localized excesses in bound charges, real excesses caused by polarization.

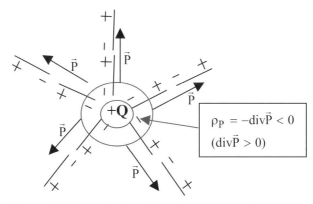

Figure 2.9. *Schematization of the divergence operator.*

As shown in Figure 2.9, when a dielectric is not deliberately charged in its volume (with real charges), the resultant for bound charges over the volume is zero and $\rho_P = 0$. So that ρ_P becomes nonzero, the divergence polarization vector also is nonzero (as $\rho_P = - \text{div } \vec{P}$). In the presence of real charges in the volume, the charge +Q indicated in Figure 2.9, which gives a physical representation of the divergence operator, localized polarization charges appear, so that $\rho_P = - \text{div } \vec{P} \neq 0$.

If the material is neutral prior to polarization it must remain neutral following polarization as it is only the existing, bound charges which are displaced. This property can be verified by considering a volume (V) of dielectric delimited by a surface (S). Here, the total of polarized charges is given by:

$$\iiint_V \rho_P \; dV + \oiint_S \sigma_P \; dS = \iiint_V - \text{div } \vec{P} \; dV + \oiint_S \vec{P}.\vec{n} \; dS = -\oiint_S \vec{P}. \; \overrightarrow{dS} + \oiint_S \vec{P}. \; \overrightarrow{dS} = 0$$

The equation shows how the algebraic sum of the charges equals zero. We also can see in Figure 2.8, where $\rho_P = 0$ and the surface polarization negative charges of density σ_P facing the positive electrode are exactly compensated for by the surface polarization, positive charges facing the negative electrode.

2.2.3.   Important comment: Under dynamic regimes the polarization charges are the origin of polarization currents

Even though this section deals with static states, it is worth making the occasional *sortie* into dynamic systems.

In effect, for the results obtained from static systems to be acceptable for their dynamic counterparts, with polarization values in particular, the variations in magnitude should be negligible with respect to a given macroscopic domain. A reasonable estimate for this is to assume, for example, that a polarization is constant over a distance (d) of around 10 nm (the "atomic" dimensions being of an order of less than 1 nm). So if the length of a polarization signal wave (λ) is such that $\lambda > 10d$, we can assume that for a "macroscopic" domain of size d the signal remains more or less constant and therefore definable. Figure 2.10 gives an example of this where the frequency (v) under consideration is $v < \dfrac{c}{100 \text{ nm}} \approx 3 \times 10^{15}$ Hz .

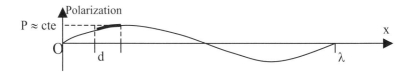

Figure 2.10. *Distance d over which P remains more or less constant.*

Above a frequency of this order, it is necessary to take into account the variation with time for a polarization in the macroscopic domain. If the bound charge carriers (q_i) are vibrated with an average velocity $\vec{v}_i = \dfrac{d}{dt}(\vec{\delta}_i)$, there is a corresponding current termed a "polarization current", which has a density $\vec{j}_i = n_i\,q_i\,\vec{v}_i$. The average resultant current density is given by

$$\vec{J}_P = \sum_i n_i\,q_i\,\vec{v}_i = \sum_i n_i\,q_i\,\dfrac{d\vec{\delta}_i}{dt}, \text{ so that with } \vec{P} = \sum_i n_i\,q_i\,\vec{\delta}_i,$$

$$\boxed{\vec{J}_P = \dfrac{\partial \vec{P}}{\partial T}}$$

This current density is of a macroscopic scale and represents an average over the microscopic currents associated with the slight displacements of bound charges. The current therefore is not imaginary and gives rise to the same magnetic effects as conduction currents. Its specificity is that it cannot leave the dielectric material since it is tied to bound charges and therefore cannot be measured or used in an external circuit.

2.3. Vectors for an Electric field ($\vec{E}$) and an Electric Displacement ($\vec{D}$) : Characteristics at Interfaces

Section 2.2.1.1 showed that the potential formed at a point (P) in space by a polarized dielectric is given by an equation that brings in charge densities. The electric field at P can be determined by the often used equation $\vec{E} = -\,\overline{grad}_P\,V$. The inconvenience though is that it is assumed that the polarization vector is known.

In a more classic method from the Anglo-Saxon school of thought, it is interesting to reason in terms of an electric field in a vacuum on which is imposed a limiting law in order to take into account the characteristics of bound charges associated with a dielectric medium.

2.3.1 Vectors for an electric field ($\vec{E}$) and an electric displacement ($\vec{D}$): electric potentials

2.3.1.1 Coulomb's law applied to dielectrics

In general terms, if q and q' are two electric charges placed in a linear, homogeneous and isotropic (l.h.i.) dielectric at two points M and P, which are such that $\vec{r} = \overrightarrow{MP}$, the force that should appear between the two charges is given by Coulomb's law, i.e.:

$$\vec{F} = \dfrac{q\,q'}{4\,\pi\,\varepsilon_0\,\varepsilon_r\,r^3}\,\vec{r}\,.$$

where ε_r is the relative dielectric permittivity characteristic to each dielectric material. The quantity given by $\varepsilon = \varepsilon_0\varepsilon_r$ is the absolute dielectric permittivity and in the electrostatic system unity (e.s.u.), $\varepsilon_0 = 1$, so that $\varepsilon = \varepsilon_r$ and is the magnitude generally called the dielectric constant.

2.3.1.2 Electric field

In simple terms an electric field can be defined over all points P, each defined in respect to M, where there is a charge q by $\vec{r} = \overrightarrow{MP}$, just as if there were a force exerted on the charge unit (unit here is q' = 1), which is assumed here to be very small, such that its dimensions tend toward zero in order to limit the singularity at P. This gives:

$$\vec{E}(P) = \frac{q}{4\pi\varepsilon_0\varepsilon_r}\frac{\vec{r}}{r^3}.$$

For a group of charges (q_i), we have $\displaystyle \vec{E}(P) = \frac{1}{4\pi\varepsilon_0\varepsilon_r}\sum_i q_i\frac{\vec{r_i}}{r_i^3}$.

2.3.1.3. Electrical potential

Just as for the system in a vacuum, detailed in Section 1.2.2.1, the potential (V) is always such that $\vec{E} = -\overrightarrow{grad}V$. Therefore, for a dielectric,

$$V = \frac{q}{4\pi\varepsilon_0\varepsilon_r\, r}.$$

2.3.1.4. Electric displacement vector

By definition, the vector, which is also termed electrical induction, is given by:

$$\vec{D} = \varepsilon\vec{E} = \varepsilon_0\varepsilon_r\vec{E}$$

and for a linear homogeneous and isotropic (lhi) is dielectric, as detailed in Section 2.4.3.1.

2.3.2. Gauss's theorem

The calculation is identical as that carried out for a vacuum, with the exception that here the absolute permittivity of the medium $(\varepsilon = \varepsilon_0\varepsilon_r)$ is substituted in place of that of a vacuum (ε_0). We therefore have:

1. for a discrete distribution of charges, where q_i represents the internal real charges at the surface S:

$$\Phi = \oiint_S \vec{E}.\overrightarrow{dS} = \frac{\sum_i q_i}{\varepsilon_0\varepsilon_r} ;$$

2. for a continuous distribution of charges with a volume density (ρ_ℓ) of deliberately added free charges

$$\Phi = \oiint_S \vec{E} \cdot \overrightarrow{dS} = \frac{1}{\varepsilon_0 \varepsilon_r} \iiint_V \rho_\ell \, d\tau = \frac{1}{\varepsilon} \iiint_V \rho_\ell d\tau$$

Use of Ostrogradsky's theory, $\Phi = \oiint_S \vec{E} \cdot \overrightarrow{dS} = \iiint \operatorname{div} \vec{E} \, d\tau$, gives rise to a localized Gauss's theorem:

$$\operatorname{div} \vec{E} = \frac{\rho_\ell}{\varepsilon_0 \varepsilon_r} = \frac{\rho_\ell}{\varepsilon}$$

With $\vec{D} = \varepsilon \vec{E}$, we also have:

$$\operatorname{div} \vec{D} = \rho_\ell$$

For its part, the integral form of Gauss's theorem with respect to the flux from $\vec{D}$ is written:

1. for a discrete distribution of charges, $\boxed{\Phi_D = \oiint_S \vec{D}.\overrightarrow{dS} = \sum_i q_i}$ and

2. for a continuous distribution $\boxed{\Phi_D = \oiint_S \vec{D}.\overrightarrow{dS} = \iiint \rho_\ell \, d\tau}$.

2.3.3.   *Conditions under which $\vec{E}$ and $\vec{D}$ move between two dielectric materials*

2.3.3.1. The continuity of potential

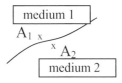

Figure 2.11. *Potential at an interface.*

Figure 2.11 details two points (A_1 and A_2) each situated in, respectively, medium 1 and medium 2 and both close to the interface. If these two points are sufficiently close to one another so that $\vec{E}$ is a constant between them, then

$$V_{A1} - V_{A2} = \int_{A_1}^{A_2} \vec{E}.\vec{dl} \approx \vec{E}.\,\overrightarrow{A_1 A_2}$$

At the immediate neighborhood of the interface, A_1 tends toward A_2 so that $\overrightarrow{A_1 A_2} \approx 0$, and therefore $V_{A_1} = V_{A_2}$ so that the potential can be known at the interface.

2.3.3.2. Continuity of the tangential component of $\vec{E}$

For two trajectories, $A_1 B_1$ (in medium 1) and $A_2 B_2$ (in medium 2), which are equal neighbors around the interface surface (S) described in Figure 2.12a, given the continuity of the potential at the interface, we have: $V_{A_1} - V_{B_1} = V_{A_2} - V_{B_2}$ so that $E_{1t} A_1 B_1 = E_{2t} A_2 B_2$. From also knowing that $A_1 B_1 = A_2 B_2$, we can determine that:

$$\boxed{E_{1t} = E_{2t}}\ .$$

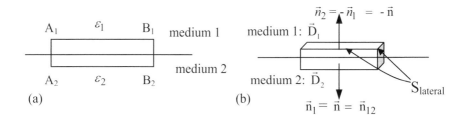

Figure 2.12. *Setup to study continuity of (a) E_t and (b) D_n.*

2.3.3.3. Continuity of the normal component of $\vec{D}$

2.3.3.3.1. No real charges at the interface

The construction of a small parallelepipedal element around the interface with infinitely small lateral sides makes it possible to study the properties of the displacement vector $\vec{D}$ in the neighborhood of the interface where $S_{lateral} \to 0$. For this system shown in Figure 2.12, we have

$$\Phi_D = \oiint_{S_{parallelepiped}} \vec{D}.\,\overrightarrow{dS} = \iiint \rho_\ell \,.\, d\tau\,.$$

As there are no real charges inside the parallelepiped, we have:

$$\Phi_D = \oiint_{S_{parallelepiped}} \vec{D} \cdot \vec{dS} = 0$$

$$= \iint_{S_{superior}} \vec{D} \cdot \vec{dS} + \iint_{S_{inferior}} \vec{D} \cdot \vec{dS} + \iint_{S\,lateral} \vec{D} \cdot \vec{dS} .$$

With the upper and lower faces being opposite one another and identical, we have $S_{superior} = S_{inferior} = S_{si}$. In addition, $S_{lateral} \approx 0$.

By making $\vec{n} = \vec{n}_1 = -\vec{n}_2$, and by stating that the displacement vectors are such that $\vec{D}_1$ is in medium 1, and $\vec{D}_2$ is in medium 2, we then have:

$$\Phi_D = \iint_{S_{superior}} \vec{D} \cdot \vec{dS} + \iint_{S_{inferior}} \vec{D} \cdot \vec{dS}$$

$$= \iint_{S_{si}} (\vec{D}_2 . \vec{n} dS + \vec{D}_1 . (-\vec{n}) dS) = \iint_{S_{si}} (\vec{D}_2 . \vec{n} - \vec{D}_1 . \vec{n}) dS = 0$$

where $\overrightarrow{n_1}$ is normal to the exterior of medium 1 and $\overrightarrow{n_2}$ is normal exterior to medium 2. From this can be determined that $\vec{D}_2 . \vec{n} - \vec{D}_1 . \vec{n} = 0$, and thus

$$\boxed{D_{1n} = D_{2n}} .$$

2.3.3.3.2. Polarization charges and no real charges at the interface

There are not always real charges in the Gaussian volume (the parallelepiped), which means that $\Phi_D = 0$ and that the preceding result, $D_{1n} = D_{2n}$, again is valid.

2.3.3.3.3. Distribution (σ_ℓ) of real charges at the interface (surface layer of real

charges)

For Gauss's theorem applied in Figure 2.12, there now is a parallelepiped that contains a total interior charge $\iint_{S_{interface}} \sigma_\ell dS$, which can be written as:

$$\Phi_D = \iint_{S_{si}} (\vec{D}_2 . \vec{n}_1 dS + \vec{D}_1 . \vec{n}_2 dS) = \iint_{S_{interface}} \sigma_\ell dS .$$

Where $S_{interface} = S_{si}$, and with the same notations as above,

$$\iint_{S_{si}} (\vec{D}_2 . \vec{n} - \vec{D}_1 . \vec{n}) dS = \iint_{S_{si}} \sigma_\ell dS, \text{ which is to say}$$

$$\boxed{D_{2n} - D_{1n} = \sigma_\ell} .$$

2.3.4. Refraction of field or induction lines

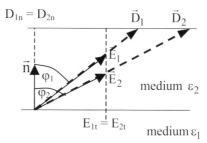

Figure 2.13. *Field and induction lines refracted.*

It is assumed here that there are no real charges at the interface, so that we can go on to write the two equations of continuity (Figure 2.13):

$$E_{1t} = E_{2t}, \text{ so that } E_1 \sin \varphi_1 = E_2 \sin \varphi_2 \quad (1)$$

$$D_{1n} = D_{2n}, \text{ so that } D_1 \cos \varphi_1 = D_2 \cos \varphi_2 \quad (2)$$

Dividing Eq. (1) by Eq. (2) we obtain $\dfrac{E_1}{D_1} \tan \varphi_1 = \dfrac{E_2}{D_2} \tan \varphi_2$, which also gives:

$$\boxed{\frac{\tan \varphi_1}{\tan \varphi_2} = \frac{\varepsilon_1}{\varepsilon_2}},$$

and with $\varepsilon_1 < \varepsilon_2$, we find also that $\varphi_1 < \varphi_2$.

2.4. Relations between Displacement and Polarization Vectors

2.4.1. Coulomb's theorem

The theorem concerns the conductor-dielectric interface. A condenser with an armature of a known surface (S) carries a total charge (Q_T) so that $Q_T = \sigma_T S$ (Figure 2.14).

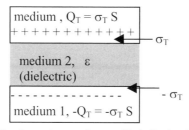

Figure 2.14. *Configuration used to establish Coulomb's theorem.*

With the condenser in equilibrium, on the inside of its metallic electrodes, which as such are equipotential, we have:

$$\vec{E}_1 = 0 \quad \text{and} \quad \vec{D}_1 = \varepsilon_{metal}\overline{E_1} = 0 \,.$$

At the interface between the two materials, we have $D_{2n} - D_{1n} = \sigma_T$, so that here, with $\vec{D}_1 = 0$, $D_{2n} = \sigma_T$ which also can be written as $E_{2n} = \dfrac{\sigma_T}{\varepsilon}$.

The continuity of the tangential component of the electric field allows us to see that $E_{2t} = 0$ (as $E_{1t} = 0$), so that in the dielectric the field is normal to the armature of the condenser. The field that acts between the armatures (effective field) therefore is equal to

$$E_a = \frac{\sigma_T}{\varepsilon} \,. \quad (3)$$

2.4.2. Representation of the dielectric-armature system and $\vec{P} = (\varepsilon - \varepsilon_0)\vec{E}_a$

As previously mentioned in this text, a material with a dielectric permittivity (ε) can also be seen (on a microscopic scale) as a vacuum in which dipolar charges sit (attached to atoms making up the material). Therefore, regarding the armatures, the surface density of dipolar charges ($\mp\sigma_P$), detailed in Section 2.1.4, annihilate an equivalent number of charges (with an opposing sign $\pm\sigma_P$) carried by the armature surfaces, as shown in Figure 2.15a. Charges belonging to the upper armature not canceled out by the dipolar charges thus have a density σ_0 such that

$$\sigma_T = \sigma_P + \sigma_0 \quad (4).$$

The resulting problem with respect to the charges shown in Figure 2.15b is that of a condenser with arms carrying charges of density $\pm\sigma_0$, while there is a vacuum with a permittivity ε_0 separating the armatures. The effective electric field between the armatures therefore is:

$$E_a = \frac{\sigma_0}{\varepsilon_0} \quad (5).$$

From Eqs. (3) and (4), we have $\sigma_T = \sigma_P + \sigma_0 = \varepsilon E_a$, as from Eq. (5), $\sigma_0 = \varepsilon_0 E_a$, we also find that

$$\sigma_P = \left|\vec{P}\right| = \sigma_T - \sigma_0 = \varepsilon E_a - \varepsilon_0 E_a = (\varepsilon - \varepsilon_0)E_a \,.$$

In terms of vectors and for $\vec{P} \,// \, \vec{E}_a$ (detailed in Section 2.1.4) we arrive at

$$\vec{P} = (\varepsilon - \varepsilon_0)\vec{E}_a \,.$$

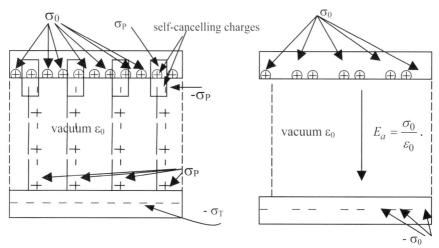

Figure 2.15. *Microscopic analysis of charges in (a) a polarized dielectric and (b) a vacuum.*

2.4.3. Linear, homogeneous, and isotropic dielectrics and the relation $\vec{D} = \varepsilon_0 \vec{E}_a + \vec{P}$

2.4.3.1. Lhi dielectrics

A lhi dielectric is one that gives a response to either a displacement ($\vec{D}$) or polarization ($\vec{P}$) vector in which the excitation of an effective electric field ($\vec{E}_a$) can be described by a relation:

1. $\vec{D} = \varepsilon\vec{E}_a$ where ε is a real number in a linear system and the dielectric can be called perfect, which is to say it has no dielectric losses associated with leak currents through the material (as detailed later on, when there are losses, ε becomes a complex number);

2. which is true at any point in the dielectric, *i.e.* it is homogeneous; and

3. which is verifiable in all directions throughout the dielectric (isotropic).

This restates the working hypothesis set out in Section 2.3.1.4 with the notation $\vec{E} \equiv \vec{E}_a$.

The relation between $\vec{P}$ and $\vec{E}_a$ is $\vec{P} = (\varepsilon - \varepsilon_0)\vec{E}_a$, and this can be rewritten as:

$\vec{P} = (\varepsilon - \varepsilon_0)\vec{E}_a = \varepsilon_0(\varepsilon_r - 1)\vec{E}_a$, which makes:

$$\boxed{\chi_d = (\varepsilon_r - 1)}$$

where χ_d is the dielectric susceptibility, and leads to:

$$\boxed{\vec{P} = \varepsilon_0 \chi_d \vec{E}_a}.$$

2.4.3.2. The relation $\vec{D} = \varepsilon_0 \vec{E}_a + \vec{P}$

Starting from $\vec{P} = (\varepsilon - \varepsilon_0)\vec{E}_a$, we can write $\vec{P} = \varepsilon\vec{E}_a - \varepsilon_0\vec{E}_a = \vec{D} - \varepsilon_0\vec{E}_a$, so that:

$$\boxed{\vec{D} = \varepsilon_0\vec{E}_a + \vec{P}}.$$

2.4.4. Comments

Comment 1. As indicated in Section 2.1.4 and recalled in Section 2.4.3.1, when an electric field ($\vec{E}$) is denoted without any further precision, it is the effective field which is under discussion.

In addition, we have $\vec{D} = \varepsilon\vec{E}$, $\vec{P} = \varepsilon_0\chi_d\vec{E}$, and $\vec{D} = \varepsilon_0\vec{E} + \vec{P}$.

Comment 2. The latter relationship ($\vec{D} = \varepsilon_0\vec{E} + \vec{P}$) shows how the electric induction comes from the polarization ($\vec{P}$) of the dielectric, on which is superimposed an effect ($\varepsilon_0\vec{E}$) of the corresponding space vacuum, which would mean that a vacuum can be seen as a dielectric with zero polarization.

Comment 3. Poisson's equation and dielectrics

On substituting $\vec{E} = -\overrightarrow{\text{grad}}\, V$ into $\text{div}\vec{E} = \dfrac{\rho_\ell}{\varepsilon}$, we directly have $\Delta V + \dfrac{\rho_\ell}{\varepsilon} = 0$. By replacing ε_0 by ε, we have the same relation for a vacuum.

Comment 4. Gauss's equation and dielectric media without real volume charges where $\rho_\ell = 0$

Gauss's theorem, which was written for a dielectric of permittivity ε containing a distribution of real charges (ρ_ℓ) in the form

$$\Phi = \oiint_S \vec{E} \cdot \overrightarrow{dS} = \iiint \text{div}\,\vec{E}\, d\tau = \iiint \frac{\rho_\ell}{\varepsilon}\, d\tau.$$ This can be rewritten to take into account the equivalence of a dielectric to a vacuum in which the volume densities of real charges (ρ_ℓ) with a polarization ρ_P are found distributed so that

$$\Phi = \oiint_S \vec{E} \cdot \overrightarrow{dS} = \iiint \text{div}\,\vec{E}\, d\tau = \iiint \frac{\rho_\ell + \rho_P}{\varepsilon_0}\, d\tau = \iiint \frac{\rho_\ell - \text{div}\,\vec{P}}{\varepsilon_0}\, d\tau.$$

The localized form can be written as

$$\mathrm{div}\,\vec{E} = \frac{\rho_\ell - \mathrm{div}\,\vec{P}}{\varepsilon_0}$$.

However, if the dielectric does not contain real polarization charges, then:

1. $\mathrm{div}\,\vec{E} = \dfrac{\rho_\ell}{\varepsilon} = 0$ (if $\rho_\ell = 0$)

2. and $\mathrm{div}\,\vec{E} = \dfrac{\rho_\ell - \mathrm{div}\,\vec{P}}{\varepsilon_0} = \dfrac{-\mathrm{div}\,\vec{P}}{\varepsilon_0}$ (if $\rho_\ell = 0$)

$$\Rightarrow$$

$$\mathrm{div}\,\vec{P} = 0 \quad \text{if } \rho_\ell = 0 .$$

As $\rho_P = -\mathrm{div}\vec{P}$, therefore $\rho_P = 0$ when $\rho_\ell = 0$, and we return to the result already given in Section 2.2.2.4. The volume charges equivalent to the polarization are zero when the dielectric contains no real volume charges, and so polarization charges exist only at the surface of the dielectric ($\sigma_n = P_n$).

2.4.5.   Linear, inhomogeneous, and nonisotropic dielectrics

For this particular case, within a Cartesian framework, the Dx, Dy, and Dz components of $\vec{D}$, and the Ex, Ey, and Ez components of $\vec{E}$ are all related by the relation:

$$\begin{bmatrix} D_x \\ D_y \\ D_z \end{bmatrix} = \begin{pmatrix} \varepsilon_{xx} & \varepsilon_{xy} & \varepsilon_{xz} \\ \varepsilon_{yx} & \varepsilon_{yy} & \varepsilon_{yz} \\ \varepsilon_{zx} & \varepsilon_{zy} & \varepsilon_{zz} \end{pmatrix} \begin{bmatrix} E_x \\ E_y \\ E_z \end{bmatrix}$$

The matrix for permittivities is symmetric in that $\varepsilon_{ij} = \varepsilon_{ji}$, so the matrix is diagonal for a system with axes OX, OY and OZ, which are the electric axes for the media, making it possible to state that: $D_X = \varepsilon_{XX} E_X$, $D_Y = \varepsilon_{YY} E_Y$ and $D_Z = \varepsilon_{ZZ} E_Z$.

2.5. Problems: Lorentz field

2.5.1. Dielectric sphere

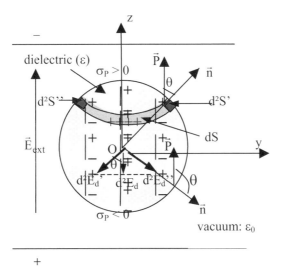

A uniform external field ($\vec{E}_{ext}$) is applied on a dielectric sphere with absolute permittivity ε. The sphere is assumed to be of small dimensions with respect to the distance between the armatures producing $\vec{E}_{ext}$, so it is worth noting that the above scheme is not to scale.

1. Determine the charges equivalent to the polarization.

2. Show that at the center (O) of the sphere, the depolarizing field is $\vec{E}_d = -\dfrac{\vec{P}}{3\varepsilon_0}$.

3. Calculate the resultant field at O. Give the expression for the polarization vector.

---

1. The polarization vector ($\vec{P}$) is parallel to $\vec{E}_{ext}$, and the surface polarization charge densities on the upper half of the sphere are $\sigma_p = P\cos\theta > 0$ as in this zone $0 < \theta < \dfrac{\pi}{2}$. On the lower half, $\dfrac{\pi}{2} < \theta < \pi$ and $\sigma_p < 0$.

These signs are evidently in accordance with the orientation of the dipoles due to the electric field.

2. The field produced by these polarization charges, placed in a vacuum, at O can be calculated by considering an element of the surface (d^2S') giving rise to an elementary field $d^2E_d' = \dfrac{\sigma_P \, d^2S'}{4\pi\varepsilon_0 R^2}$, while an element of a symmetric surface d^2S''

results in a elementary field $d^2E_d'' = \dfrac{\sigma_P \, d^2S''}{4\pi\varepsilon_0 R^2}$ (see figure). Given the symmetry of

the problem, the resultant contribution is directed along Oz. The effective

contribution (d^2E_d) of an elementary contribution such that d^2E_d is directed along

Oz is such that:

$$d^2E_d = d^2E_d' \cos\theta = \frac{\sigma_P \cos\theta \, d^2S'}{4\pi\varepsilon_0 R^2} = \frac{P \cos^2\theta \, d^2S'}{4\pi\varepsilon_0 R^2} = \frac{P \cos^2\theta \sin\theta \, d\theta \, d\varphi}{4\pi\varepsilon_0} \text{ (with}$$

$d^2S' = R^2 \sin\theta \, d\theta \, d\varphi$) so that

$$dE_d = \int_{\varphi=0}^{\varphi=2\pi} \frac{P \cos^2\theta \sin\theta \, d\theta}{4\pi\varepsilon_0} d\varphi = \frac{P \cos^2\theta \sin\theta \, d\theta}{2\varepsilon_0} .$$

Finally, we reach

$$E_d = \int_{\theta=0}^{\theta=\pi} dE_d = \int_{\theta=0}^{\theta=\pi} \frac{P \cos^2\theta \sin\theta \, d\theta}{2\varepsilon_0} = \int_{\theta=\pi}^{\theta=0} \frac{P \cos^2\theta \, d(\cos\theta)}{2\varepsilon_0} = \frac{P}{3\varepsilon_0} \text{ and given}$$

the orientations indicated in the Figure ($\vec{d^2E_d}$ and therefore also $\vec{E}_d$ antiparallel to

$\vec{P}$), we have:

$$\boxed{\vec{E}_d = -\frac{\vec{P}}{3\varepsilon_0}}.$$

3) The active field resulting locally at O is:

$$\boxed{\vec{E}_{al} = \vec{E}_{ext} - \frac{\vec{P}}{3\varepsilon_0}}.$$

2.5.2. Empty spherical cavity

Into a dielectric of permittivity ε is placed a spherical cavity that has a center O and

a radius R. The cavity is subject to a field ($\vec{E}_a$) which acts in the dielectric, which is

assumed to be uniformly polarized.

1. Calculate the field created at O by the polarization charges, which is the Lorentz

field : $\vec{E}_L = \dfrac{\vec{P}}{3\varepsilon_0}$.

2. Determine the resulting field ($\vec{E}_{al}$) at the center of the cavity which can be expressed as a function of ε_r and $\vec{E}_a$.

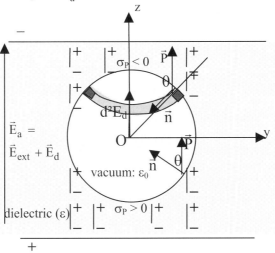

1. Considering the sphere surrounded by dielectric material, the polarization charges appear at the exterior of the sphere (and also near the armatures). As shown in the scheme above, the sign of these polarization charges is inverted with respect to the preceding problem and the normals being external to the dielectric are now directed toward the interior of the sphere, which causes a variation in the values of $\theta = (\vec{P}, \vec{n})$. $\overline{d^2\vec{E}_d}$ and therefore also $\vec{E}_d$ are parallel $\vec{P}$, and $\vec{E}_d = \vec{E}_L = +\dfrac{\vec{P}}{3\varepsilon_0}$.

2. The expression for the resulting field at O is therefore:

$$\vec{E}_{al} = \vec{E}_a + \frac{\vec{P}}{3\varepsilon_0} = \vec{E}_a + \frac{\varepsilon_0(\varepsilon_r - 1)}{3\varepsilon_0}\vec{E}_a = \frac{\varepsilon_r + 2}{3}\vec{E}_a.$$

The problem above is repeated using a parallelepipedal cavity at the end of this chapter.

2.6. Mechanism of Dielectric Polarization: Response to a Static Electric Field

2.6.1. Induced polarization and orientation

2.6.1.1. Induced polarization

When a field is applied to a dielectric, it acts on the charges attached to the atoms making up the molecules of the material. The positive and negative electric charges are then displaced so that the initially unpolarized molecules become polarized, and those that were polarized have their polarization modified. As shown in Figure 2.16, different types of polarization are possible.

2.6.1.1.1. Electronic and atomic polarizations

Under the action of an electric field, the charges in an atom—electrons and nucleus—can be moved. Initially their center of gravities coincide in a nonpolar molecule, but following the application of an electric field, they are no longer the same.

As a first approximation (linear regime), the dipole moment ($\vec{\mu}_i$), which appears once the electric field is applied, is proportional to the field ($\vec{E}_{al}$) locally acting on the molecules, so that $\vec{\mu}_i = \alpha_m \vec{E}_{al}$ where $\vec{\mu}_i$ is the induced dipole moment and α_m is the polarizability of the molecule that is particular to the molecule and expresses its ability to deform to yield $\vec{\mu}_i$. In fact, α_m can be considered a composite of two essential terms, such that $\alpha_m = \alpha_e + \alpha_N$ where:

- α_e is the electronic polarizability expressing the capability of the electronic cloud to deform when exposed to an electric field; and
- α_N is the nucleus polarizability, which also is often inappropriately termed the atomic polarizability, and translates the ability of a nucleus to move under the effect of an electric field.

All atoms can undergo this type of polarization, which occurs in a very short period, corresponding to high frequencies from around 10^{12} Hz for nuclei polarizations up to around 10^{15} Hz for the electron polarizations, which also are called optical polarizations as these frequencies are of the order of the optical domain.

2.6.1.1.2. Ionic polarizability

Ionic polarizability (α_{ionic}), found in ionic crystals, results from the opposing movements of opposite charges. With ions being relatively difficult to move, as they are part of a lattice, this polarization occurs over longer time frames than those mentioned above, compatible with frequencies from the hyperfrequencies to the infra-red.

2.6.1.1.3. Interface polarizability ($\alpha_{interface}$)

Interface polarizability is the result of an applied field on residual charges found in macroscopic domains in a dielectric, which can be found, for example, in heterogeneous dielectrics, which have joints between aggregates, in domains associated with dislocation defaults or even in particulates near interfaces. With the charge carriers being rather slow, as dielectrics are poor conductors, the polarization also is slow in being established—taking up to several minutes—so in general $\nu \approx 10^{-2}$ Hz.

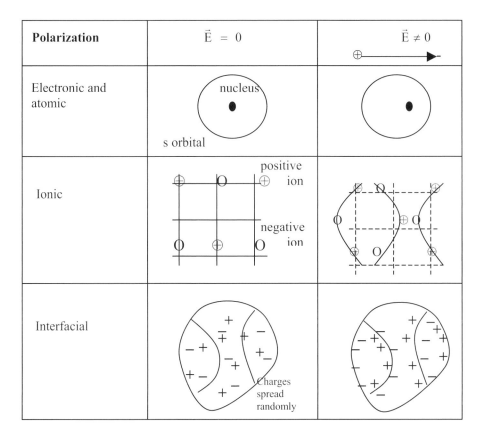

Figure 2.16. *Mechanisms for polarization by induction.*

2.6.1.2.   *Orientation polarization*

Generally speaking, a molecule formed from several different atoms has a spontaneous dipole moment, also commonly called *permanent* ($\vec{\mu}_p$), which exists even in the absence of an applied electric field. Nevertheless, and as we have seen, these particular moments generally are orientated in a random manner due to thermal vibrations so that there is no observable macroscopic polarization. However, in the presence of a field, these moments tend to orientate themselves in the direction of the field. An equilibrium is then struck up between the concurrent effects of thermal disorientation and field orientation. This phenomenon, termed polarization by orientation, can be observed at frequencies typically between 1 kHz and 1 MHz (for materials in the liquid state, these frequencies can be higher).

2.6.1.3. *Conclusion*

Different types of polarization may be observed depending on what the dielectric is. In addition, with a weak polar molecule exhibiting a permanent moment the induced polarization is not necessarily negligible with respect to $\vec{\mu}_p$. In all cases, the resulting moment is the sum of all the possible dipole moments, whatever their origin.

2.6.2. *Study of the polarization induced in a molecule*

We have seen that for a molecule, which here will be assumed to be nonpolar, the induced polarization is of the form: $\vec{\mu}_i = \alpha_m \vec{E}_{al}$. The local field that acts upon the molecule denoted by $\vec{E}_{al}$ is different from the applied external field from the armatures. Using Debye's approximation, we can assume that each molecule can be thought of as in a small spherical cavity of free space. The polarization charges appear on the surface of this cavity and result in the so – called Lorentz local field ($\vec{E}_L$) at the center where it is assumed the molecule lies. From Section 2.4.6.2, we have $\vec{E}_L = \dfrac{\vec{P}}{3\varepsilon_0}$. This field is superimposed over the external applied field, an effective field that as elsewhere is denoted $\vec{E}_a$, is the resultant of the external and the depolarizing fields, and is the actual field measured coming from the charges exhibited on the armatures of the condenser. We therefore have:

$$\vec{E}_{al} = \vec{E}_a + \frac{\vec{P}}{3\varepsilon_0}$$

For a certain number (n) of nonpolar molecules per unit volume, we have:

$$\left.\begin{array}{l}\vec{P} = n\vec{\mu}_i = n\alpha_m\vec{E}_{a\ell} = n\alpha_m(\vec{E}_a + \dfrac{\vec{P}}{3\varepsilon_0}) \\[3mm] \vec{P} = (\varepsilon - \varepsilon_0)\vec{E}_a \end{array}\right\} \Rightarrow$$

$n\alpha_m(\vec{E}_a + \dfrac{[\varepsilon - \varepsilon_0]\,\vec{E}_a}{3\varepsilon_0}) = (\varepsilon - \varepsilon_0)\vec{E}_a$, from which can be determined that

$n\alpha_m(\dfrac{3\varepsilon_0 + [\varepsilon - \varepsilon_0]}{3\varepsilon_0}) = (\varepsilon - \varepsilon_0)$ so that

$$\alpha_m = \frac{3\varepsilon_0}{n}\left(\frac{\varepsilon_r - 1}{\varepsilon_r + 2}\right).$$

If $n = \dfrac{N}{V} = N\dfrac{\rho}{M_a}$ where N is Avogadro's number, V the molar volume, ρ the

volume mass, and M_a the atomic mass, then:

$$\boxed{\dfrac{N\alpha_m}{3\varepsilon_0} = \dfrac{M_a}{\rho}\dfrac{\varepsilon_r - 1}{\varepsilon_r + 2} = P_M}\ .$$

where P_M is the induced molar polarizability.

The above relation is called the Clausius-Mossotti equation and relates the microscopic polarizability (α_m) of a molecule with the macroscopic characteristic of permittivity (ε_r), which is measured via a capacitance ratio $\varepsilon_r = \dfrac{C}{C_0}$, where C and C_0 are capacities of the dielectric and of a vacuum, respectively.

2.6.3.   Study of polarization by orientation

2.6.3.1.   The principle and the result

For a polar molecule with a permanent dipole ($\vec{\mu}_p$) placed in an electric field, the whole molecule turns so that its dipole is in the same sense as the field. Thermal vibrations though limit the effectiveness of the orientation. If the applied field is constant, then the phenomenon is added to that of induced polarization and the molar polarizability is then such that:

$$\boxed{P_M = \dfrac{M_a}{\rho}\dfrac{\varepsilon_r - 1}{\varepsilon_r + 2} = \dfrac{N}{3\varepsilon_0}\left(\alpha_m + \dfrac{\mu_p^2}{3kT}\right) = \dfrac{N}{3\varepsilon_0}\left(\alpha_m + \alpha_{or}\right)}$$

where $\alpha_{0r} = \dfrac{\mu_p^2}{3kT}$.

A demonstration of the above equation is given below. If an alternating field is applied, the dipoles oscillate and rub against each other, resulting in dielectric losses. This phenomenon will be described in more detail in the second volume under the subject of wave-dielectric interactions.

2.6.3.2. The Langevin function

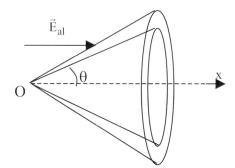

Figure 2.17. *Relative position of an applied field and a solid angle in which are distributed dipoles.*

A collection of identical polar molecules, with moment $\vec{\mu}_p = q\,\vec{1}$, prior to any application of an electric field, are randomly distributed by thermal agitation so that overall they give rise to a zero resultant moment. Once the field is applied, each molecule is subject to a localized effective field ($\vec{E}_{al}$), which tends to orientate the dipole associated with the molecule in the same direction and sense. For each molecule to have a minimum energy (E_p), then the following relation also must reach a minimum:

$$Ep = -\vec{\mu}_p \cdot \vec{E}_{al} = -\mu_p\,E_{al}\,\cos\theta,$$

where $\theta = 0\ (2\pi)$. The alignment of the dipoles along the sense of the applied field nevertheless is limited by thermal agitation so that the orientation angle θ described in Figure 2.17 is nonzero. It is interesting therefore to determine the contribution of each dipolar moment ($\vec{\mu}_p$) to the resultant polarization.

The number of molecules (dN) with moments belonging to a given solid angle (dΩ) is such that $dN = A'\,d\Omega$, i.e., the greater dΩ is, the more molecules present. With Boltzmann's distribution law, the coefficient A' is such that:

$$A' = A\exp(-\frac{E_P}{KT}),\ \text{so that}$$

$$dN = A\exp(+\frac{\mu_p E_{al}\,\cos\theta}{kT})d\Omega.$$

Given that the space dΩ has Ox as an axis of symmetry, the resultant dipolar moment for each molecule in that space with respect to Ox is $\mu_p\cos\theta$. The contribution of dN molecules situated in dΩ with respect to the same axis is therefore:

$$dm = \mu_p\,\cos\theta\,dN.$$

The resultant dipolar moment along Ox for the molecules thus can be written as:

$$M = \int dm = \int_{\theta=0}^{\theta=\pi} \mu_p \cos\theta \, dN .$$

As each dipole makes a contribution, we then have:

$$\overline{\mu} = \frac{M}{N} = \frac{\displaystyle\int_{\theta=0}^{\theta=\pi} \mu_p \cos\theta dN}{\displaystyle\int_{\theta=0}^{\theta=\pi} dN} = \frac{\displaystyle\int_{\theta=0}^{\theta=\pi} \mu_p \cos\theta A \, e^{\frac{\mu_p E_{al} \cos\theta}{kT}} d\Omega}{\displaystyle\int_{\theta=0}^{\theta=\pi} A \, e^{\frac{\mu_p E_{al} \cos\theta}{kT}} d\Omega} .$$

With $\Omega = 2\pi \, (1 - \cos\theta)$, or rather $d\Omega = 2\pi \sin\theta \, d\theta$, we have:

$$\overline{\mu} = \frac{\displaystyle \mu_p \int_{\theta=0}^{\theta=\pi} \cos\theta \, e^{\frac{\mu_p E_{al} \cos\theta}{kT}} \sin\theta \, d\theta}{\displaystyle \int_{\theta=0}^{\theta=\pi} e^{\frac{\mu_p E_{al} \cos\theta}{kT}} \sin\theta \, d\theta} = \mu_p <\cos\theta> .$$

By making $x = \cos\theta$, so that $dx = - \sin\theta \, d\theta$ and $\beta = \dfrac{\mu_p E_{al}}{kT}$, we have:

$$\overline{\mu} = \mu_p <\cos\theta> = \mu_p \frac{\displaystyle\int_{x=-1}^{x=+1} x \, e^{\beta x} dx}{\displaystyle\int_{x=-1}^{x=+1} e^{\beta x} dx} . \qquad (1)$$

The integration in parts of the numerator gives (making $u = x$ and $dv = e^{\beta x} dx$):

$$\int_{x=-1}^{x=+1} x \, e^{\beta x} dx = \left[\frac{x e^{\beta x}}{\beta} \right]_{-1}^{+1} - \int_{-1}^{+1} \frac{1}{\beta} e^{\beta x} dx = \frac{1}{\beta}\left(e^\beta + e^{-\beta} \right) - \frac{1}{\beta^2}\left(e^\beta - e^{-\beta} \right)$$

The denominator directly gives:

$$\int_{x=-1}^{x=+1} e^{\beta x} dx = \frac{1}{\beta}\left(e^\beta - e^{-\beta} \right),$$

from which can be determined that

$$<\cos\theta> = \frac{(e^\beta + e^{-\beta}) - \frac{1}{\beta}(e^\beta - e^{-\beta})}{(e^\beta - e^{-\beta})} = \coth\beta - \frac{1}{\beta} = L(\beta) ,$$

where L is Langevin's function.

If β is large $(\beta \to \infty)$, $\frac{1}{\beta} \to 0$, then:

$$<\cos\theta> \approx \coth\beta = \frac{e^\beta + e^{-\beta}}{e^\beta - e^{-\beta}} = \frac{e^{2\beta} + 1}{e^{2\beta} - 1} \approx \frac{e^{2\beta}}{e^{2\beta}} = 1.$$ The upshot is that as $\beta \to \infty$,

$L(\beta) \to 1$.

If β is small $(\beta \approx 0)$, using the following relation obtained from Eq. (1):

$$<\cos\theta> = \frac{\int\limits_{x=-1}^{x=+1} x \, e^{\beta x} dx}{\int\limits_{x=-1}^{x=+1} e^{\beta x} dx} ,$$ we can write that $e^{\beta x} \approx 1 + \beta x$, so that

$$<\cos\theta> \approx \frac{\int\limits_{x=-1}^{x=+1} x \, (1 + \beta x) \, dx}{\int\limits_{x=-1}^{x=+1} (1 + \beta x) \, dx} = \frac{\left[\frac{x^2}{2} + \frac{\beta x^3}{3}\right]_{-1}^{+1}}{\left[x + \frac{\beta x^2}{2}\right]_{-1}^{+1}} = \frac{\frac{2\beta}{3}}{2} = \frac{\beta}{3} .$$

Accordingly, as $\beta \to 0$, $L(\beta) = <\cos\theta> \approx \frac{\beta}{3} = \frac{\mu_p E_{al}}{3kT}$.

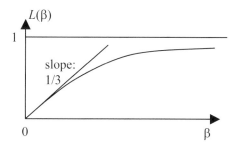

Figure 2.18. *Representation of Langevin's function L(β).*

2.6.3.3. Molar polarizability

In practical terms and under normal conditions, $\beta \leq 10^{-2}$. This gives

$$\bar{\mu} = \mu_p <\cos\theta> \simeq \mu_p \frac{\beta}{3} = \frac{\mu_p^2 E_{al}}{3kT}.$$

The orientation polarization (dipolar moment per unit volume) is thus:

$P_{or} = n\,\bar{\mu} = n\dfrac{\mu_p^2 E_{al}}{3kT}$. This is superimposed on the induced polarization,

$P_i = n\,\alpha_m\,E_{al}$, and the resultant polarization for a polar molecules is

$P_T = n(\,\alpha_m + \dfrac{\mu_p^2}{3kT})E_{al}$ (this equation also can be written in terms of vectors).

The molar polarization therefore can be written as :

$$P_M = \frac{M_a}{\rho}\frac{\varepsilon_r - 1}{\varepsilon_r + 2} = \frac{N}{3\varepsilon_0}\left(\alpha_m + \frac{\mu_p^2}{3kT}\right) \quad \text{(Debye's formula)},$$

and is the summation of the induced molar and orientation polarizations.

2.6.3.4. Application: determination of permanent moments and the polarizability α_m

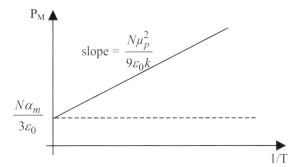

Figure 2.19. *Determination of μ_p and α_m using the Clausius-Mossotti equation.*

On knowing M_a, ρ, and ε_r (by capacitance measurements), it is possible to trace $P_M = f(\dfrac{1}{T})$ which gives a straight line, as it is of the form $P_M = A(\dfrac{1}{T}) + B$, and then determine from the point where the ordinate crosses the y axis the value of

$B = \dfrac{N\alpha_m}{3\varepsilon_0}$, and hence the value of α_m . From the slope, as indicated in Figure 2.19,

which has the value $A = \dfrac{N\mu_p^2}{9\varepsilon_0 k}$, one can determine the value of μ_p .

The commonly used unit for the dipole moment is the Debye (equal to 3×10^{-30} MKS) and is equivalent to a unit charge of 1 u.e.s and a distance of 0.1 nm (1 Å) apart. The dipole moment of HCl thus is 1.1 Debye.

2.7. Problems
2.7.1. Electric field in a small cubic cavity found within a dielectric
Using a large flat condenser, a uniform electric field ($\vec{E}$) is generated in a dielectric with permittivity (ε) which is placed between its electrodes so that $\vec{E} \parallel \overrightarrow{Oz}$. A small empty cubic cavity is placed in the dielectric so that it presents both its upper and lower surfaces, shown in the figure below, parallel to the electrodes.

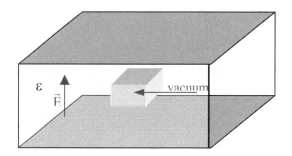

1
(a) Indicate the direction and sense of the polarization vector.
(b) Algebraically calculate the charge densities equivalent to the polarization, and schematically show the position in space of these charges.
2 The origin (O) of the trihedral is at the center of gravity of the cavity, and the electric field ($\vec{E}_P$) due to the polarization charges may be calculated at that point.

(a) Given the symmetry of the problem, indicate the sense and direction of the resultant field $\vec{E}_P$ by considering in succession the effect of polarization charges at the lower and then upper interfaces.
(b) For an element (dS_P) of the flat surface of one of the interfaces, indicate the expression of the electric field produced by dS_P, specifying the useful component of the electric field as a function of a solid angle from which the point O can be observed dS_p.
(c) Give the value of the solid angle through which all the space is observed, and then from this determine the value of the solid angle through which one may observe

one and then two faces of the cube. Give the modulus of the electric field generated by the charges equivalent to the polarization.

(d) Give the vectorial expression for the resultant $\vec{E}_P$ at O produced by charges equivalent to the polarization.

Answers

1.

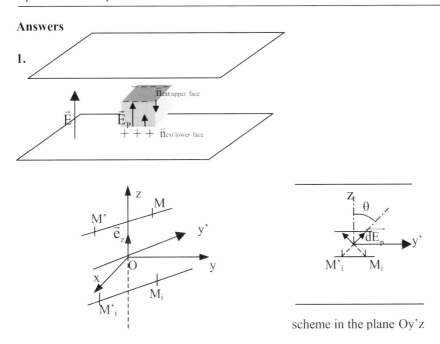

scheme in the plane Oy'z

(a) The external electric field ($\vec{E}$) is assumed to be uniform outside of the cavity, which is supposed to be of negligible size with respect to the size of the armatures of the condenser. Given the geometry of the exercise, we have $\vec{E} \parallel \overline{Oz}$. For its part, the electric field ($\vec{E}_{cav}$) inside the cavity is also assumed to be uniform. The polarization vector in the dielectric (outside of the cavity) is defined by $\vec{P} = (\varepsilon - \varepsilon_0)\vec{E}$, and therefore is directed in the same sense and direction as the field $\vec{E}$, with $\varepsilon > \varepsilon_0$.

(b) The polarization charges can be of two types: volume (ρ_P) or surface (σ_P). As $\rho_P = -\mathrm{div}\vec{P}$, so that $\rho_P = -(\varepsilon\text{-}\varepsilon_0)\mathrm{div}\,\vec{E}$, with $\vec{E}$ and the localized form of Gauss's theorem, $\mathrm{div}\vec{E} = \dfrac{\rho_\ell}{\varepsilon}$, we have $\rho_P = -\dfrac{(\varepsilon\text{-}\varepsilon_0)}{\varepsilon}\rho_\ell$, so that $\rho_P = 0$, as the volume density of real charges in the dielectric (ρ_ℓ) is zero.

The surface charges, $\sigma_P = \vec{P}.\vec{n}_{ext}$, use external normals to the dielectric. Given the directions they follow, which are Oy at the two lateral interfaces of the cavity (parallel to the Oxz plane) and Ox at the two sides behind and in front (parallel to the plane Oyz), only surface charge densities at the upper and lower sides (parallel to the plane Oxy with normals along Oz, in the direction of the polarization) are zero. Therefore:

• on the upper face, $\sigma_P = \vec{P}.\vec{n}_{ext \ upper \ face} = \vec{P}.(-\vec{e}_z) = -P < 0$; and

• on the lower side $\sigma_P = \vec{P}.\vec{n}_{ext \ lower \ face} = \vec{P}.(\vec{e}_z) = P > 0$.

2

(a)   Any point M on either the upper or lower side can be associated with a symmetrically equivalent point M' through Oz. The components following Oy' for an electric field ($\overrightarrow{dE_P}$) generated by the polarization charges at these two points are opposed so that the resultant component of the electric field is directed along Oz. In addition, given the sign of the polarization charges, the sense of the electric field is the same as $\overrightarrow{Oz}$ whether the points are respectively on upper or lower sides. As a result, we have $\overrightarrow{E_P} \parallel \overrightarrow{Oz}$.

(b)   We have $dE_p = \dfrac{1}{4\pi\varepsilon_0} \dfrac{\sigma_P dS_P}{r^2}$, for which the projection along Oz is:

$dE_z = dE_P \cos\theta$. With $|\sigma_P| = P$, we also find: $dE_z = \dfrac{1}{4\pi\varepsilon_0} \dfrac{P \cos\theta \ dS_P}{r^2}$.

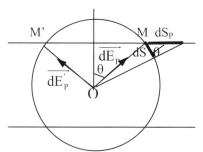

As the solid angle through which a surface element dS of the sphere is $d\Omega = \dfrac{dS}{r^2}$ where dS and dS_P are such that $dS = dS_P \cos\theta$, from which

$dE_z = \dfrac{1}{4\pi\varepsilon_0} \dfrac{P \ dS}{r^2} = \dfrac{P}{4\pi\varepsilon_0} d\Omega$ and $d\Omega = \dfrac{dS}{r^2} = \dfrac{dS_P \cos\theta}{r^2}$ represents the solid angle through which the surface dS_p is seen from O.

Finally, we have $E_Z = \dfrac{P}{4\pi\varepsilon_0}\int d\Omega$ where the integral must be taken over the solid angles that cover the charge-carrying upper and lower sides of the cavity seen from O.

(c)   The solid angle through which the whole of the cube can be seen from the point O, and therefore all of its six sides, and is a solid angle through all space is 4π steradians. With the six sides all being equivalent, 1 side can be seen through the solid angle $\dfrac{4\pi}{6}$, and two sides thus are seen through $\dfrac{4\pi}{6} \times 2 = \dfrac{4\pi}{3}$ steradians. We therefore find that $E_Z = \dfrac{P}{4\pi\varepsilon_0}\int d\Omega = \dfrac{P}{4\pi\varepsilon_0}\dfrac{4\pi}{3} = \dfrac{P}{3\varepsilon_0}$.

(d)   In terms of vectors, and given the answers in 2 (a) and (b), we can state that:

$$\vec{E}_Z = \frac{P}{3\varepsilon_0}\vec{e}_z = \frac{\vec{P}}{3\varepsilon_0}.$$

2.7.2. Polarization of a dielectric strip

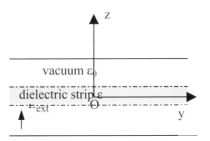

A thin dielectric strip with an absolute dielectric permittivity equal to ε is placed parallel to and in between condenser armatures which generate an uniform "external" field denoted $\vec{E}_{ext}$, as shown in the figure above. It is assumed that the dielectric, like the armatures, has infinite dimensions and a center of gravity called O. Its thickness is such that the field external to the strip is not modified by polarization charges, and that the field inside the strip is uniform.

1.   Algebraically indicate the value of the polarization charges and indicate their position on a figure. The value of the charges should be expressed as a function of the dielectric polarization (P).

2.   With the help of an equation for the continuity for the dielectric strip-vacuum interface and as a function of $\vec{E}_{ext}$ and ε_r, directly find the expression for the

resultant field in the strip ($\vec{E}_L$) and then that of the depolarising field and polarization vector.

Answers

1. The polarization vector is in the same direction and sense as the electric field and follows Oz. The external normal at the upper face of the strip goes along Oz, but at the lower face goes along –Oz. The result is that the superficial charge density equivalent to the polarization $\sigma_P = \vec{P}.\vec{n}_{ext}$ is positive on the upper side and negative on the bottom. These polarities are realized simply by considering the orientation of the permanent dipoles in the strip when subject to an external field.

With the strip not being charged in its volume, then $\rho_P = 0$.

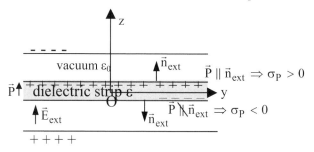

2. By taking the upper interface as an example, the continuity of the component normal to the electric (displacement) induction can be used. As there is no real charge surface density ($\sigma_\ell = 0$), we have $D_n - D_{n0} = 0$ where $D_n = \varepsilon E_L$ (induction in the dielectric for a field denoted E_L) and $D_{n0} = \varepsilon_0 E_{ext}$. The result is $\varepsilon E_L - \varepsilon_0 E_{ext} = 0$, and given the same sense of the two fields,

$$\vec{E}_L = \frac{\varepsilon_0}{\varepsilon}\vec{E}_{ext} = \frac{\vec{E}_{ext}}{\varepsilon_r}.$$

The depolarizing field ($\vec{E}_d$) is such that $\vec{E}_L = \vec{E}_{ext} + \vec{E}_d$, where $\vec{E}_d = -\frac{\varepsilon_r - 1}{\varepsilon_r}\vec{E}_{ext}$. When $\varepsilon_r > 1$, we have $\left(-\frac{\varepsilon_r - 1}{\varepsilon_r}\right) < 0$ and it is confirmed that $\vec{E}_d$ is antiparallel to $\vec{E}_{ext}$. In the dielectric strip, the polarization vector given by $\vec{P} = (\varepsilon - \varepsilon_0)\vec{E}_a$ is such that the effective field $\vec{E}_a = \vec{E}_L$, so that $\vec{E}_a = \frac{\vec{E}_{ext}}{\varepsilon_r}$ and

$$\vec{P} = \frac{\varepsilon_0(\varepsilon_r - 1)}{\varepsilon_r}\vec{E}_{ext}.$$

2.7.3. Dielectric planes and charge distribution (electric images)

In a trihedral about Oxyz, the plane xOz separates a vacuum (permittivity ε_0) in the region $y > 0$ from a dielectric (permittivity ε) which extends through the lower half of the space where $y < 0$. At S, which has the coordinates $(0, a, 0)$, there is a charge (q). An additional point of reference is S' which is symmetrical to S about O and therefore has the coordinates $(0, -a, 0)$; we also have $r = SP$ and $r' = S'P$. This problem concerns the Oxy plane and a point P with coordinates $(x, y, 0)$ from an electrical point of view.

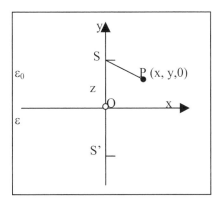

1. In order to deal with this problem, it is worth showing that at the point P $(x, y, 0)$, the scalar potential (V) is such that:

• when $y > 0$: $V_1(P) = \dfrac{1}{4\pi\varepsilon_0}\left[\dfrac{q}{r} + \dfrac{q_1}{r'}\right]$

• when $y < 0$: $V_2(P) = \dfrac{1}{4\pi\varepsilon}\dfrac{q_2}{r}$

where q_1 and q_2 are constants.

(a) What do the potentials $\dfrac{1}{4\pi\varepsilon_0}\dfrac{q}{r}$, $\dfrac{1}{4\pi\varepsilon_0}\dfrac{q_1}{r'}$, $\dfrac{1}{4\pi\varepsilon}\dfrac{q_2}{r}$ represent? What actual physical origin might they have?

(b) Calculate as a function of x and y the Cartesian components for the field vectors $\vec{E}_1(P)$ and $\vec{E}_2(P)$ which are derived from $V_1(P)$ and $V_2(P)$, respectively.

(c) Determine the two constant q_1 and q_2.

2.

(a) Show that the charges equivalent to the polarization are only at the surface.

(b) Determine the density of the surface charges and the total charge through the plane xOz, noting that the distribution is around the axis Oy.

Answers

1.

(a) $V^{(1)} = \dfrac{1}{4\pi\varepsilon_0}\dfrac{q}{r}$ represents the potential generated at P by the charge q at S. The dielectric in the region $y < 0$ is polarized by q. The overall polarization of the dielectric can be taken into account by using an intermediate charge (q_1), which remains to be determined, at S'. The potential $V'^{(1)} = \dfrac{1}{4\pi\varepsilon_0}\dfrac{q_1}{r'}$ is a representation of the problem using a reflected electric image for which q_1 must be determined. This representation is more acceptable when seen as an analogy of an optical system, where it is possible to imagine a mirror, for example, reflecting an image of an object of a similar size and shape to q to give q_1, much as a fisherman might see his float (F) dangling in the air taking on the form F_1 as an image in the water. The total potential in the vacuum is therefore:

$$V_1(P) = V^{(1)} + V'^{(1)} = \frac{1}{4\pi\varepsilon_0}\left[\frac{q}{r} + \frac{q_1}{r'}\right].$$

This equation also can be interpreted by imagining the fisherman sitting on the bank in the air (permittivity equal to ε_0), and him being able to see $B + B_1$, both his float and its reflection.

The potential $V^{(2)} = \dfrac{1}{4\pi\varepsilon}\dfrac{q_2}{r}$ represents the potential at point P (which is in the dielectric), generated by a charge (q_2) at S. Carrying on with the analogy, here it is as if the charge q at S is being observed in the shape of q_2 by the fisherman who has dived into the water to be at P ($y < 0$). Once again though, this representative is only valid if q_2 can be determined.

(b) Bringing in the unit vectors $\vec{u}_{\overrightarrow{SP}}$ and $\vec{u}_{\overrightarrow{S'P}}$ in the directions SP and S'P gives:

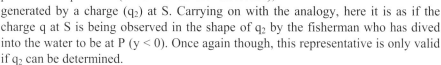

$$\vec{u}_{\overrightarrow{SP}} = \frac{\overrightarrow{SP}}{|\overrightarrow{SP}|} = \left\{\begin{array}{c}\dfrac{x}{r}\\[2mm]\dfrac{(y-a)}{r}\\[2mm]0\end{array}\right. \quad\text{and}\quad \vec{u}_{\overrightarrow{S'P}} = \frac{\overrightarrow{S'P}}{|\overrightarrow{S'P}|} = \left\{\begin{array}{c}\dfrac{x}{r'}\\[2mm]\dfrac{(y+a)}{r'}\\[2mm]0\end{array}\right.$$

The result is that the fields $\vec{E}_1$ and $\vec{E}_2$ are from the same charges generating $V_1(P)$ and $V_2(P)$ and are such that:

$$\vec{E}_1 \begin{cases} \dfrac{x}{4\pi\varepsilon_0}\left[\dfrac{q}{r^3}+\dfrac{q_1}{r'^3}\right] \\[2mm] \dfrac{1}{4\pi\varepsilon_0}\left[\dfrac{q(y-a)}{r^3}+\dfrac{q_1(y+a)}{r'^3}\right] \\[2mm] 0 \end{cases}$$
and
$$\vec{E}_2 \begin{cases} \dfrac{1}{4\pi\varepsilon}\dfrac{q_2 x}{r^3} \\[2mm] \dfrac{1}{4\pi\varepsilon}\dfrac{q_2(y-a)}{r^3} \\[2mm] 0 \end{cases}$$

(c) There remain two constants to determine, namely q_1 and q_2. To do this we need two equations with two unknowns, and these can be derived from the limiting conditions at the interface $y = 0$. At this plane there is the continuity of the tangential component of the electric field, and in the absence of a real superficial charge a continuity of the component normal to D. At the interface where $r = r'$, $y = 0$, and 'x = x', we thus have:

$E_{1x} = E_{2x}$ and $\varepsilon_0 E_{1y} = \varepsilon E_{2y}$, which give, respectively:

$$\dfrac{x}{4\pi\varepsilon_0}\left[\dfrac{q}{r^3}+\dfrac{q_1}{r^3}\right] = \dfrac{1}{4\pi\varepsilon}\dfrac{q_2 x}{r^3} \quad \text{and} \quad \dfrac{1}{4\pi}\left[\dfrac{-q\,a}{r^3}+\dfrac{q_1\,a}{r^3}\right] = \dfrac{1}{4\pi}\dfrac{-q_2\,a}{r^3}.$$

From which we also find:

$$\dfrac{q+q_1}{\varepsilon_0}=\dfrac{q_2}{\varepsilon} \quad \text{and} \quad q-q_1=q_2.$$

From this can be deduced that $q_1 = \dfrac{\varepsilon_0 - \varepsilon}{\varepsilon + \varepsilon_0}q$, and that $q_2 = \dfrac{2\varepsilon}{\varepsilon + \varepsilon_0}q$.

2
(a) There are no volume charges in a dielectric, so $\rho_\ell = 0$ with the result that $\rho_P = 0$ (see course work).
Only surface charges equivalent to the polarization can be present.

(b) We have $\sigma_P = \vec{P}.\vec{n}$ where $\vec{n}$ is the normal outside of the dielectric as shown in the figure. The result is that $\sigma_P = P\cos(\pi - \theta) = - P\cos\theta$.

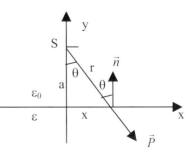

To calculate the polarization, $\vec{P} = (\varepsilon - \varepsilon_0)\vec{E}_2$ wherein the term $\vec{E}_2$ intervenes as the field in the dielectric.

Deriving the potential, $V_2(P) = \dfrac{1}{4\pi\varepsilon}\dfrac{q_2}{r}$, we have for the field $\vec{E}_2$,

$$\vec{E}_2 = \frac{1}{4\pi\varepsilon}q_2\frac{\vec{r}}{r^3} = \frac{1}{2\pi}q\frac{1}{\varepsilon + \varepsilon_0}\frac{\vec{r}}{r^3}$$

from which

$$\vec{P} = \frac{1}{2\pi}q\frac{\varepsilon - \varepsilon_0}{\varepsilon + \varepsilon_0}\frac{\vec{r}}{r^3}.$$

From this can be derived $\sigma_P = -P\cos\theta = -\dfrac{q}{2\pi}\dfrac{\varepsilon - \varepsilon_0}{\varepsilon + \varepsilon_0}\dfrac{1}{r^2}\cos\theta$, and with $\cos\theta = \dfrac{a}{r}$

from which $\dfrac{1}{r^2} = \dfrac{\cos^2\theta}{a^2}$. By making $A = q\dfrac{\varepsilon - \varepsilon_0}{\varepsilon + \varepsilon_0}$, we find that

$$\sigma_P = -\frac{A}{2\pi a^2}\cos^3\theta.$$

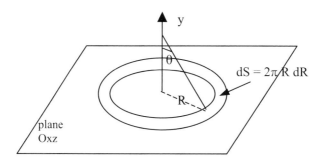

In the Oxz plane, the resultant of charges equivalent to the polarization is:

$q_p = \displaystyle\int\limits_{planOxz} \sigma_P dS$, where $dS = 2\pi R\,dR$, and $R = a\tan\theta$ and $dR = \dfrac{a}{\cos^2\theta}d\theta$.

This gives $q_p = -A\displaystyle\int\limits_{\theta=0}^{\pi/2}\sin\theta\,d\theta = -A = -q\dfrac{\varepsilon - \varepsilon_0}{\varepsilon + \varepsilon_0}$.

The angle θ in the figure varies from 0 to $\pi/2$ so as to cover the whole surface when integrating over the whole surface of the Oxz plane.

2.7.4. Atomic polarizability using J.J. Thomson's model

To represent an atom of hydrogen, J.J. Thomson used the following model. A fixed sphere with a radius R, a center O contains a charge (+q) uniformly spread throughout its volume. The electron charge (-q) is thought of as a point charge and moves in the sphere, assuring the overall neutrality of the atom. All charges are considered to be moving in a vacuum.

1. Calculate the internal electric field ($\vec{E}_{int}$) generated by +q for a point M inside the sphere and located by the vector $\vec{OM} = \vec{r} = r\,\vec{e}_r$.

2. Determine the attractive force between +q and the electron (-q) at M. Detail the equilibrium position of minimum energy for the electron.

3. An external field ($\vec{E}_{ext}$) is now applied along $\vec{e}_r$ to the system. Determine the value of $\vec{d}$ which $\vec{r}$ must take so that the system is in equilibrium.

4. For the equilibrium position, derive the equation for the dipole moment induced by $\vec{E}_{ext}$ and then from this the polarizability of the atom.

Answers

1. Gauss's theorem applied to sphere with radius r (Gaussian surface) gives:

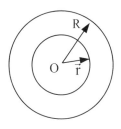

$$\Phi = \iint \vec{E}_{int}.\overrightarrow{dS} = \frac{1}{\varepsilon_0} \iiint_{\text{int érieur}}^{\text{Volume}} \rho\, d\tau \text{ , from which}$$

$$4\pi\, r^2\, E_{int} = \frac{\rho}{\varepsilon_0}\frac{4}{3}\pi r^3. \text{ Then with } \rho = \frac{q}{(4/3)\pi R^3} \text{ , we can}$$

derive that $E_{int} = \dfrac{q}{4\pi\varepsilon_0}\dfrac{r}{R^3}$, and as q > 0, in terms of vectors gives:

$$\vec{E}_{int} = \frac{q}{4\pi\varepsilon_0}\frac{r}{R^3}\vec{e}_r \text{ where } \vec{e}_r = \frac{\vec{r}}{r}.$$

2. The force exerted on an electron at M, such that $\vec{OM} = \vec{r} = r\,\vec{e}_r$, is given by

$$\vec{F} = -q\vec{E}_{int} = \frac{q^2}{4\pi\varepsilon_0}\frac{r}{R^3}\vec{e}_r. \text{ The energy of the electron } (W(r)) \text{ is such that}$$

$$\vec{F} = -\overrightarrow{grad}W, \text{ so that } \int_0^r dW = -\int_0^r \vec{F}.\overrightarrow{dr}, \text{ from which can be deduced that}$$

$$W(r) = \frac{q^2r^2}{8\pi\varepsilon_0 R^3} + \text{constant. Thus } W(r) \text{ is at a minimum when } r = 0, \text{ and therefore}$$

also the force $\vec{F}$ is canceled out at this position, which would make it an equilibrium position for the electron.

3. In the presence of $\vec{E}_{ext}$ applied along $\vec{e}_r$, the resultant force on the electron is $\vec{F}_T = -q\left(\vec{E}_{int} + \vec{E}_{ext}\right)$. The equilibrium position, the value of $\vec{r}$ equal to $\vec{d}$, again corresponds to when $\vec{F}_T = 0$ in that $\left(\vec{E}_{int}\right)_{\vec{r}=\vec{d}} = -\vec{E}_{ext}$. From this can be determined that $\vec{d} = -\dfrac{4\pi\varepsilon_0 R^3}{q}\vec{E}_{ext}$ and that $\vec{d}$ is antiparallel to $\vec{E}_{ext}$.

4. $\vec{E}_{ext}$ moves $-q$ by $\vec{d}$. This in return induces a dipole moment, which can be defined by $\vec{\mu}_i = -q\,\vec{d} = 4\pi\varepsilon_0 R^3 \vec{E}_{ext}$. The polarizability ($\alpha$), as given by the relation $\vec{\mu}_i = \alpha\,\vec{E}_{ext}$, therefore has a value of $\alpha = 4\pi\varepsilon_0 R^3$.

2.7.5. The field in a molecular- sized cavity

Preliminary comments

Debye's theory

The calculation for the effective localized field in a spherical cavity above was carried out with the assumption that the field was the result of a superposition of the applied external field and Lorentz's field, the latter being generated by polarization charges at the surface of a dielectric. Assuming that inside the cavity the molecules are distributed with a symmetry such that the local field they generate is zero, then the resultant field in the cavity experienced by the molecules is the Debye field ($\vec{E}_D$), given by $\vec{E}_D = \vec{E}_{ext} + \dfrac{\vec{P}}{3\varepsilon_0}$.

Onsager's theory

Onsager's theory was introduced to account for localised interactions between molecules. The method used is to superimpose two steps to give the result, respectively, given below in 1, 2, and 3.

1. *Internal field ($\vec{G}$) formed by external field in a small cavity ($\vec{E} \neq 0$ and $\mu = 0$)*

An external field is applied to a small, molecule-sized cavity that does not contain molecules. First the internal field (G) in the cavity is calculated assuming that it is of the same form as the external field except modified by refraction at the dielectric interface.

A molecular-sized spherical cavity, of radius a and a center that is taken as the origin, filled with a dielectric of known absolute permittivity (ε_1), is placed in a dielectric medium, with an absolute permittivity denoted by ε_2. If a field ($\vec{E}$) is applied such that it is parallel to the Oz axis and far from the cavity, as shown in the figure, the sphere distorts the field lines in its neighboring volume. As there is a symmetry around the axis Oz, the problem is not reliant on an azimuthal angle φ. Given this symmetry, for a point P with spherical coordinates $[r,\theta,\varphi]$, we can use the Oxz plane where $\varphi = 0$ so that the potential can be written as:

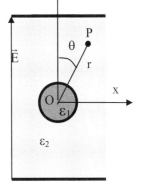

- in the volume outside of the cavity $((r > a)$,

$$V_2 = - E\, r\, \cos\theta + \frac{A\, \cos\theta}{r^2}\; ;\text{ and}$$

- in the cavity $(r < a)$, $V_1 = B\, r\, \cos\theta$,

where A and B are constants.

(**a**) Give a physical justification for the form of the two potentials.
(**b**) Determine the constants A and B.
(**c**) Give an expression for the internal electric field ($\vec{G}$),

which can be expressed as a function of $\vec{E}$ and $\varepsilon_r = \dfrac{\varepsilon_2}{\varepsilon_1}$.

N.B. It also is given that $(\mathrm{grad}V)_r = \dfrac{dV}{dr}$ and $(\mathrm{grad}V)_\theta = \dfrac{1}{r}\dfrac{dV}{d\theta}$

2. *Reaction field ($\vec{R}$) in the absence of an external field, due to polarization charges induced by dipole in the cavity ($\vec{E}=0$, dipole moment (μ) is non zero)*
A pinpoint dipole with an axis in the direction Oz is now at the point O. This dipole causes a polarization of the spherical surface. It is supposed that the potentials are similar to those found in Section 1 just above.

(**a**) Show how the potentials now can be described by:

- outside the cavity $(r > a)$, $U_2 = \dfrac{C\, \cos\theta}{r^2}$; and

- in the cavity $(r < a)$, $U_1 = D\, r\, \cos\theta + \dfrac{D'\cos\theta}{r^2}$

where C, D and D' are constants to be determined.
(**b**) By using the limiting conditions of the interface, determine C and D as a function of D'. Considering a limiting case, give D' as a function of the total moment (μ). Give the final form of U_1 and U_2.
(**c**) Given that the reaction field ($\vec{R}$) is that which is induced by a dipole at the surface of the cavity, find $\vec{R}$ in terms of the corresponding component of U_1.

3. Onsager's field

(a) Show how the internal field acting on the molecule (Onsager's field) takes on the form $\vec{F} = g\vec{E} + \dfrac{r_0\vec{\mu}}{a^3}$. Detail the constants g and r_0 as a function of ε_r and ε_1 (often assumed equal to ε_0).

(b) Formulate the expression of the resulting dipole moment ($\vec{\mu}$) for the molecule located at the center of the cavity that exhibits a permanent moment ($\vec{m}$) and a polarizability (α) in the presence of the electric field $\vec{E}$.

Answers

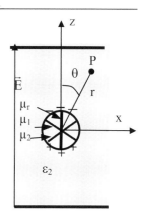

1. (a) Given the direction of $\vec{E}$, it can be supposed that the polarization charges at the interface of the two media will give rise to dipoles with a resultant directed along Oz. As the sphere is of molecular dimensions, the distance between dipoles is around the same as that as the dimension of a molecule—which well generates dipoles. In effect, any given dipole (μ_1) will have a symmetrical equivalent (μ_2) so that combined their resultant (μ_r) will follow Oz, as shown in the adjacent figure. The sense of direction of the dipoles will depend in the polarity of the polarization charges, and hence also the respective values of ε_1 and ε_2. By consequence, for a point P outside of the cavity, the potential due to $\vec{E}$ is such that $V_P - V_O = -\int_O^P \vec{E}.\vec{dl} = -\vec{E}.\overrightarrow{OP} = -\,E\,r\cos\theta$,

so that $V_P = -\,E\,r\cos\theta$ (at potentials with respect to the origin O), to which must be added the potential due to the resultant dipole μ_r which is of the form:

$$V_{dP} = -\frac{1}{4\pi\varepsilon_2}\vec{\mu}_r.\overrightarrow{\mathrm{grad}}_P\frac{1}{r} = \frac{1}{4\pi\varepsilon_2}\frac{\vec{\mu}_r.\vec{r}}{r^3} = \frac{1}{4\pi\varepsilon_2}\frac{\mu_r r\cos\theta}{r^3}\text{ , with}$$

$$A = \frac{\mu_r}{4\pi\varepsilon_2} \Rightarrow V_{dP} = \frac{A\cos\theta}{r^2}.$$

The resultant potential in the medium outside the cavity with permittivity ε_2 is therefore:

$$V_2 = V_P + V_{dP} = -\,E\,r\cos\theta + \frac{A\cos\theta}{r^2} \quad (\,Q.E.D.\,).$$

In the cavity the field E is modified by the refraction of field lines so that form of the first term for V_2 is modified to become $B\,r\cos\theta$ where B is unknown for the present moment. The second term will disappear inside the cavity: If it were kept, when $r \to 0$ the term would diverge and so that this does not happen A has to go to zero, thus removing the component. In the cavity (permittivity ε_1), the potential is thus in the form $V_1 = Br \cos\theta$.

(b) The two equations of continuity for the electric field and induction can be used to find the unknown constant A and B. For $r = a$, the equations are $E_{1t} = E_{2t}$ and $D_{1n} = D_{2n}$. The "tangential" and "normal" directions following, respectively, $\vec{e}_\theta$ and $\vec{e}_r$, can be described by $(E_{1\theta} = E_{2\theta})_{r=a}$ and $(D_{1r} = D_{2r})_{r=a}$, so that

$$\frac{1}{a}\left(\frac{\partial V_1}{\partial \theta}\right)_{r=a} = \frac{1}{a}\left(\frac{\partial V_2}{\partial \theta}\right)_{r=a} \quad \text{and} \quad \varepsilon_1\left(\frac{\partial V_1}{\partial r}\right)_{r=a} = \varepsilon_2\left(\frac{\partial V_2}{\partial r}\right)_{r=a}.$$

From these can be determined that

$$B = \frac{A}{a^3} - E \quad \text{and} \quad B = -\frac{\varepsilon_2}{\varepsilon_1}E - \frac{2}{a^3}\frac{\varepsilon_2}{\varepsilon_1}A \quad \text{so that}$$

$$A = \frac{\varepsilon_1 - \varepsilon_2}{\varepsilon_1 + 2\varepsilon_2}a^3E \quad \text{and} \quad B = -\frac{3\varepsilon_2}{\varepsilon_1 + 2\varepsilon_2}E.$$

Therefore, the potentials V_1 and V_2 are:

$$V_1 = -\frac{3\varepsilon_2}{\varepsilon_1 + 2\varepsilon_2}Er \cos\theta, \quad \text{and} \quad V_2 = -Er \cos\theta - \frac{\varepsilon_2 - \varepsilon_1}{\varepsilon_1 + 2\varepsilon_2}\frac{E\cos\theta}{r^2}a^3.$$

(c) The field $\vec{G}$ in the cavity is such that $\vec{G} = -\overline{\text{grad}V_1}$, where $V_1 = Br \cos\theta$, so that with $z = r \cos\theta$ we have $V_1 = B z$. The result is that the only component of $\vec{G}$ is:

$$G_z = -\frac{\partial V_1}{\partial z} = -B = \frac{3\varepsilon_2}{\varepsilon_1 + 2\varepsilon_2}E, \quad \text{which in terms of vectors gives}$$

$$\vec{G} = \frac{3\varepsilon_r}{2\varepsilon_r + 1}\vec{E}.$$

2. (a) In general terms, the potentials can be written as:

$$U_2 = -Er \cos\theta + \frac{C\cos\theta}{r^2},$$

so that with $E = 0$, we have $U_2 = \frac{C\cos\theta}{r^2}$.

And $U_1 = D r \cos\theta + \dfrac{D'\cos\theta}{r^2}$ where O is a singularity (a pinpoint dipole with zero moment) so that r can no longer go towards zero.

(b) The limiting conditions

$$\frac{1}{a}\left(\frac{\partial U_1}{\partial\theta}\right)_{r=a} = \frac{1}{a}\left(\frac{\partial U_2}{\partial\theta}\right)_{r=a} \quad\text{and}\quad \varepsilon_1\left(\frac{\partial U_1}{\partial r}\right)_{r=a} = \varepsilon_2\left(\frac{\partial U_2}{\partial r}\right)_{r=a} \quad\text{give}$$

$$C = \frac{3\varepsilon_1}{2\varepsilon_2 + \varepsilon_1} D' \quad\text{and}\quad D = -\frac{2(\varepsilon_2 - \varepsilon_1)}{2\varepsilon_2 + \varepsilon_1} \frac{D'}{a^3}.$$

The potentials are therefore:

$$U_1 = -\frac{2(\varepsilon_2 - \varepsilon_1)}{2\varepsilon_2 + \varepsilon_1} \frac{D' r \cos\theta}{a^3} + \frac{D'\cos\theta}{r^2} \quad\text{and}\quad U_2 = \frac{3\varepsilon_1}{2\varepsilon_2 + \varepsilon_1} \frac{D'\cos\theta}{r^2}.$$

If it is proposed that the surface of the spherical surface can become quite large, the potential (U_1) will be that generated by an isolated dipole of total moment μ in a medium of permittivity ε_1, with a value at a coordinate point (r,θ) given by

$$U_d = \frac{\mu\cos\theta}{4\pi\varepsilon_1 r^2}$$ so that when the surface increases, and $1/a^2 \to 0$, we have

$$\frac{\mu\cos\theta}{4\pi\varepsilon_1 r^2} = U_d \approx (U_1)_{\frac{1}{a^2}\to 0} = \frac{D'\cos\theta}{r^2}, \text{ from which by identification } D' = \frac{\mu}{4\pi\varepsilon_1}.$$

Finally, $U_1 = -\dfrac{\mu}{4\pi\varepsilon_1} \dfrac{2(\varepsilon_2 - \varepsilon_1)}{2\varepsilon_2 + \varepsilon_1} \dfrac{r\cos\theta}{a^3} + \dfrac{\mu\cos\theta}{4\pi\varepsilon_1 r^2}$,

and

$$U_2 = \frac{3}{2\varepsilon_2 + \varepsilon_1} \frac{\mu\cos\theta}{4\pi r^2}.$$

(c) In the equation for U_1, the potential inside the sphere, the second term is simply due to the dipole itself and the first term is the potential caused by charges at the surface of the sphere induced by the dipole. The resulting field is the reaction field ($\vec{R}$).

With $z = r\cos\theta$, $\vec{R}$ can be derived from the potential

$$U = -\frac{\mu}{4\pi\varepsilon_1} \frac{2(\varepsilon_2 - \varepsilon_1)}{2\varepsilon_2 + \varepsilon_1} \frac{z}{a^3}, \text{ so that } R = -\frac{\partial U}{\partial z} = \frac{2(\varepsilon_2 - \varepsilon_1)}{2\varepsilon_2 + \varepsilon_1} \frac{\mu}{4\pi\varepsilon_1 a^3} \text{ and in}$$

terms of vectors, $\vec{R} = \dfrac{2(\varepsilon_2 - \varepsilon_1)}{2\varepsilon_2 + \varepsilon_1} \dfrac{\vec{\mu}}{4\pi\varepsilon_1 a^3} = \dfrac{2(\varepsilon_r - 1)}{2\varepsilon_r + 1} \dfrac{\vec{\mu}}{4\pi\varepsilon_1 a^3}.$

3.

(a) The internal field acting on the molecule is given by $\vec{F} = \vec{G} + \vec{R}$, and thus

$$\vec{F} = \frac{3\varepsilon_r}{2\varepsilon_r + 1}\vec{E} + \frac{2(\varepsilon_r - 1)}{2\varepsilon_r + 1}\frac{\vec{\mu}}{4\pi\varepsilon_1 a^3}.$$

This can be written then as $\vec{F} = g\vec{E} + \dfrac{r_o \vec{\mu}}{a^3}$, where

$$g = \frac{3\varepsilon_r}{2\varepsilon_r + 1} \text{ and } r_o = \frac{1}{4\pi\varepsilon_1}\frac{2(\varepsilon_r - 1)}{2\varepsilon_r + 1}.$$

(b) The total moment ($\vec{\mu}$) of the molecule is the sum of its permanent moment ($\vec{m}$) and its induced moment ($\alpha\vec{F}$), so that $\vec{\mu} = \vec{m} + \alpha\vec{F} = \vec{m} + \alpha g\vec{E} + \dfrac{r_o \alpha \vec{\mu}}{a^3}$, from which

we have $\vec{\mu} = \dfrac{\vec{m} + \alpha g\vec{E}}{1 - \dfrac{r_o \alpha}{a^3}}.$

This equation shows how the reaction field modifies the molecule's moment.

Chapter 3

Magnetic Properties of Materials

3.1. Magnetic Moment

3.1.1. Preliminary remarks on how a magnetic field cannot be derived from a uniform scalar potential

Ampere's theorem written for a vacuum, $\overrightarrow{\mathrm{rot}}_P \vec{B} = \mu_0 \vec{j}$, shows how $\overrightarrow{\mathrm{rot}}_P \vec{B}$ is only zero at points (P) which are without current, and therefore is nonzero elsewhere. The $\vec{B}$ therefore cannot be derived from a uniform scalar potential (V), and in effect, if we could write for all P that $\vec{B} = \overrightarrow{\mathrm{grad}}_P V$, we will end up with $\overrightarrow{\mathrm{rot}}_P \vec{B} = 0$, which is not true as we have just seen. Nevertheless, we can define a pseudoscalar potential (V*) such that $\vec{B} = -\overrightarrow{\mathrm{grad}}_D V^*$ where D represents points without current, as at these points $\overrightarrow{\mathrm{rot}}_D \vec{B} = 0$ (a pseudoscalar is a scalar that is defined by its being limited to certain points).

3.1.2. The vector potential and magnetic field at a long distance from a closed circuit

3.1.2.1. Form of the vector potential

Here we will calculate the vector potential $\vec{A}$ and the induction $\vec{B}$ at a distance $|\vec{r}_M|$ far from a current in a closed coil that is considerably smaller than the distance shown in Figure 3.1. For a point P in a vacuum, for which $\vec{r}_M = \overrightarrow{MP}$, the vector $\vec{A}$ given by the coil (C) in which a current of intensity i moves is given by:

$$\vec{A} = \frac{\mu_0 i}{4\pi} \oint \frac{\overrightarrow{dl}}{r_M}.$$

This also is referred to in Section 1.4.2.

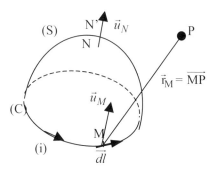

Figure 3.1. *Coiled current.*

If for M we consider a unit vector ($\vec{u}_M$) such that $\vec{u}_M = \text{constant} = \vec{u}$ whatever

the position used, the scalar $\vec{u}_M . \vec{A} = \oint \frac{\mu_0 i}{4\pi} \frac{\vec{u}_M}{r_M} \overrightarrow{dl}$ can correspond to the circulation

around C of the vector $\vec{G}_M = \frac{\mu_0 i}{4\pi} \frac{\vec{u}_M}{r_M}$.

Stokes theory written for the vector $\vec{G}_M$ gives:

$$\vec{u}_M . \vec{A} = \oint_C \vec{G}_M . \overrightarrow{dl} = \iint_S \left(\overrightarrow{rot}_N \vec{G}_N \right) . \overrightarrow{dS} ,$$

where N is on the surface (S) through C.

So now with $\vec{G}_N = \frac{\mu_0 i}{4\pi} \frac{\vec{u}_N}{r_N}$ and $\vec{r}_N = \overrightarrow{NP}$, we find

$$\overrightarrow{rot}_N \vec{G}_N = \frac{\mu_0 i}{4\pi r_N} \overrightarrow{rot}_N \vec{u}_N + \overrightarrow{grad}_N \left(\frac{\mu_0 i}{4\pi r_N} \right) \times \vec{u}_N .$$

Yet with the vector $\vec{u}_N$ (detailed in Figure 3.1), the components

$(x_{N'} - x_N)$, $(y_{N'} - y_N)$, $(z_{N'} - z_N)$, which allow $|\vec{u}_N| = 1$, are such that $\overrightarrow{rot}_N \vec{u}_N = 0$

(as for example $\frac{\partial(z_{N'} - z_N)}{\partial y_N} = 0$).

We therefore have:

$$\vec{u}_M . \vec{A} = \iint_S \left(\overrightarrow{rot}_N \vec{G}_N \right) . \overrightarrow{dS} = \iint_S \frac{\mu_0 i}{4\pi} \left(\overrightarrow{grad}_N \frac{1}{r_N} \times \vec{u}_N \right) . \overrightarrow{dS}$$

$$\vec{u}_M \cdot \vec{A} = \frac{\mu_0 i}{4\pi} \iint \left(\overrightarrow{dS} \times \overrightarrow{grad}_N \frac{1}{r_N} \right) \cdot \vec{u}_N$$

$$= \vec{u}_N \cdot \left[\frac{\mu_0 i}{4\pi} \iint \left(\overrightarrow{dS} \times \overrightarrow{grad}_N \frac{1}{r_N} \right) \right].$$

From $\vec{u}_M \cdot \vec{A} = \vec{u}_N \cdot \vec{A} = A \cos(\vec{u}, \vec{A})$ it can be inferred that $\vec{A}$ can be written as:

$$\vec{A} = \frac{\mu_0 i}{4\pi} \iint_S \left(\overrightarrow{dS} \times \overrightarrow{grad}_N \frac{1}{r_N} \right). \qquad (1)$$

3.1.2.2. The approximation r >> C and the magnetic moment

3.1.2.2.1. The nature of the approximation
We can justify the approximation by recognizing that in real applications the coils through which the current passes are extremely small with respect to the distances over which the magnetic effects can be dealt with. More specifically, the approximation states that $\overrightarrow{grad}_N \dfrac{1}{r_N}$ is constant for each and every point (N) over the surface S, which can be whatever size, although it depends directly on the size of C, and that $\left| \vec{r}_N \right| = \left| \overrightarrow{NP} \right| \approx$ constant, with respect to the position of point N. In other words, from a position P outside of the coil, we would see that any part of the coil would be a distance r away, and indeed, the coil would seem miniscule from P. We also can see that there is an analogy with the electric doublet considered in the section on electrostatics, in which it is also supposed that $r \approx$ constant.

3.1.2.2.2. Magnetic moment

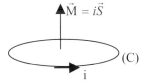

Figure 3.2. *The magnetic moment from a circular current.*

The magnetic moment is introduced by definition as a vector:

$$\vec{M} = i \iint_S \overrightarrow{dS} \qquad (2)$$

thus giving it its name – spin angular magnetic moment – as schematized in Figure 3.2. Given that only C has been determined and that S is dependent on C, the magnitude of Eq. (2) results from Eq. (1).

Taking into account Stokes relation, the magnetic moment therefore depends only on the size of C. If C is a flat coil, then S can be simply that defined by the coil C, and in which case $\vec{M}$ is normal to the plan of the circuit so that its modulus is iS. The direction of $\vec{M}$ is defined by the corkscrew rule, which gives the rotational direction of i through the coil. (Stokes theory states $\vec{S}$ that the direction is given by that of the current i $\overrightarrow{dl}$.)

Carrying Eq. (2) into Eq. (1), we can write (when $\overrightarrow{grad}_N \dfrac{1}{r_N} \approx$ constant

$= \overrightarrow{grad}_N \dfrac{1}{r}$) that

$$\vec{A} = \frac{\mu_0}{4\pi} \vec{M} \times \overrightarrow{grad}_N \frac{1}{r} .$$

This expression can be turned around:

$$\vec{A} = -\frac{\mu_0}{4\pi} \vec{M} \times \overrightarrow{grad}_P \frac{1}{r} . \qquad (3)$$

Given that the circuit is fixed in space, and that $\overrightarrow{rot}_P \vec{M} = 0$ as $\vec{M}$ is independent of P:

$$\overrightarrow{rot}_P \left(\frac{1}{r} \frac{\mu_0}{4\pi} \vec{M} \right) = \frac{1}{r} \frac{\mu_0}{4\pi} \overrightarrow{rot}_P \vec{M} + \left(\frac{\mu_0}{4\pi} \overrightarrow{grad}_P \frac{1}{r} \right) \times (\vec{M}) = -\frac{\mu_0}{4\pi} \vec{M} \times \overrightarrow{grad}_P \frac{1}{r} .$$

we therefore arrive at:

$$\vec{A} = \overrightarrow{rot}_P \left(\frac{\mu_0 \vec{M}}{4\pi r} \right) . \qquad (4)$$

3.1.2.2.3. Magnetic field

We know that $\vec{B}$, for a point P, is given by $\vec{B} = \overrightarrow{rot}_P \vec{A} = \overrightarrow{rot}_P \overrightarrow{rot}_P \left(\frac{\mu_0 \vec{M}}{4\pi r} \right)$,

so (with $\overrightarrow{rot} \ \overrightarrow{rot} = \overrightarrow{grad} \ div - \vec{\Delta}$) we have $\vec{B} = \frac{\mu_0}{4\pi} \left[\overrightarrow{grad}_P \ div_P \frac{\vec{M}}{r} - \vec{\Delta}_P \left(\frac{\vec{M}}{r} \right) \right]$.

However, with $\vec{M}$ fixed and independent from P, we find that:

$$\vec{\Delta}_P\left(\frac{\vec{M}}{r}\right) = \vec{i}\left[\Delta_P\left(\frac{M_x}{r}\right)\right] + \vec{j}\left[\Delta_P\left(\frac{M_y}{r}\right)\right] + \vec{k}\left[\Delta_P\left(\frac{M_z}{r}\right)\right]$$

$$= \left(\vec{i}\ M_x\ +\ \vec{j}\ M_y\ +\ \vec{k}\ M_z\ \right)\Delta_P\left(\frac{1}{r}\right),$$

so, with $\Delta_P\left(\frac{1}{r}\right) = 0$, then $\vec{\Delta}_P\left(\frac{\vec{M}}{r}\right) = 0$.

This finally gives us:

$$\vec{B} = \frac{\mu_0}{4\pi}\overrightarrow{grad}_P\ div_P\ \frac{\vec{M}}{r}. \qquad (5)$$

Therefore, at a large distance from a coil through which moves i, the vector $\vec{B}$ is derived from a (pseudo) scalar potential (V^*) and that

$$V^* = -\frac{\mu_0}{4\pi}div_P\ \frac{\vec{M}}{r}. \qquad (6)$$

The vector $\vec{B}$ is such that

$$\vec{B} = -\overrightarrow{grad}_P\ V^*. \qquad (7)$$

3.1.3. *The analogy of the magnetic moment to the (dipolar) electric moment and the justification of the term magnetic doublet for* $\vec{M}\ =\ I\ \iint_S \vec{dS}$

We can recall briefly that the scalar potential generated by an electric dipole of moment $\vec{\mu}\ =\ q\ \vec{l}$ is such that

$$V = -\frac{1}{4\pi\varepsilon_0}\vec{\mu}\ .\ \overrightarrow{grad}_P\ \frac{1}{r}. \qquad (8)$$

On comparing Eqs. (3) and (8), we rapidly can see that the potentials take on the same form, the scalar product (in the expression for the scalar potential) can be substituted by the vectorial product (in the expression for the potential vector $\vec{A}$), and the vector $\vec{M}\ =\ i\ \iint_S \vec{dS}$ can take the place of the electric dipolar moment ($\vec{\mu}$).

Therefore, by analogy, the term $\vec{M}$ is called the magnetic dipole moment.

At the level of the pseudoscalar potential given by Eq. (6), it also is possible to write:

$$V^* = -\frac{\mu_0}{4\pi}\mathrm{div_P}\,\frac{\vec{M}}{r} = -\frac{\mu_0}{4\pi}\frac{1}{r}\,\mathrm{div_P}\,\vec{M} - \frac{\mu_0}{4\pi}\vec{M}.\overrightarrow{\mathrm{grad_P}}\,\frac{1}{r} = -\frac{\mu_0}{4\pi}\vec{M}.\overrightarrow{\mathrm{grad_P}}\,\frac{1}{r}\ \ (6'),$$

because $\mathrm{div_P}\,\vec{M} = 0$ ($\vec{M}$ is independent of P). We can see in the expression for the magnetic pseudoscalar a resemblance to the relationship for electric potential when exchanging $\vec{M}$ for $\vec{\mu}$. There also is the coefficient, $\frac{\mu_0}{4\pi}$ to change to $\frac{1}{4\pi\varepsilon_0}$ for the electrostatic potential, and μ_0 which takes on the role of $\frac{1}{\varepsilon_0}$ (as detailed in Chapter 1 in the formulation of Poisson's equation). This equivalence again justifies calling $\vec{M}$ the magnetic dipole moment.

To conclude this section, we have seen that for a very small coil with respect to the distance $|\vec{r}|$ to where the magnetic effects are observed, the vector potential, the pseudoscalar potential, and the magnetic field are given by expressions such as Eqs. (3), (6) or (6'), and (7). These relationships are analogous to those found in dielectrics—which are supposed as vacuums in which "sit" electrostatic dipoles (Section 2.2.1.1). Similarly, a small closed circuit can be associated with a magnetic dipole, and outside of the volume defined by the circuit the magnetic effects can be described by potentials expressed with the help of the magnetic moment $\vec{M} = i\iint_S\overrightarrow{dS}$. This study is carried over to Section 3.2.2.

Interestingly enough, in material, the same closed currents can be tied to the movement of electrons in their orbitals.

3.1.4. Characteristics of magnetic moments $\vec{M} = i\iint_S\overrightarrow{dS}$

3.1.4.1. "The right hand rule": magnetic moments are positive when the rotational sense of the current corresponds to the north pole

For electrons in their orbitals, $\rho = - ne$, and as $\vec{i} = \rho\vec{v}$ where $\rho < 0$, we find $\vec{i}$ antiparallel to $\vec{v}$, so that from the "north face", as defined in Figure 3.3(a), $\vec{M}//\overrightarrow{Oz}$, while the kinetic moment $\vec{l} = r\times m\vec{v}$ is such that $\vec{l}$ is antiparallel to $\overrightarrow{Oz}$, and therefore $\vec{M}$ is antiparallel to $\vec{l}$.

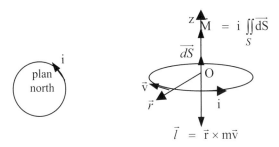

Figure 3.3(a). *Northern plan and the direction of the magnetic moment.*

With respect to the "south face" shown in Figure 3.3(b), $\vec{M}$ is antiparallel to $\overrightarrow{Oz}$, although $\vec{M}$ remains antiparallel to $\vec{1}$.

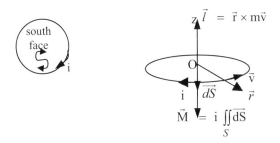

Figure 3.3(b). South face and the direction of the magnetic moment.

3.1.4.2. *The energy of a magnetic dipole in a magnetic field*

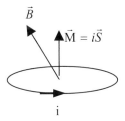

Figure 3.4. *Magnetic dipole in a magnetic field.*

If the dipole is very small, then it can be assumed that $\vec{B}$ is constant with respect to the rest of the dipole, a supposition that indeed is correct if $\vec{B}$ is uniform. If $\vec{S}$ is the surface of the dipole, the flux from $\vec{B}$ going through the dipole is:

$$\Phi = \iint_S \vec{B}.\overrightarrow{dS} = \vec{B} \iint_S \overrightarrow{dS}.$$

From $\vec{M} = i \iint_S \overrightarrow{dS}$, we can carry forward $\iint_S \overrightarrow{dS} = \dfrac{\vec{M}}{i}$ into Φ to obtain

$$\Phi = \vec{B}\dfrac{\vec{M}}{i}.$$

In addition, the potential energy of the closed loop circuit (magnetic dipole) traversed by i and placed within the flux Φ is $E_p = -i\,\Phi$, so that with the preceding expression for Φ, we have

$$E_p = -\vec{M}.\vec{B}.$$

3.1.4.3. Forces and couples active on $\vec{M}$ placed in $\vec{B}$

3.1.4.3.1. Forces

In general terms, we have $\vec{F} = -\overrightarrow{grad}\,E_p$, so that more specifically $\vec{F} = \overrightarrow{grad}\left(\vec{M}.\vec{B}\right)$.

3.1.4.3.2. Couples

While moving through $\overrightarrow{dl}$, $\overrightarrow{d\Omega}$ of a dipole, if $\vec{F}$ and $\vec{\Gamma}$ are the elements of reduction around O for the torsor of forces acting upon the dipole, we have
$$dW = -dE_p = \vec{M}\,d\vec{B} + \vec{B}d\,\vec{M}.$$

Performing the calculation $\vec{M}\,d\vec{B} = \vec{M}\left(\dfrac{\partial \vec{B}}{\partial x}dx + \dfrac{\partial \vec{B}}{\partial y}dy + \dfrac{\partial \vec{B}}{\partial y}dy\right)$, recognizing that $\vec{M}$ is independent of the position of the dipole, means that

$$\vec{M}\,d\vec{B} = \dfrac{\partial}{\partial x}\left(\vec{M}.\vec{B}\right)dx + \dfrac{\partial}{\partial y}\left(\vec{M}.\vec{B}\right)dy + \dfrac{\partial}{\partial z}\left(\vec{M}.\vec{B}\right)dz = \left[grad\left(\vec{M}.\vec{B}\right)\right].\overrightarrow{dl}.$$

It is also possible to derive the expression $\vec{B}.d\vec{M} = \vec{B}.\overrightarrow{d\Omega} \times \vec{M} = \overrightarrow{d\Omega}.(\vec{M} \times \vec{B})$, given that $\overrightarrow{d\Omega}$ is defined by $d\vec{M} = \overrightarrow{d\Omega} \times \vec{M}$.

Finally, from $dW = -dE_p = \vec{M}\,d\vec{B} + \vec{B}\,d\vec{M} = \left[\text{grad}\left(\vec{M}.\vec{B}\right)\right].\overrightarrow{dl} + (\vec{M}\times\vec{B}).\overrightarrow{d\Omega}$

$$= \vec{F}.\overrightarrow{dl} + \vec{\Gamma}\overrightarrow{d\Omega},$$

which by identity gives

$$\boxed{\vec{F} = \text{grad}\left(\vec{M}.\vec{B}\right) \text{and } \vec{\Gamma} = \vec{M}\times\vec{B}} \quad .$$

Comment: If $\vec{B}$ is uniform, then $\vec{F} = 0$ and $\vec{\Gamma} = \vec{M}\times\vec{B}$. In effect, the dipole is subject to a single couple with moment $\vec{\Gamma} = \vec{M}\times\vec{B}$.

3.1.5. *Magnetic moments in materials*

3.1.5.1. Introduction to the elementary terms

The kinetic moment ($\vec{\sigma}$) of a particle with a known mass (m) is the moment of a known movement such that $\vec{\sigma} = \vec{r}\times m\vec{v} = \vec{r}\times\vec{p}$.

The moment of a force is defined by: $\vec{\Gamma} = \vec{r}\times\vec{F}$.

The theory for kinetic moment states that $\dfrac{d\vec{\sigma}}{dt} = \dfrac{d\vec{r}}{dt}\times\vec{p} + \vec{r}\times\dfrac{d\vec{p}}{dt} = \vec{r}\times\vec{F} = \vec{\Gamma}$, and

therefore if $\vec{\Gamma}$, $\vec{\sigma}$ is constant.

3.1.5.2. Atomic magnetic moments

For every kinetic moment there is an associated magnetic moment. Similarly, for each orbital kinetic moment there is an orbital magnetic moment, and for a spin kinetic moment a spin magnetic moment.

Concerning the orbital kinetic and magnetic moments, the orbital kinetic moment is given by $\vec{l} = r\times m\vec{v}$ and the related magnetic moment is $\vec{\mu}_l = i\vec{S}$, as schematized in Figure 3.4.

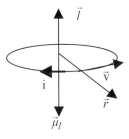

Figure 3.5. *Kinetic moment and the atomic orbital.*

The current intensity is given by $i = -ne$ where n is the number or orbital rotations (about a radius r) made by an electron of charge $-e$ and speed v. This then gives $n = \dfrac{v}{2\pi r}$, and $\vec{\mu}_1 = -\dfrac{v}{2\pi r}\, e\, \pi\, r^2 = -e\, \dfrac{r\, v}{2}$, so that in terms of vectors,

$$\vec{\mu}_1 = -e\, \frac{\vec{r} \times m\,\vec{v}}{2\,m} = -\frac{e}{2\,m}\,\vec{l}\, .$$

If we make $\beta = \dfrac{e}{2m} = \mu_B$, where μ_B is the Bohr magneton, we have:

$$\vec{\mu}_1 = -\beta\,\vec{l}\, .$$

Concerning the spin kinetic and magnetic moments, the same reasoning can be repeated; however, the speed of the system is so much greater as the distance involved is so considerably smaller that the calculation must be carried out using relativity. So, in relativistic quantum mechanics, the resulting spin magnetic moment ($\vec{\mu}_s$) is $\vec{\mu}_s = -2\,\beta\,\vec{s}$, where $\vec{s}$ is the spin kinetic moment.

For the atomic magnetic moment then, the resulting moment is

$$\vec{\mu}_T = -\beta(\vec{l} + 2\,\vec{s})\, .$$

For an atom with more than one peripheral electron and within the spin-orbit coupling approximation we have:

$$\vec{\mu}_T = -\beta(\vec{L} + 2\,\vec{S})\, .$$

3.1.6. Precession and magnetic moments

3.1.6.1. A magnetic field and the gyroscopic effect and the Larmor precession

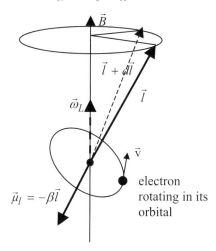

Figure 3.6. *Magnetic field effect in a magnetic moment.*

A field $\vec{B}$ acting on the movement of an electron in its orbital can be schematized as in Figure 3.6 as a moment couple due to $\vec{B}$ on $\vec{\mu}_l$.

So
$$\left.\begin{array}{l} \vec{\Gamma} = \vec{\mu}_l \times \vec{B} = \dfrac{d\vec{l}}{dt} \\ = -\beta\vec{l} \times \vec{B} = \beta\vec{B} \times \vec{l} \end{array}\right\} \Rightarrow \dfrac{d\vec{l}}{dt} = \beta\vec{B} \times \vec{l} = \vec{\omega}_L \times \vec{l},$$

and then placing $\vec{\omega}_L = \beta\vec{B} = \dfrac{e}{2m}\vec{B}$ where $\vec{\omega}_L$ is the rotational vector, here collinear to $\vec{B}$.

So the vector $\vec{l}$ undergoes a rotation around $\vec{B}$ with an angular frequency, or rather a Larmor frequency given by $\omega_L = \beta B = \dfrac{eB}{2m}$.

The movement of an electron in the plane of its orbit is not altered by $\vec{B}$ (as $\left|\vec{l}\right|$ remains constant during the rotation), but the plane of the orbit goes through a rotation, with a rotational vector $\vec{\omega}_L$, around $\vec{B}$ in what is called the Larmor precession. The movement of precession resembles that of a gyroscope.

Comment If we multiply the equation $\dfrac{d\vec{l}}{dt} = \vec{\omega}_L \times \vec{l}$ by β, the direct result is $\dfrac{d\vec{\mu}_l}{dt} = \vec{\omega}_L \times \vec{\mu}_l$. In effect, the moment $\vec{\mu}_l$ also goes through a rotation about $\vec{B}$, characterized by the same rotational vector $\vec{\omega}_L = \beta\vec{B} = \dfrac{e}{2m}\vec{B}$.

3.1.6.2. Precession and the coupling $\vec{L}, \vec{S}$: internal precession

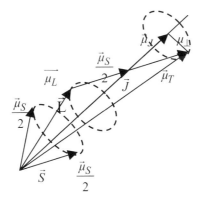

Figure 3.7. *Precession of the couple L,S.*

When there is a couple such as $\vec{L}, \vec{S}$, first the orbital kinetic moments are coupled, then the spin moments; this gives rise to a resultant kinetic moment of $\vec{J} = \vec{L} + \vec{S}$. In the absence of an external field, $\vec{\Gamma} = \vec{\mu} \times \vec{B} = 0 = \dfrac{d\vec{J}}{dt}$ and $\vec{J}$ is a constant of movement, so that a fixed direction is retained in space, as shown in Figure 3.7. The figure also shows the kinetic and magnetic moments as, by design, $\beta \equiv -1$, so that we also can see $\mu_L = L$.

However, $\overrightarrow{\mu_L}$ and $\overrightarrow{\mu_S}$ display an interaction of energy $\Delta E \propto \vec{L}.\vec{S} \cos(\vec{L}, \vec{S})$. At a given energy, $\cos(\vec{L}, \vec{S})$ must be constant and $\vec{L}$ and $\vec{S}$ can only conduct one precession movement (their direction may change) about $\vec{J}$, which retains a fixed direction in the absence of an external field. The result of this is that $\vec{\mu}_T = -\dfrac{e}{2m}(\vec{L} + 2\vec{S})$, which is invariably tied to the triangle (L,S,J), which also goes through a precession about $\vec{J}$.

In addition, $\vec{\mu}_T$ can be broken down into a component $\vec{\mu}_J$ along $\vec{J}$, and a perpendicular component $\vec{\mu}_\perp$ that turns around $\vec{J}$ at the precession rate of motion. If we consider the average value of $\vec{\mu}_\perp$ over an interval of time much greater than the time required for one rotation, it is actually zero and the effect of $\vec{\mu}_T$ is the same as that of $\vec{\mu}_J$ (the apparent magnetic moment).

It is possible then to write $\vec{\mu}_J = -\dfrac{e}{2m}g\vec{J} = -\beta g\vec{J}$ where g is the Landé factor. We also can state that $\gamma = \dfrac{e}{2m}g$, where the factor γ, such that $\vec{\mu}_J = -\gamma\vec{J}$ is called, for its part, the gyromagnetic ratio.

The factor g can be calculated easily. For example, by calculating $\left|\vec{\mu}_J\right|$ and $\left|\vec{J}\right|$, we obtain:

$$g = 1 + \frac{J(J+1) + S(S+1) - L(L+1)}{2J(J+1)} \ .$$

3.2. Magnetic Fields in Materials

3.2.1. *Magnetization intensity*
Here we suppose that there is a bar of material of a volume $d\tau$ that is equivalent to having a volume of atoms and/or molecules sitting in a vacuum each having an elementary magnetic moment associated with the movement of electrons in their orbits. This setup is comparable to that used for dielectrics.

A priori, the moments are randomly orientated by thermal agitation when there is no external applied field. The dipoles, however, will tend to orientate themselves along the lines of an applied field ($\vec{B}$) so that they assume a minimum potential energy. This will confer on each element in the volume of the material a nonzero dipole moment equal to $d\vec{M}$. In addition, this corresponds to the hypothesis given by Ampere for molecular currents.

By analogy with dielectrics, where the polarization vector is defined by $\vec{P} = \dfrac{d\vec{\mu}}{d\tau}$, we define the vector for the magnetization intensity at a point in the magnetic medium such that:

$$\vec{I} = \frac{d\vec{M}}{d\tau} \ .$$

This is quite often termed a vector of "magnetic polarization".

3.2.2. Potential vector due to a piece of magnetic material (magnetized and characterized by $\vec{I}$) and Amperian currents

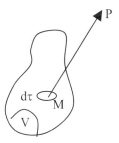

Figure 3.8. *Calculation for a potential vector through P.*

In physical terms, the potential vector is the sum of all the potential vectors generated by the electronic atomic or molecular currents in a vacuum that give rise to a magnetization intensity ($\vec{I}$) from the points (M) in the volume (M). In other words, the potential vector at a point P is the sum over all the elementary potentials ($d\vec{A}$) due to the elementary magnetic moments ($d\vec{M} = \vec{I}d\tau$) carried by the elements in the volume $d\tau$. Using Eq. (3) in Section 3.1.2. we can state that:

$$d\vec{A} = \frac{\mu_0}{4\pi}d\vec{M} \times \overrightarrow{\text{grad}}_M\left(\frac{1}{r}\right) = \frac{\mu_0}{4\pi}\frac{d\vec{M}}{d\tau} \times \overrightarrow{\text{grad}}_M\left(\frac{1}{r}\right)d\tau = \frac{\mu_0}{4\pi}\vec{I} \times \overrightarrow{\text{grad}}_M\left(\frac{1}{r}\right)d\tau \ ,$$

from which can be determined that expression for the resultant potential vector ($\vec{A}$):

$$\vec{A} = \frac{\mu_0}{4\pi} \int_V \vec{I} \times \overrightarrow{grad}_M \left(\frac{1}{r}\right) d\tau .$$

Again using the relationship $\overrightarrow{rot} (a\vec{T}) = a \overrightarrow{rot} \vec{T} + (\overrightarrow{grad} a) \times \vec{T}$ where now $a = 1/r$

and $\vec{T} = \vec{I}$, we can say that $\overrightarrow{rot}_M \left(\dfrac{\vec{I}}{r}\right) = \dfrac{1}{r} \overrightarrow{rot}_M \vec{I} - \vec{I} \times (\overrightarrow{grad}_M \dfrac{1}{r})$, from which

can be derived:

$$\vec{A} = \frac{\mu_0}{4\pi} \int_V \frac{1}{r}\overrightarrow{rot}_M \vec{I} \, d\tau - \frac{\mu_0}{4\pi} \int_V \overrightarrow{rot}_M \frac{\vec{I}}{r} \, d\tau .$$

Taking into account the equation for a rotational in Section 1.1, $\int_V \overrightarrow{rot} \vec{T} \, d\tau = \int_S \overrightarrow{dS} \times \vec{T}$, the second integral of the equation giving $\vec{A}$ can be

transformed with $\vec{T} = \dfrac{\vec{I}}{r}$, so that $\vec{A} = \dfrac{\mu_0}{4\pi} \int_V \dfrac{1}{r}\overrightarrow{rot}_M \vec{I} \, d\tau - \dfrac{\mu_0}{4\pi} \int_S \dfrac{\overrightarrow{dS} \times \vec{I}}{r}$, giving

$$\vec{A} = \frac{\mu_0}{4\pi} \int_V \frac{1}{r}\overrightarrow{rot}_M \vec{I} \, d\tau + \frac{\mu_0}{4\pi} \int_S \frac{\vec{I} \times \vec{n}}{r} dS .$$

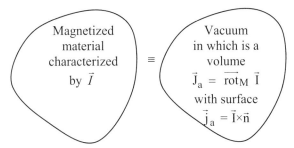

Figure 3.9. *Equivalence of magnetized material to Amperian current densities sitting in a vacuum.*

To conclude, the preceding equation shows how the potential vector due to a volume V of magnetized material, characterized by a magnetization intensity vector $\vec{I}$, is equal to the potential vector formed in a vacuum by so-called Amperian (or molecular) currents distributed—as also shown in Figure 3.9—so that:

- there is one part characterized by the volume current density vector, such that
$$\vec{J}_a = \overrightarrow{rot}_M \vec{I} ;$$

- there is another paper at the surface (limited of course by V) characterized by the superficial (or surface) current density vector, such that

$$\vec{j}_a = \vec{I} \times \vec{n}.$$

These calculations can be used to determine the internal or external vector potential due to the magnetized material.

Example: Determine the Amperian currents equivalent to a uniformly magnetized cylinder and then the magnetization intensity through its axis, as shown in Figure 3.10.

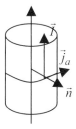

Figure 3.10. *Cylinder magnetized along its axis.*

We have $\vec{J}_a = \overrightarrow{\text{rot}}\ \vec{I} = 0$ (as $\vec{I}$ is uniform throughout the volume).

$$\vec{j}_a = \begin{cases} \bullet \text{ at the ends, } \vec{I} \text{ and } \vec{n} \text{ are parallel, and } \vec{j}_a = \vec{I} \times \vec{n} = 0\ ; \\ \bullet \text{ on the sides, } \vec{I} \text{ and } \vec{n} \text{ are perpendicular, and } \vec{j}_a = \vec{I} \times \vec{n} = I\vec{e}_\theta\ , \text{ where} \\ \quad \vec{e}_\theta \text{ is at a tangent to the surface of the cylinder and normal to its axis.} \end{cases}$$

3.2.3. Physical representation of the magnetization of material and the Amperian currents

The Amperian currents, defined above, are sometimes simply called magnetization currents. However, as they are the indirect result of a calculation, and therefore seem to have little bearing in reality, they are also often called imaginary currents. In effect, they cannot really be termed macroscopic and in addition are not associated with the movement of charge through a material, as is the case for a current density vector. Amperian currents are not then what one would call "your normal type of current".

Yet they are not fictional, as for example in orbital magnetism (spin magnetism is even more complex) they are the charge carriers tied to their nuclei, which give rise to localized, individual currents that generate the magnetic moments and result

in $\vec{I} = \dfrac{d\vec{M}}{d\tau}$.

We can use two simple examples with geometrical representations to show how Amperian currents can be seen as averages over space of individual current densities, which also qualify as microscopic currents.

3.2.3.1. A uniformly magnetized material perpendicular to the plane $\left|\vec{I}\right| = I_z$

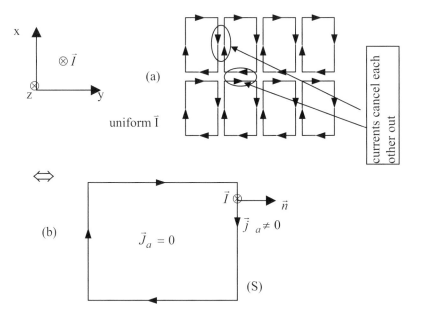

Figure 3.11.   *Equivalence of (a) a uniformly magnetized material and (b) a material crossed by surface Amperian currents.*

In a parallelepipedal volume uniformly magnetized along Oz and in a plane perpendicular to the Oz axis, there are ring currents associated with the movement of electron orbitals. In order to detail these currents in geometrical terms, the orbitals are assimilated into rectangular trajectories as indicated in Figure 3.11 a. It can be observed in the figure that, given the direction that the currents take, they cancel one another out in the volume of the material but not on the surface, and hence the representation of the overall current in Figure 3.11b. This result is in perfect agreement with the calculation of the density of Amperian volume currents ($\vec{J}_a = \overrightarrow{\text{rot}}_M \, \vec{I} = 0$ as $\vec{I}$ is uniform and independent of the points M in the material). For its part, the calculation $\vec{j}_a = \vec{I} \times \vec{n}$ indicates the correct direction and sense of the resultant current, already seen using the geometrical argument detailed above.

3.2.3.2. A nonuniformly magnetized material with $\vec{I}$ along Oz and such that
$$\frac{\partial I}{\partial y} > 0.$$

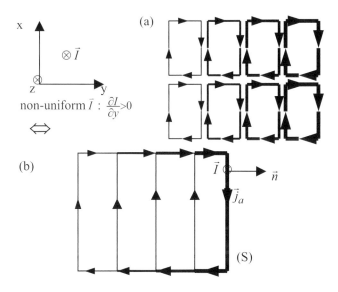

Figure 3.12. *Equivalence of (a) a nonuniformly magnetized material and (b) a material crossed by surface and volume Amperian currents.*

For a nonuniform I crossing toward positive values of y as shown in Figure 3.12(a), the geometric resultant of the volume currents is along Ox as detailed in Figure 3.12 b. The calculations for the Amperian volume currents, however, give the same result, as:

$$\vec{J}_a = \overrightarrow{\mathrm{rot}\vec{I}} = \begin{vmatrix} \dfrac{\partial}{\partial x} & \dfrac{\partial}{\partial y} & \dfrac{\partial}{\partial z} \\ 0 & 0 & I \end{vmatrix} = \begin{cases} \dfrac{\partial I}{\partial y} > 0 \ \ \text{along Ox} \\[2mm] -\dfrac{\partial I}{\partial y} > 0 \ \ \text{along Oy} \\[2mm] 0 \ \text{along Oz} \end{cases}$$

The geometric result of the Amperian surface currents also is reported in Figure 3.12 b, where the nonuniform density, the direction, and the sense of $\vec{J}_a$ are all in agreement with the calculation $\vec{j}_a = \vec{I} \times \vec{n}$.

3.2.4. *Definition of vectors* $\vec{B}$ *and* $\vec{H}$ *in materials*

3.2.4.1. *Definition of* $\vec{B}$

The vector $\vec{B}$ is defined by the relation $\vec{B} = \overrightarrow{rot}\,\vec{A}$, as detailed in Section 3.1.2.2.3. From this definition, the result is:

$$\boxed{\text{div}\vec{B} = 0}\,,$$

and therefore in a vacuum, as in the medium of a magnetic material, the flux vector $\vec{B}$ retains the same value (through conservation of flux).

For its part, the vector $\vec{A}$ can be calculated as if in a vacuum and subject to magnetic forces, with the proviso that all currents be brought into play, notably the volume ($\vec{J}_a = \overrightarrow{rot}\,\vec{I}$) and surface ($\vec{j}_a = \vec{I} \times \vec{n}$) density molecular currents.

Assuming that a certain volume of magnetic material is identical to the same volume of Amperian currents (and real currents when present) distributed in a vacuum, Ampere's theory can be written for the magnetic material as

$$\overrightarrow{rot}\,\vec{B} = \mu_0\vec{J}_T,$$

under which classic form only the volume currents ($\vec{J}$) appear, so that here

$$\vec{J}_T = \vec{J}_\ell + \vec{J}_a\,,$$

where $\vec{J}_\ell$ is the volume density of real currents (that is to say free currents, deliberately applied in the material).

3.2.4.2. *Definition of the excitation vector* $\vec{H}$

The relation $\overrightarrow{rot}\,\vec{B} = \mu_0\vec{J}_T = \mu_0(\vec{J}_\ell + \vec{J}_a)$ equally can be written:

$$\overrightarrow{rot}\,\frac{\vec{B}}{\mu_0} = \vec{J}_T = \vec{J}_\ell + \vec{J}_a = \vec{J}_\ell + \overrightarrow{rot}\,\vec{I}\text{ , so also, }\overrightarrow{rot}\left(\frac{\vec{B}}{\mu_0} - \vec{I}\right) = \vec{J}_\ell.$$

Finally, the introduction of the vector $\vec{H} = \left(\dfrac{\vec{B}}{\mu_0} - \vec{I}\right)$

gives

$$\overrightarrow{rot}\,\vec{H} = \vec{J}_\ell.$$

This last equation shows us that the real current density ($\vec{J}_\ell$) (deliberately applied) is the source current for the vector $\vec{H}$, generally called the magnetic excitation vector. For this vector, Ampere's theory is identical whether for a a vacuum or a material, which is $\overrightarrow{rot}\,\vec{H} = \vec{J}_\ell$.

In the material then, we have

$$\vec{B} = \mu_0\left(\vec{H} + \vec{I}\right),$$

(while in a vacuum, $\vec{B} = \mu_0\vec{H}$).

3.2.5. Conditions imposed on moving between two magnetic media

3.2.5.1. Continuity of the normal component of $\vec{B}$

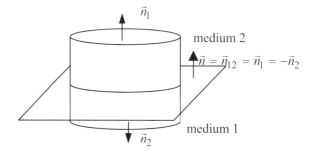

Figure 3.13. *Continuity of B_n at the interface between two magnetic media.*

In order to simplify the calculation, a cylindrical form is used of which the volume (V) is made up of two halves called medium 1 and medium 2. The height of the cylinder is infinitely small so that we can concentrate on the characteristics of the region about the interface—the sides are negligible in size with respect to the base of the volume.

With $\vec{n} = \vec{n}_{12} = \vec{n}_1 = -\vec{n}_2$, $\vec{B}$ can be written as:

$$\oiint_S \vec{B}.\overrightarrow{dS} = \iiint_V \text{div }\vec{B}\, d\tau = 0$$

$$= \iint_{\substack{\text{Sbase} \\ \text{superior}}} \vec{B}.\overrightarrow{dS} + \iint_{\text{S lower base}} \vec{B}.\overrightarrow{dS} + \iint_{\text{Sside}\to 0} \vec{B}.\overrightarrow{dS} = \iint_{\text{Sbase}} \left(\overrightarrow{B_2}.\vec{n} + \overrightarrow{B_1}.\overrightarrow{(-n)}\right)dS$$

$$\Rightarrow$$

$$\overrightarrow{B_1}.\overrightarrow{n} = \overrightarrow{B_2}.\overrightarrow{n} \text{ , which is stated as}$$

$$\boxed{B_{1n} = B_{2n}}.$$

This result indicates that there is a continuity of the normal component of B in the region of separating surfaces between the two media.

3.2.5.2. Relation between the tangential components of the vector $\vec{H}$ crossing a layer of current traversed by a superficial current of surface density $\vec{j}_s$

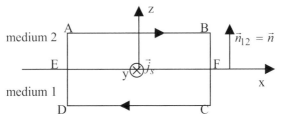

Figure 3.14. *Study of the continuities across a layer of current ($\vec{j}_s$).*

Figure 3.14 describes the outline of the form ABCDA drawn at the vicinity of an interface that contains a superficial current $\vec{j}_s$, which is perpendicular to the plane of the figure (intersection between ABCDA and the layer of current gives EF then traversed by the current $\vec{j}_s$).

Ampere's theory may be written as $\displaystyle \oint_{ABCDA} \vec{H}.\vec{dl} = I = \vec{j}_s.\vec{e}_y\, EF = j_s EF$,

because by construction $\vec{j}_s \parallel \vec{e}_y$. Therefore, with AD = BC ≈ 0 and by designating the component of $\vec{H}$ tangential to the current layer as H_t, then

$$\overrightarrow{H_{t2}}.\overrightarrow{AB} + \overrightarrow{H_{t1}}.\overrightarrow{CD} = j_s EF.$$

If $H_{t\perp}$ is the component of $\vec{H}$ perpendicular to $\vec{j}_s$ and collinear with the vector $\overrightarrow{AB}$ (itself perpendicular to $\vec{j}_s$), then with $\overrightarrow{AB} = -\overrightarrow{CD}$ we have $\left(H_{t2\perp} - H_{t1\perp}\right)AB = j_s EF$. As $AB = EF$, we can immediately see that $\left(H_{t2\perp} - H_{t1\perp}\right) = j_s$.

In effect, this relation can be written in the very general vectorial form,

$$\overrightarrow{H_{t2}} - \overrightarrow{H_{t1}} = \vec{j}_s \times \vec{n}_{12}.$$

If $j_s = 0$, the formula simplifies to $H_{t2} = H_{t1}$.

Comment: In the same way in which the current $\vec{J}_\ell$, which appears in Ampere's theory ($\overrightarrow{rot}\, \vec{H} = \vec{J}_\ell$), is a real current (deliberately applied), the superficial current ($\vec{j}_s$), which is such that $I = j_s EF$ and introduced above, also is a real current (deliberately applied) and in no circumstances contains the Amperian current component due to the Amperian surface current density (j_a).

3.2.6. Linear, homogeneous, and isotropic (l.h.i) magnetic media

3.2.6.1. Definition

These materials are such that at any point in the magnetic media (is therefore homogeneous) and in whatever direction (therefore isotropic) there is a linear relationship between the excitation ($\vec{H}$) and the response, which takes on the form

$$\vec{I} = \chi_m \vec{H}$$

where χ_m is the magnetic susceptibility of the medium.

The upshot is that

$$\vec{B} = \mu_0 \left(\vec{H} + \vec{I} \right) = \mu_0 \left(1 + \chi_m \right) \vec{H} = \mu \vec{H}$$

with $\mu = \mu_0 \left(1 + \chi_m \right)$ and $\mu_r = \dfrac{\mu}{\mu_0} = 1 + \chi_m$, where μ_r is relative permittivity and μ is the (magnetic) absolute permittivity.

3.2.6.2. Result 1

From Ampere's theorem, $\overrightarrow{rot}\vec{H} = \vec{J}_\ell$, we can determine that with $\vec{B} = \mu \vec{H}$:

$$\overrightarrow{rot}\vec{B} = \mu \vec{J}_\ell \ ,$$

so that $\mu \vec{J}_\ell = \overrightarrow{rot}\overrightarrow{rot}\vec{A} = \overrightarrow{graddiv}\vec{A} - \Delta\vec{A}$ and, finally, as $div\vec{A} = 0$:

$$\Delta\vec{A} + \mu \vec{J}_\ell = 0 \ .$$

3.2.6.3. Result 2

If the deliberately applied current density $\vec{J}_\ell = 0$, the result is that the volume density of the Amperian volumes $\vec{J}_A = 0$. In effect, as $\overrightarrow{rot}\,\vec{H} = \vec{J}_\ell$, if $\vec{J}_\ell = 0$, $\overrightarrow{rot}\,\vec{H} = 0$, so that with $\vec{I} = \chi_m \vec{H}$ we have $\dfrac{1}{\chi_m}\overrightarrow{rot}\,\vec{I} = 0 \Rightarrow \vec{J}_A = \overrightarrow{rot}\,\vec{I} = 0$.

3.2.7. Comment on the analogy between dielectric and magnetic media, terms for $\vec{H}$ and $\vec{B}$, and magnetic masses defined as calculable equivalents

3.2.7.1. The well-used analogy based on the equations $\vec{D} = \varepsilon_0 \vec{E} + \vec{P}$ *and*
$\vec{B} = \mu_0 (\vec{H} + \vec{I}) = \mu_0 \,\vec{H} + \mu_0 \vec{I}$

The direct transposition from the above equations would seem to correspond $\vec{D}$ with $\vec{B}$, $\vec{E}$ with $\vec{H}$, and $\vec{P}$ with $\mu_0\vec{I}$. As $\vec{D}$ is called the electric induction vector, $\vec{B}$ is generally called the magnetic induction vector. With $\vec{E}$ being the electric field vector, in the same way, $\vec{H}$ is called the electric field vector.

3.2.7.2. *Analogy based on the "source" equations of Gauss's and Ampere's theories*

These equations are:

$$\mathrm{div}\vec{D} = \rho_\ell \qquad \text{and} \qquad \overrightarrow{\mathrm{rot}}\vec{H} = \vec{J}_\ell \; ; \text{and}$$

$$\mathrm{div}\vec{E} = \frac{\rho_\ell + \rho_P}{\varepsilon_0} \qquad \text{and} \qquad \overrightarrow{\mathrm{rot}}\, \vec{B} = \mu_0(\vec{J}_\ell + \vec{J}_a).$$

In this case, it is the vector $\vec{H}$ that generally is termed the magnetic excitation (or induction) vector by analogy with the electric excitation (or induction) vector $\vec{D}$, as the two cases share the same type of real source (deliberately applied).

The vector $\vec{B}$ really should be called the magnetic field vector through an analogy with the electric field vector ($\vec{E}$), as the sources in both cases are both real and equivalent to the polarization/magnetization, all placed in a vacuum (of permittivity ε_0 or permeability μ_0, which intervene in the Gaussian or Amperian equations).

3.2.7.3. *Conclusion*
These two arguments driving toward two different sets of names cannot be anything but admissible and it seems reasonable to accept both terms. In order to make the problem easier, it seems nevertheless more simple to call the two magnitudes "vector H" for $\vec{H}$ and "vector B" for $\vec{B}$. In any case, in mathematical terms, $\vec{H}$ just like $\vec{B}$ corresponds to a field of vectors each determined by their three components in the trihedral reference grid.

3.2.7.4. *Comment. Magnetic masses as intermediates in calculations for permanent magnets*

As $\vec{B} = \overrightarrow{\mathrm{rot}}\vec{A}$, $\mathrm{div}\vec{B} = 0 = \mu_0\mathrm{div}(\vec{H} + \vec{I})$, where $\mathrm{div}\vec{H} = -\mathrm{div}\vec{I}$.

By analogy with electrostatics, where the charge volume density for polarization is $\rho_p = -\mathrm{div}\vec{P}$, in magnetism the volume density of magnetic masses equivalent to the magnetization can expressed as $\rho^* = -\mathrm{div}\vec{I}$. Similarly, the surface density of magnetic mass can be characterized through $\sigma^* = I_n$ (by analogy with $\sigma_p = P_n$).

So we have $\text{div}\vec{H} = -\text{div}\vec{I} = \rho^*$. The result is that

$$\Phi_H = \oiint \vec{H}.\overrightarrow{dS} = \iiint \text{div}\,\vec{H}\,d\tau = \iiint \rho^* d\tau,$$

which is Gauss's theory for magnetism, and is a relation that is equivalent to Gauss's theory for electrostatics:

$$\Phi_E = \oiint \vec{E}.\overrightarrow{dS} = \frac{1}{\varepsilon}\iiint \rho_\ell d\tau.$$

Placing $m^* = \iiint \rho^* d\tau$, we also find $\Phi_H = \oiint \vec{H}.\overrightarrow{dS} = m^*$ (the flux calculated over all space, that is $\Omega = 4\pi$), and for a solid angle $d\Omega$, the elementary flux is:

$$d\Phi_H = \frac{m^*}{4\pi}d\Omega = \frac{m^*}{4\pi}\frac{\overrightarrow{dS}.\vec{u}}{r^2} = \vec{H}.\overrightarrow{dS}, \text{ where by identification } \vec{H} = \frac{1}{4\pi}\frac{m^*}{r^2}\vec{u} \text{ and is}$$

the magnetic field due to a magnetic mass m*.

It is worth noting that the magnetic masses introduced here are only equivalents used in calculations and as such are not physically real, in contrast to the Amperian currents. Experience tells us that the negative and positive masses cannot be separated; indeed, the two poles are "inseparable" and can be physically attached only to two faces, north and south, of a closed current, as in Amperian currents.

3.3. Problems

3.3.1. Magnetic moment associated with a surface charged sphere turning around its own axis

A sphere of radius R turns around its axis Oz at an angular velocity ω. The sphere is uniformly charged at its surface with a charge density σ_ℓ.

1. Directly obtain the expression for the charge carried by an elementary part of the surface defined by $dS = 2\pi R^2 \sin\theta d\theta$. Calculate the intensity equivalent to this amount of charge rotating at the angular velocity ω.

2. Given that the closed loop of current of radius $R\sin\theta$ is traversed by the above current dI, determine the magnetic moment of the rotating sphere.

Answers

1. The charge carried by the surface element $dS = 2\pi R^2 \sin\theta d\theta$ is $dq = 2\pi\sigma_\ell R^2 \sin\theta d\theta$. The intensity dI can be thought of as the ratio of the quantity of charge that goes through the section dS with each complete turn of the sphere (i.e., dq), over the time (dt) for the charge to pass through this section. This time is that which the sphere takes to make one turn, which is $dt = T = \dfrac{2\pi}{\omega}$. The result is therefore $dI = \dfrac{dq}{dt} = \dfrac{2\pi\sigma_\ell R^2 \sin\theta d\theta}{2\pi/\omega} = \sigma_\ell R^2 \omega \sin\theta d\theta$.

2. The magnetic moment $(d\mu)$ that corresponds current ring crossed by dI is by definition such that $d\mu = dI.S$, where S is the surface limited by the closed circuit. Given that the ring of current has a radius $R\sin\theta$, its surface S is therefore:

$S = \pi (R \sin\theta)^2$, where $d\mu = dI.S = \sigma_\ell R^2 \omega \sin\theta d\theta \, \pi (R \sin\theta)^2 = \pi\sigma_\ell \omega R^4 \sin^3\theta d\theta$.

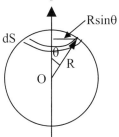

The resultant magnetic moment over the whole sphere is obtained by summing over all the elementary magnetic moments associated with the circuits of radius $R\sin\theta$, where θ varies from 0 to π:

$$\mu = \int_{\theta=0}^{\pi} d\mu = \int_{\theta=0}^{\pi} \pi\sigma_\ell \omega R^4 \sin^3\theta d\theta = \pi\sigma_\ell \omega R^4 \int_{\theta=0}^{\pi} \sin^3\theta d\theta$$

, where

$$= \pi\sigma_\ell \omega R^4 \int_{\theta=0}^{\pi} (1-\cos^2\theta)d(-\cos\theta)$$

$$\mu = \pi\sigma_\ell \omega R^4 \left(\left[-\cos\theta\right]_0^\pi + \left[\dfrac{\cos^3\theta}{3}\right]_0^\pi \right) = \dfrac{4}{3}\pi\sigma_\ell \omega R^4 .$$

3.3.2 *Magnetic field in a cavity deposited in a magnetic medium*

A magnetic medium with permeability μ and uniform magnetic intensity $\vec{I}$ parallel to Oz contains an empty spherical cavity of radius R.

1. Determine the Amperian currents $\vec{J}_A$ and $\vec{j}_A$ equivalent to the magnetization.

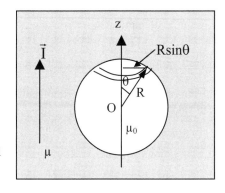

2. Calculate the elementary induction *dB* at the center O of the cavity created by a coil of size $Rd\theta$ carrying a current dI associated with a layer of surface current of density $\vec{j}_A$.

3. From this, determine the total induction formed at O only by the magnetized material. *N.B.* It is worth remembering that the field created by a coil of radius r carrying an intensity i at a point on the axis that observes the coil from an angle θ is given by $H = \dfrac{i\sin^3\theta}{2r}$.

Answers

1. We have $\vec{J}_A = \overrightarrow{rot\,I} = 0$ as $\vec{I}$ is uniform. In addition, we have $\vec{j}_A = \vec{I} \times \vec{n}_{ext}$, where $\vec{n}_{ext}$ is the normal external to the magnetic medium, as shown in the figure on the right. Therefore $\vec{j}_A$ is perpendicular to the plane of the figure and corresponds to a surface current tangential to the sphere's surface, which can be seen as a layer of current. Its modulus is $j_A = I\sin(\pi - \theta) = I\sin\theta$.

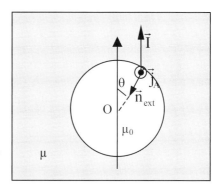

2. As noted in the question, for a coil with a radius r, the field H formed at a point M at a distance z from the coil is directed along the axis of the coil and has a modulus of

$$H = \frac{i\,r^2}{2(r^2+z^2)^{3/2}} = \frac{i\sin^3\theta}{2r}.$$

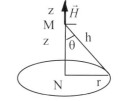

A coil carrying a layer of current with current density equal to j_A is traversed by the intensity

$di = j_A \times dl = j_A \times Rd\theta = R \times I\sin\theta d\theta$.

The result is the formation at O of an elementary induction $\overrightarrow{dB}$ of sense opposite to that of $\vec{I}$ and such that $dB = \dfrac{\mu_0}{2r}dI\sin^3\theta$.

With $r = R\sin\theta$, we have

$$dB = \frac{\mu_0}{2R\sin\theta}R\,I\sin^4\theta\,d\theta = \frac{\mu_0 I}{2}\sin^3\theta\,d\theta.$$

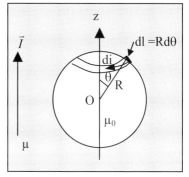

The total induction (B) is therefore:

$$B = \frac{\mu_0 I}{2} \int\limits_{\theta=0}^{\pi} \sin^3\theta\, d\theta$$

$$= \frac{\mu_0 I}{2} \int\limits_0^{\pi}(1-\cos^2\theta)d(-\cos\theta) = \frac{\mu_0 I}{2}\frac{4}{3}, \text{ that is to say:}$$

$$B = \frac{2}{3}\mu_0 I.$$

3.3.3. *A cylinder carrying surface currents*

A cylinder of radius a has a surface revolving around an axis Oz, a direction through which the cylinder is infinitely long. The surface is carrying superficial currents that have a value for each current point M(a,θ,z) of the form $\vec{k} = k_0 \sin\theta\, \vec{e}_z$.

1. Show that the distribution of the current is equivalent to a uniformly magnetized cylinder with a magnetization intensity vector $\vec{I}$ to be determined.
2. Take the problem in the inverse direction and show that its resolution is actually faster.
N.B. For a cylinder,

$$\overrightarrow{\text{rot}\vec{I}} = \left[\frac{1}{r}\frac{\partial I_z}{\partial\theta} - \frac{\partial I_\theta}{\partial z}\right]\vec{e}_r + \left[\frac{\partial I_r}{\partial z} - \frac{\partial I_z}{\partial r}\right]\vec{e}_\theta + \left[\frac{1}{r}\frac{\partial}{\partial r}(rI_\theta) - \frac{1}{r}\frac{\partial I_r}{\partial\theta}\right]\vec{e}_z.$$

Answers

1.   Given the equivalence of "magnetized material with a magnetization intensity $\vec{I}$ ", and "material traversed by currents of surface density $\vec{k}$ and volume density $\vec{K}$ ", we can write that $\vec{k} = \vec{j}_a = \vec{I}\times\vec{n}$ and $\vec{K} = \vec{J}_a = \overrightarrow{\text{rot}\vec{I}}$. As here, $\vec{k} = k_0 \sin\theta\, \vec{e}_z$ and $\vec{K} = 0$ (as no volume current is indicated in the question), the vector $\vec{I}$ should verify on one hand $k_0 \sin\theta\, \vec{e}_z = \vec{I}\times\vec{n}$, and on the other $\overrightarrow{\text{rot}\vec{I}} = 0$.

Given the symmetry of the problem, which would tend to invite the use of cylindrical coordinates, we have $\vec{I}\begin{pmatrix}I_r \\ I_\theta \\ I_z\end{pmatrix}\times\vec{n}\begin{pmatrix}1 \\ 0 \\ 0\end{pmatrix} = \begin{pmatrix}0 \\ I_z \\ -I_\theta\end{pmatrix} = \vec{k}\begin{pmatrix}0 \\ 0 \\ k_0 \sin\theta\end{pmatrix}$, where, by identification, $\vec{I} \equiv \begin{pmatrix} I_r = ? \\ I_\theta = -k_0 \sin\theta \\ I_z = 0 \end{pmatrix}$.

In order to determine I_r, we are obliged to use the second equation, $\overrightarrow{\text{rot}}\vec{I} = 0$, for which the three components give three equations (taking into account that $I_z = 0$)

so $-\dfrac{\partial I_\theta}{\partial z} = 0$, $\dfrac{\partial I_r}{\partial z} = 0$, $\dfrac{1}{r}\dfrac{\partial}{\partial r}(rI_\theta) - \dfrac{1}{r}\dfrac{\partial I_r}{\partial\theta} = 0$.

As $I_\theta = -k_0 \sin\theta$ ($=$ constant with respect to r), the last equation gives $I_\theta - \dfrac{\partial I_r}{\partial\theta} = 0$,

where $\dfrac{\partial I_r}{\partial\theta} = I_\theta = -k_0 \sin\theta$, or rather $I_r = k_0 \cos\theta$.

Finally, $I = \sqrt{I_r^2 + I_\theta^2} = \sqrt{k_0^2(\cos^2\theta + \sin^2\theta)} = k_0$.

As $\dfrac{|I_\theta|}{|I_r|} = \tan\theta$, we have the geometrical

representation which details how $\vec{I} // \vec{e}_x$, where

$\vec{I} = k_0 \vec{e}_x$.

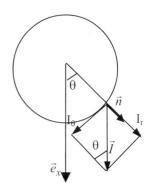

2. Turning question 1 on its head implies finding the Amperian currents equivalent to a magnetic cylinder with a magnetization intensity of $\vec{I} = k_0 \vec{e}_x$. As this magnetization intensity is uniform and has $k_0 =$ constant and an orientation at all points following Ox, the volume Amperian currents are zero and all that remains to determine are the surface Amperian currents $\vec{j}_a = \vec{I} \times \vec{n}$. Taking into account that $\vec{I}$ is orientated following Ox, we are more than obliged to use Cartesian coordinates, and thus directly find:

$$\vec{j}_a = \vec{I} \times \vec{n} = \begin{pmatrix} k_0 \\ 0 \\ 0 \end{pmatrix} \times \begin{pmatrix} n_x = \cos\theta \\ n_y = \sin\theta \\ 0 \end{pmatrix} = \begin{pmatrix} 0 \\ 0 \\ k_0\sin\theta \end{pmatrix},$$

so that

$$\vec{j}_a = k_0 \sin\theta\, \vec{e}_z.$$

The resolution of the first question is greatly facilitated by using Cartesian coordinates, rather than the cylindrical coordinates that the symmetry of the problem otherwise would suggest when $\vec{I}$ is unknown!

3.3.4 Virtual current

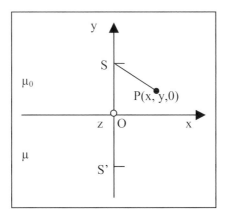

In the trihedral indicated by Oxyz, the plane xOz separates a vacuum (of permeability μ_0) corresponding to the region $y > 0$, from a magnetic medium (of permeability μ) that extends through the inferior part of the defined space where $y < 0$. An infinitely long conducting wire, both rectilinear and parallel to Oz, is placed at S, which has coordinates (0, a, 0). A current of intensity I goes through the wire in sense $\vec{u}$, a unit vector along the Oz axis. An additional point, S' which is used in the problem, has coordinates (0, -a, 0).

It is possible to show that for any point P with coordinates (x, y, 0), the vector $\vec{B}$ is in the form

- when $y > 0$, $\vec{B}_0 = \dfrac{\mu_0 I}{2\pi} \vec{u} \times (\dfrac{\vec{r}}{r^2} + K \dfrac{\vec{r'}}{r'^2})$; and

- when $y < 0$, $\vec{B} = \dfrac{\mu I}{2\pi} \vec{u} \times \dfrac{L\vec{r}}{r^2}$,

noting that in both cases, $\vec{r} = \overline{SP}$ and $\vec{r'} = \overline{S'P}$. Also, K and L are two constant that are given by $K = \dfrac{\mu - \mu_0}{\mu + \mu_0}$ and $L = \dfrac{2\mu_0}{\mu + \mu_0}$, respectively.

1. Recall the method by which the constants K and L can be determined.

2. Directly calculate the expression for the magnetizing field $\overrightarrow{H_e(P)}$ formed by the current I.

3. As a function of $\overrightarrow{H_e(P)}$, give the expressions for:

    (a) the field $\overrightarrow{H(P)}$ in the medium ($y < 0$);

(b) the demagnetizing field $\overline{H_d(P)}$; and

(c) the magnetization intensity vector $\overline{J(P)}$.

4. Determine the Amperian currents equivalent to the magnetization.

5. Calculate for the point S the expression for $\vec{B}$ via two different routes:

(a) directly from the above derived equations for $\vec{B}$ expressed as a function
of predetermined constants; and

(b) by bringing in to bear on the calculation all the currents, both real and Amperian.

Answers

1. Ampere's theory states that $H\ 2\pi r = I$, so at point P,

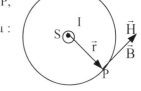

$\vec{H} = \dfrac{I}{2\pi}\vec{u}\times\dfrac{\vec{r}}{r^2}$, and in the medium with permeability μ :

$\vec{B} = \dfrac{\mu\,I}{2\pi}\vec{u}\times\dfrac{\vec{r}}{r^2}$.

At a point P in the medium with permeability μ_0, the field results from the current of intensity I (in the medium where the permeability is μ_0) and the current KI (K remains to be determined) at S' (where the permeability is μ). S' is symmetric to S about the origin and KI is the "reflected" (virtual) current.

When P is in the medium with permeability μ, the field to which it is subject is of a "perturbed" current—by the presence of the two different media—in the form LI where L has yet to be determined.

The problem therefore is to find two unknowns, K and L, which can be determined by introducing two equations, namely for the continuity at the interface (plane xOz) of the tangential component of H which gives $[1 - K] = L$, and for the normal component of B which gives $K[\mu+\mu_0] = \mu-\mu_0$.

2. The calculation was performed in (1), where $\vec{H}_e(P) = \dfrac{I}{2\pi}\vec{u}\times\dfrac{\vec{r}}{r^2}$, with $\vec{r} = \overline{SP}$.

3. $\vec{H} = \vec{H}_e + \vec{H}_d$ for a point inside the material and given that the fields are:

(a) in the material $\vec{H} = \dfrac{\vec{B}}{\mu} = \dfrac{L\,I}{2\pi}\vec{u}\times\dfrac{\vec{r}}{r^2} = L\,\vec{H}_e$

(b) $\vec{H}_d = \vec{H} - \vec{H}_e = (L-1)\,\vec{H}_e = - K\,\vec{H}_e$

(c) $\vec{B} = \mu_0\left(\vec{H} + \vec{I}_a\right) = \mu\,\vec{H}$ where $\vec{I}_a$ is the magnetization intensity, we find

$$\Rightarrow \vec{I}_a = \frac{\vec{B} - \mu_0 \vec{H}}{\mu_0} = \frac{\mu - \mu_0}{\mu_0} \frac{2\mu_0}{\mu + \mu_0} \vec{H}_e = 2\frac{\mu - \mu_0}{\mu + \mu_0} \vec{H}_e = 2K \, \vec{H}_e$$

(where $\vec{I}_a = \chi \vec{H} = (\mu_r - 1) L \, \vec{H}_e = 2\dfrac{\mu_r - 1}{\mu_r + 1} \vec{H}_e$).

4. $\vec{J}_A = \overrightarrow{\text{rot}} \, \vec{I}_a = (\mu_r - 1) \overrightarrow{\text{rot}} \, \vec{H}$, while $\overrightarrow{\text{rot}} \, \vec{H} = \vec{j}_\ell = 0$ (no current source in the material) $\Rightarrow \vec{J}_A = 0$.

$$\vec{j}_A = \vec{I}_a \times \vec{n} = 2K \, \vec{H}_e \times \vec{n} = \frac{2KI}{2\pi r^2} \vec{u} \times \vec{r} \times \vec{n} = \frac{KI}{\pi r^2} \, a \, \vec{u}$$

5. We have

(a) $\vec{B}(r = 2a) = \dfrac{\mu \, I}{2\pi} \vec{u} \times \dfrac{L \, \vec{r}}{r^2} = \dfrac{2\mu_0}{\mu + \mu_0} \dfrac{I}{2\pi} \dfrac{1}{2a} \vec{e}_x$;

(b) $\vec{B} = \vec{B}(\text{produced by I in vacuum}) + \vec{B}(\text{produced by } \vec{j}_A \text{ in vacuum})$.

B(produced in S' by I located in S in vacuum) $= \dfrac{\mu_0 I}{4\pi a}$.

$dB_{\text{ampérien}} = \dfrac{\mu_0 j_A \, dx}{2\pi \, r}$; the relevant component, given its symmetry, is:

$(dB_{\text{Amperian}})_{\text{useful}} = dB_{\text{Amperian}} \cos\theta = \dfrac{\mu_0 j_A \, dx}{2\pi \, r} \cos\theta$ (see Figure below).

With $x = a \, \tan\theta$ and $\cos\theta = \dfrac{a}{r}$, we have:

$B_{\text{Amperian}} = \displaystyle\int_{\theta = -\pi/2}^{\theta = \pi/2} \dfrac{\mu_0 KI}{2\pi a^2} \cos^2\theta \; d\theta = \dfrac{\mu_0 KI}{4\pi a}$,

from which comes the same result for $\vec{B}$ as found in part (a).

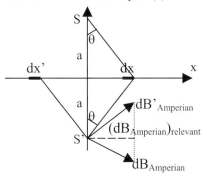

Chapter 4

Dielectric and Magnetic Materials

4.1. Dielectrics

4.1.1. Definitions

Dielectrics are in effect electrical insulators. A scale of conductivity can be divided into somewhat arbitrary but well recognized characteristics for each group of materials.

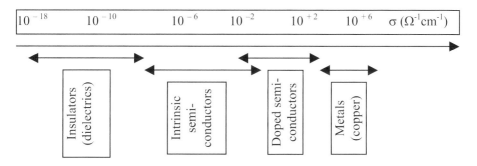

There are certain characteristics that are specific to dielectrics. These include:

- Dielectric withstand strength (Ec) (usually given in the units kV mm^{-1}). When the electric field E > Ec, the dielectric is no longer an insulator and an electrical discharge is generated.

- Breakdown potential (Uc). When the potential applied is such that U > Uc, the dielectric is no longer an insulator.

- Electrical discharge, which is the current passing through the dielectric when it breaks down. The discharge is due to the formation of a highly conductive passage between two electrodes.

4.1.2. Origins and types of breakdowns

4.1.2.1. Thermal breakdown

When a dielectric exhibits losses through dipolar absorptions or leak currents, a Joule effect ensues that generates heat. If the heat produced is greater than the heat given out by the insulator, the temperature rises and this can come about rapidly as dielectrics are often both good electrical and thermal insulators. As the conductivity (σ) is related to temperature by $\sigma = \sigma_0 e^{-(\frac{E_G}{kT})}$, where E_G is the band gap of the insulator or semiconductor, the conductivity also increases with temperature to the point where the material can no longer be termed an insulator.

4.1.2.2. Intrinsic breakdown

Such a breakdown is caused by a snowball effect rather than the Joule effect, which no longer plays a role. Once the electric field is sufficiently high, a significant number of electrons impact with and ionize the dielectric. Electron-hole pairs are then separated by the electric field, and holes (poorly mobile) tend to accumulate near the cathode. The resulting space-charge reinforces the local electric field and contributes to an increase in the number of ionizations. Field effects also can give rise to additional emissions.

4.1.2.3. Ageing and changes in Ec with time

If a dielectric contains inhomogeneities such as cavities or imperfections due to foreign particles, partial discharges can develop around these defaults and an erosion or even a localized melting of the dielectric can result. A network of more or less conducting channels may then develop resembling so closely the branches on a tree that the effect is indeed called treeing. An example is shown in Figure 4.1.

Just as mechanical strains can generate cracks, humidity and ionizing radiation, present in our everyday environment, can provoke similar disruptions in certain polymers. In addition, the shape of the electrodes (or contacts) can play a central role, so that to limit localized breakdowns, bumps, and deformities are avoided. An important example of a way in which such effects are limited is the use of π-conjugated polymers in the insulation of high tension cables, where they are used as an inner sheath around the copper core.

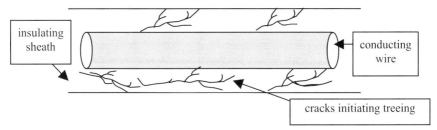

Figure 4.1. *Electric cable covered by an insulating sheath showing treeing.*

4.1.3 Insulators

4.1.3.1. Natural and inorganic insulators

Examples include:

- composites made from natural materials such as paper or cotton impregnated with oil (paraffin) and wax;
- electronegative gases, in particular those with halogen atoms, for example, SF_6, which have high electron affinities, and thus reduce discharges by reducing the density of free charges; and
- inorganic materials such as ceramics prepared at high temperatures and pressures (eliminating the need for binders) and engineering ceramics (containing titanium resulting in a high permittivities) that are used in specific applications, for example high value capacitors. Ceramics containing high amounts of aluminum facilitate metal plating, a useful property when used as substrates for electrical circuits.

4.1.3.2. Synthetic organic insulators

For the most part, these materials are based on polymers, which consist of a chain of a high number of monomers (M). The principal types of polymers are:

linear homopolymer $-M - M - M - M -$

branched polymer
$$M - M - M\begin{cases} M - M - M - M - M - \\ M - M - M - M - M - \end{cases}$$

reticulated polymer
$$\begin{array}{c} -M-M-M-M-M-\ \ M-M-M-M-\\ \quad | \qquad\qquad\qquad | \\ \quad M \qquad\qquad\qquad M \\ \quad | \qquad\qquad\qquad | \\ M-M-M-M-M-M-M-M-M- \end{array}$$

alternating copolymer $- M_1 - M_2 - M_1 - M_2 -$

The degree of crystallinity is defined by $\tau = \dfrac{\text{volume of crystalline part}}{\text{sample volume}}$.

The most common examples are:

1. Polyethylene (PE), which is based on the structure:

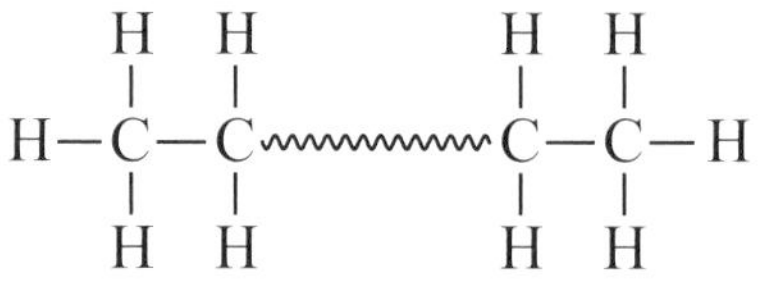

$$
\begin{array}{cccc}
\text{H} & \text{H} & & \text{H} & \text{H} \\
| & | & & | & | \\
\text{H}-\text{C}-\text{C}&\!\!\!\!\text{\scriptsize wwwwww}\!\!\!\!&\text{C}-\text{C}-\text{H} \\
| & | & & | & | \\
\text{H} & \text{H} & & \text{H} & \text{H}
\end{array}
$$

and is prepared from ethylene, which has the chemical structure $H_2C = CH_2$. High-density polyethylene (often indicated as HDPE) has a high level of crystallinity.

2. Poly(vinyl chloride) (PVC), is based on the structure:

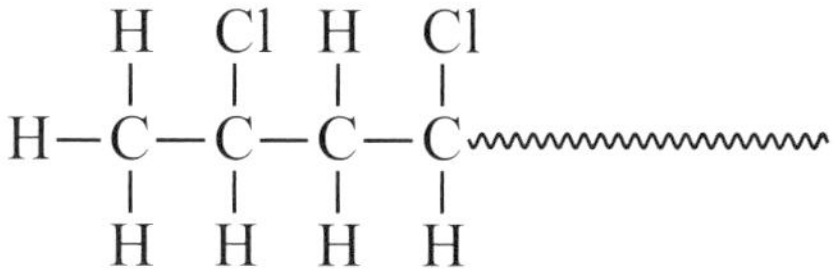

$$
\begin{array}{cccc}
\text{H} & \text{Cl} & \text{H} & \text{Cl} \\
| & | & | & | \\
\text{H}-\text{C}-\text{C}-\text{C}-\text{C}&\!\!\!\!\text{\scriptsize wwwwwwww}\\
| & | & | & | \\
\text{H} & \text{H} & \text{H} & \text{H}
\end{array}
$$

and is prepared from vinyl chloride.

3. Polystyrene (PS)

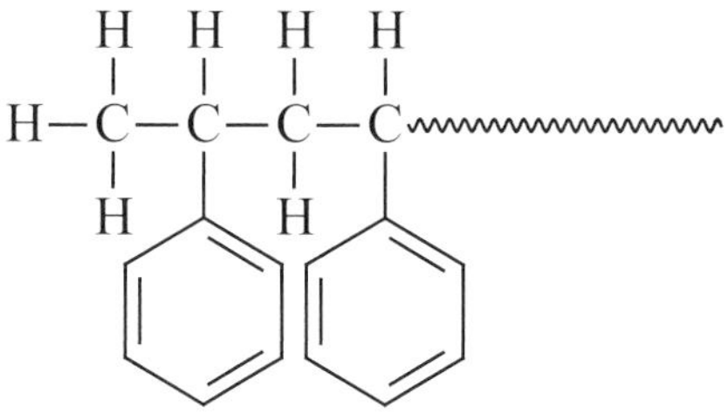

comes from the polymerization of styrene, which unlike the above-noted systems, carries aromatic phenyl rings.

4. Polypropylene (PP)

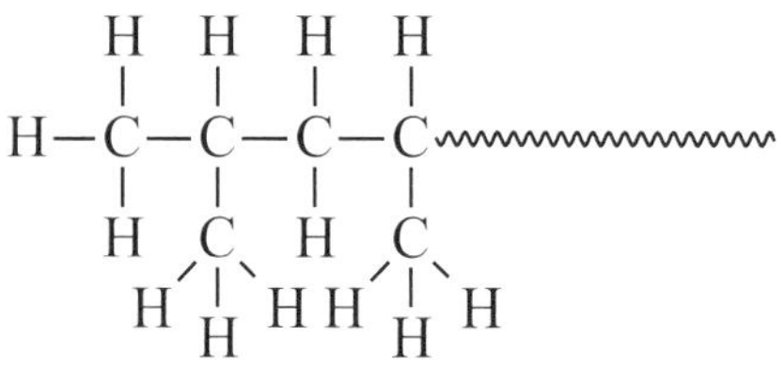

as its name indicates results from the polymerization of propylene.

5. Polytetrafluoroethylene (PTFE) is also known under its commercial name of Teflon, owned by DuPont. This is an amorphous polymer that, on carrying a "thermal history", and having undergone mechanical treatment, does not tend toward a crystalline state.

$$
\begin{array}{cccc}
\text{F} & \text{F} & \text{F} & \text{F} \\
| & | & | & | \\
\text{C} & \text{C} & \text{C} & \text{C} \\
| & | & | & | \\
\text{F} & \text{F} & \text{F} & \text{F}
\end{array}
$$

Dielectric characteristics of the polymers

	PE	PVC	PS	PP	PTFE
ε_r (50 Hz)	2.3	4	2.4	2.3	2
Tanδ (1 MHz)	$< 5 \times 10^{-4}$	10^{-1}	10^{-4}	$< 5.10^{-4}$	$< 5.10^{-4}$
$\rho = 1/\sigma$ (Ω m)	$> 10^{14}$	$10^9 - 10^4$	$> 10^{14}$	$> 10^{14}$	$> 10^{13}$
Ec (kV mm^{-1})	17-28	11-32	16-28	20-26	17-24

4.1.4 Electrets
4.1.4.1. Definition and properties
Electrets are dielectric materials that carry a quasipermanent charge and are analogous to magnets. They have a permanent polarization; however, the charges involved are relatively small. Under normal ambient conditions, ions in the atmosphere, resulting from natural ionizing radiation, can neutralize the deposited charges. In Figure 4.2, an example of the evolution of surface charge on an electret is given. Typically, after around 2 or 3 months, there is only 20 to 30 % of the initial charge.

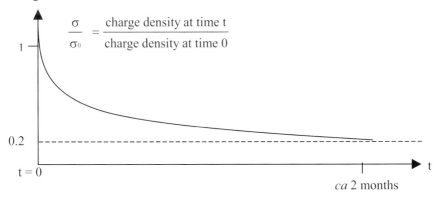

Figure 4.2. *Diminution of charge on an electret with time.*

4.1.4.2. *Operational details of the voltammeter* (devices which measure charge densities)
These devices operate on the principle of compensation. That is to say, that for a charged dielectric placed on the lower electrode, the following equation can be

written:

$$V_0 = V_A - V_B = \int_A^B \vec{E}.\vec{dl} = E\,s + E'.\,s' \qquad (1)$$

where σ is the surface charge density of the dielectric, s is its thickness, and s' is the distance from the upper electrode. E and E' are the electric fields at the dielectric and between the electret and the upper electrode, respectively.

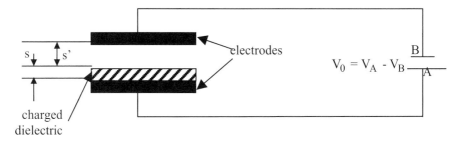

Figure 4.3. *Schematization of the operating principles of an electrostatic voltammeter.*

Imposing the condition of continuity on the component, normal to the displacing vector gives:

$$D_n - D'_n = \sigma, \text{ which in turn yields } \varepsilon E - \varepsilon_0 E' = \sigma. \qquad (2)$$

V_0 is adjusted until $E' = 0$ (by using the compensation method), so that

       (1) gives (1'): $V_0 = E\,s$,
       (2) gives (2'): $\sigma = \varepsilon E$

Determining the ratio (2')/(1') directly gives $\sigma = \dfrac{\varepsilon V_0}{s}$.

4.1.4.3. Piezoelectrets
These dielectrics become polarized when subject to an applied force. Inversely, when subject to a polarization and in the absence of any mechanical constraints, they change their dimensions. The more common piezoelectrets are quartz (SiO_2), barium titanate ($BaTiO_3$), and aluminum phosphate ($AlPO_4$).

There is a linear relationship between the causal applied force and the resulting polarization, and elasticity theory can be used to describe the phenomenon. Piezoelectricity is a property tied closely to the structure of a material. For it to exist, the centers of gravity of positive and negative charges, which coincide in the absence of any strain as shown in Figures 4.4a and 4.5a, are separated by deformation. In Figure 4.4b, the dipolar moment remains at 0 because of symmetry, so the material is not piezoelectric; however, in Figure 4.5b, the deformation in Oy leads to a nonsymmetry and a dipolar moment in the direction Ox.

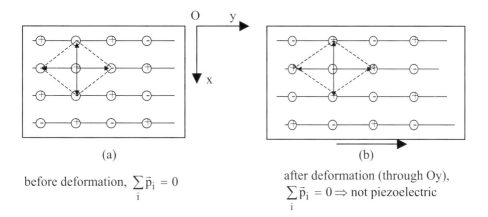

(a)

before deformation, $\sum_{i} \vec{p}_i = 0$

(b)

after deformation (through Oy),
$\sum_{i} \vec{p}_i = 0 \Rightarrow$ not piezoelectric

Figure 4.4. *Deformation yielding no piezoelectric effect.*

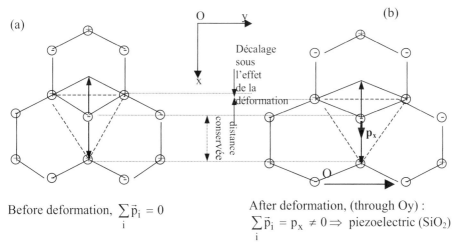

Before deformation, $\sum_{i} \vec{p}_i = 0$

After deformation, (through Oy) :
$\sum_{i} \vec{p}_i = p_x \neq 0 \Rightarrow$ piezoelectric (SiO_2)

Figure 4.5. *Deformation resulting in a piezoelectric effect (SiO_2).*

4.1.5. Ferroelectrics

4.1.5.1. Definition

Ferroelectric materials possess domains, called ferroelectric, inside which dipolar moments are coupled with each other, thus giving rise to spontaneous polarizations.

4.1.5.2. Properties

The dielectric permittivity (ε_r) of ferroelectric materials is very high and can reach values of around 10^3 to 10^4. It is for this reason that they are used in high-strength capacitors. They are very sensitive to temperature: above the so-called Curie ferroelectric temperature, the ferroelectricity disappears.

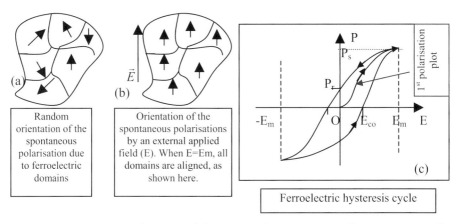

Random orientation of the spontaneous polarisation due to ferroelectric domains	Orientation of the spontaneous polarisations by an external applied field (E). When E=Em, all domains are aligned, as shown here.

Ferroelectric hysteresis cycle

Figure 4.6. *Ferroelectric domains and their (a) random orientation; (b) organization under an external field; and (c) hysteresis plot.*

A plot of polarization against the applied electric field resembles a normal cycle of hysteresis:

- At the initial zero field strength, as in Figure 4.6a, the overall polarization of the sample is zero even though each individual domain gives rise to a polarization. This polarization is due to the gradual coupling of dipolar moments up to the limits of the domain, which are structural dislocations of various origin. As the orientation of all the domains is completely random, the initial overall polarization is zero.

- When an electric field is applied, a coupling energy tends to orientate the ferroelectric domains in the directional sense of the field. This coupling energy (W) follows the equation

$$W = - p.E \cos \theta$$

and is directly proportional to E. The disorientated polarized domains orientate themselves to the field bit by bit with the increasing field, and this first polarization gives the plot shown in Figure 4.6b. Once the value E_m is attained, saturation occurs, i.e., $P = Ps$.

The completed cycle is shown in Figure 4.6c, where E is varied from $-E_m$ to $+E_m$. At $E = 0$, the permanent polarization (P_r) remains, so to return to $P = 0$, the coercive field (E_{co}) needs to be applied.

4.1.5.3. Polarization with respect to temperature
4.1.5.3.1 Conditions for a spontaneous polarization (Ps ≠0)

At a temperature $T < \theta_c$, the ferroelectric material can exhibit a nonzero spontaneous polarization (i.e., Ps ≠ 0 without an external field). To understand how this can be brought about, we shall look at a highly polar ferroelectric material in which the orientated polarization (P_{or}) dominates the other polarized components; that is to say that $P_{or} \approx P$.

For this system, and from Sections 2.5.3.2-3, we have:
$$P = n\,\bar{\mu} = n\,\mu_p <\cos\theta> = n\,\mu_p L(\beta) \qquad (1),$$

where $\beta = \dfrac{\mu_p E_{al}}{kT}$ and E_{al} is such that $\vec{E}_{al} = \vec{E}_a + \dfrac{\vec{P}}{3\varepsilon_0}$.

In the absence of an applied field, the presence of P is such that $E_a = 0$ so that

$$E_{al} = \frac{P}{3\varepsilon_0}, \quad \beta = \frac{\mu_p P}{3\varepsilon_0 kT}, \text{ and accordingly } P = \frac{3\varepsilon_0 kT}{\mu_p}\beta \qquad (2).$$

Given that Eq. (1) = Eq. (2), we find that $L(\beta) = \dfrac{3\varepsilon_0}{n}\dfrac{kT}{\mu_p^2}\beta$. This equation,

which contains the condition $P \neq 0$ when $E_a = 0$, is in fact the straight slope due to $\dfrac{3\varepsilon_0}{n}\dfrac{kT}{\mu_p^2}$.

To find the solution, which corresponds to $P \neq 0$ when $E_a = 0$, the line of

the equation $\dfrac{3\varepsilon_0}{n}\dfrac{kT}{\mu_p^2}\beta$ must intercept the Langevin function ($L(\beta)$), which is a

tangent with a slope of 1/3 and has certain aspects detailed in Chapter 2. This means

that the slope of the equation $\dfrac{3\varepsilon_0}{n}\dfrac{kT}{\mu_p^2}\beta$ must be less that the slope of the tangent to

the origin, i.e., $\dfrac{3\varepsilon_0}{n}\dfrac{kT}{\mu_p^2} < \dfrac{1}{3}$, which means that:

$$T < \theta_C = \frac{n\mu_p^2}{9\varepsilon_0 k}$$

where θ_C is the Curie ferroelectric temperature, below which point a spontaneous polarisation can appear without an applied field.

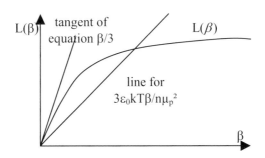

Figure 4.7. *Condition to obey to obtain a ferroelectric material.*

4.1.5.3.2. The Curie law

Given that $\vec{E}_{al} = \vec{E}_a + \dfrac{\vec{P}}{3\varepsilon_0}$, and for a strongly polar ferroelectric material

$P \approx P_{or} = n\,\mu_p L(\beta)$, with $L(\beta) \approx \dfrac{\beta}{3}$, $P = n\dfrac{\mu_p^2 E_{al}}{3kT}$. This can be used to deduce that

$E_{al} = \dfrac{3\,k\,T\,P}{n\,\mu_p^2} = E_a + \dfrac{P}{3\varepsilon_0}$, and thus $P\left(\dfrac{3kT}{n\mu_p^2} - \dfrac{1}{3\varepsilon_0}\right) = E_a$. The last equation can be rewritten as

$$P\left(1 - \dfrac{\theta_C}{T}\right) = \dfrac{C}{T}\,E_a$$

where $\theta_C = \dfrac{n\,\mu_p^2}{9\,\varepsilon_0\,k}$ and $C = \dfrac{n\,\mu_p^2}{3\,k}$ (C is the Curie constant). Again, the equation can be rewritten the form:

$$\boxed{\dfrac{P}{E_a} = \dfrac{C}{T - \theta_C}}.$$

This last equation is an expression of the Curie law and shows that as $T \to \theta_c$, $\dfrac{P}{E_a} \to \infty$. As P has a finite maximum value—it cannot go above $n\,\mu_p$—the relationship $T = \theta_c$ (and $\dfrac{P}{E_a} \to \infty$) can only be true when $E_a = 0$; that is to say at the point at which the spontaneous polarization occurs.

4.1.5.3.3. Conclusion

Ferroelectricity only occurs in small number of crystalline materials, an example of which is perovskite. The property comes about when, at low temperatures $(T < \theta_c)$, the localized dipole moments are sufficiently intense to induce a gradual alignment of dipoles. As the temperature increases, thermal motions obliterate this established order so that local polarizations are deformed, the dipolar moment is reduced to nothing, and the ferroelectricity disappears.

4.2. Magnetic Materials
4.2.1. Introduction
4.2.1.1. Field inside a bar (with permeability μ) placed in a magnetic field $\vec{H}_0$

4.2.1.1.1. Field outside and parallel to the bar $\vec{H}_0$

Given the conditions of continuity, it is possible to write:

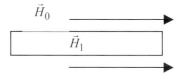

$H_0 = H_{0t} = H_{1t}$

$B_{0n} = 0 = B_{1n}$.

As $B_{1n} = \mu H_{1n}$, the last two equations indicate that $H_{1n} = 0$.

The result, from the first equation, is that $H_1 = H_{1t} = H_0$, where the field ($\vec{H}_1$) in the bar is equal to the field outside the bar ($\vec{H}_0$) and $\vec{H}_1 = \vec{H}_0$.

In addition, it is possible to state that:

$\vec{B}_1 = \mu \vec{H}_1 = \mu \vec{H}_0 = \mu_r \mu_0 \vec{H}_0 = \mu_r \vec{B}_0$ (if $\mu_r \approx 1$, $\vec{B}_1 \approx \vec{B}_0$),

While the intensity of the magnet is be given by:

$\vec{I} = \chi_m \vec{H}_1 = \chi_m \vec{H}_0 = (\mu_r - 1) \vec{H}_0$ (if $\mu_r \approx 1$, $\vec{I} \approx 0$; if $\mu_r \to \infty$, $\vec{I} \to$ *very large*).

4.2.1.1.2. Field exterior and perpendicular to the bar $\vec{H}_0$

Continuity conditions make it possible to write that

$B_0 = B_{0n} = B_{1n}$, that $H_{0t} = 0 = H_{1t} = \dfrac{B_{1t}}{\mu}$, and

with $B_{1t} = 0$, we find $B_1 = B_{1n} = B_0$.

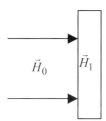

As $H_{1t} = 0$, we have $H_1 = H_{1n} = \dfrac{B_{1n}}{\mu} = \dfrac{B_0}{\mu}$, where $B_0 = \mu_0 H_0$, so $H_1 = \dfrac{H_0}{\mu_r}$.

For the magnetic intensity,

$$\vec{I} = \chi_m \vec{H}_1 = (\mu_r - 1)\, \vec{H}_1 = \dfrac{\mu_r - 1}{\mu_r} \vec{H}_0 \text{ , and if } \mu_r \approx 1,\ \vec{I} \approx 0 \text{ ,}$$

whereas if $\mu_r \to \infty$, $\vec{I} \approx \vec{H}_0$.

In general terms

- If $\chi_m > 0$, where $\mu_r > 1$, we have $H_1 = \dfrac{H_0}{\mu_r} < H_0$. On writing $\vec{H}_1$ in the form

  $\vec{H}_1 = \vec{H}_0 + \vec{h}$, the vector $\vec{h}$ needs to be antiparallel to $\vec{H}_0$. Here, $\vec{h}$ is said to be a demagnetizing field.

- If $\chi_m < 0$, where $\mu_r < 1$, we have $H_1 = \dfrac{H_0}{\mu_r} > H_0$. Once again, writing $\vec{H}_1$ in the

  form $\vec{H}_1 = \vec{H}_0 + \vec{h}$, the vector $\vec{h}$ is now parallel to $\vec{H}_0$, and $\vec{h}$ is now a magnetizing field.

4.2.1.2. General properties
As we shall see, there are two main classes of magnetic materials.

4.2.1.2.1. Linear materials
These materials have an diamagnetic intensity ($\vec{I}$) that is proportional to the magnetic field ($\vec{H}$), so that $\vec{I} = \chi_m \vec{H}$. When $\chi_m < 0$, the material is diamagnetic, and when $\chi_m > 0$, it is paramagnetic. In fact, diamagnetism is a quite general phenomenon and can be found in all materials, as it results from orbital magnetic moments, while paramagnetism can be observed only in materials with a total, resulting magnetic moment $\mu_T \neq 0$. Apart from the sign of χ_m , its constancy or variation with temperature also can be used to indicate the type of magnetism one is dealing with: in the case of diamagnetism, χ_m is independent of temperature, whereas in paramagnetism, $\chi_m(T) \propto \dfrac{1}{T}$. This relationship is observed for dilute systems, which is detailed below.

4.2.1.2.2. Nonlinear materials (essentially the ferromagnets)
In this class, the relationship $\vec{I} = \chi_m \vec{H}$ is still appropriate; however, χ_m is now a function of $\vec{H}$ such that $\chi_m = \chi_m(\vec{H})$. Similarly, $\mu = \mu(\vec{H})$, so that $\vec{B} = \mu(\vec{H}) \cdot \vec{H}$.

In addition, μ and χ_m are now not only dependent on the strength of H at any particular time t, but also on its anterior values: the system is now subject to hysteresis.

4.2.2. Diamagnetism and Langevin's theory

The Larmor precession, associated with the orbital moment, can be found even in atoms where the resultant magnetic movement is zero (such as noble gases). To understand the effect of a resultant magnetic field on an intraatomic orbital, we suppose that the field is applied as indicated by the Larmor precession (Section 3.1.6), shown in Figure 4.8.

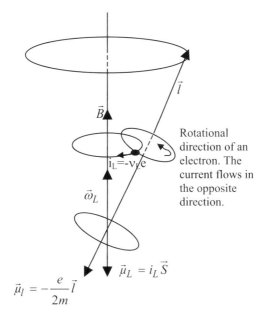

Figure 4.8. *Effect of magnetic field on the orbital magnetic moment.*

The frequency of the Larmor precession ($v_L = \dfrac{\omega_L}{2\pi}$) forms a current (i_L) which is such that $i_L = -v_L\, e$, where v_L is the number of rotations per second. We also can write $i_L = -\dfrac{\omega_L}{2\pi} e$. With $\omega_L = \beta B$, where $\beta = \dfrac{e}{2m}$, we find $i_L = -\dfrac{e^2 B}{4\pi m}$.

If $\langle \rho^2 \rangle$ is the average value of the square of the distance between the electron and the axis Oz through which the magnetic field is applied, then

$$\mu_L = i_L S = i_L \pi \langle \rho^2 \rangle = -\frac{e^2 B}{4m} \langle \rho^2 \rangle .$$

Note that μ_L is in the opposite direction to B, just as μ_l is opposite to l. In addition,

$$\begin{cases} r^2 = x^2 + y^2 + z^2 & \Rightarrow <r^2> = <x^2> + <y^2> + <z^2> = 3<x^2> \\ \rho^2 = x^2 + y^2 & \Rightarrow <\rho^2> = <x^2> + <y^2> = 2<x^2> \end{cases}$$

$$\Rightarrow <\rho^2> = \frac{2}{3} <r^2> .$$

For a number n of atoms per volume, each containing z electrons, the magnetic moment per unit volume for the precession is $nz\mu_L$. The magnetic intensity ($\vec{I}$) therefore is such that

$$\vec{I} = nz\vec{\mu}_L = -\frac{nze^2}{6m}\vec{B}\langle r^2 \rangle .$$

Given that the magnetic material is represented as a vacuum through which currents associated with orbiting electrons progress, and that the magnetic moments moving in this vacuum are such that $B = \mu_0 H$, we can write that

$$\chi_m = \frac{I}{H} = \frac{\mu_0 I}{B} = -\frac{\mu_0 nze^2 \langle r^2 \rangle}{6m} ,$$

where χ_m therefore is negative and temperature independent, and $\langle r^2 \rangle$ can be calculated for atoms or ions using quantum mechanics.

Even when the orbital magnetic moments and the spin give a resultant equal to zero, this susceptibility related only to the orbital magnetic moment is still apparent. This is because it is tied to the single Larmor precession. When the resultant is not equal to zero, then there is diamagnetism, however, its contribution to the magnetic susceptibility is less intense than that of paramagnetism. Indeed, the latter masks the former.

4.2.3. Paramagnetism

Paramagnetism appears for atoms that carry a permanent magnetic moment, such that $\mu_T \neq 0$. The effect due to diamagnetism is less than that caused by paramagnetism, and in atoms where $\mu_T \neq 0$ paramagnetic effects dominate.

4.2.3.1. Langevin's theory

Langevin's theory can be thought of as an analogy of the theory developed for dielectrics under an orientating polarization (which gives rise to Clausius Mossotti's general formula). Here we consider in terms of magnetism the distribution of magnetic dipoles, rather than the dielectric effect due to a distribution of dielectric dipoles. In the presence of an external field ($\vec{B}$) and at a certain temperature (T), the dipoles are subject to:

- on one hand, an orientation due to $\vec{B}$, for which the coupling energy in the stable state is in the form $(E_p)_{min} = - \overrightarrow{\mu_T} \cdot \vec{B} = - \mu_T.B \cos \theta = - \mu_T.B$, so that $\theta = 0$ (2π) gives $\overrightarrow{\mu_T}//\vec{B}$; and

- on the other hand, a disorientation with respect to the direction OO' of the applied field $\vec{B}$ due to a thermal agitation of energy kT, where k is Boltzmann's constant.

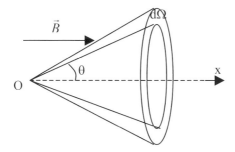

Figure 4.9. *Spatial distribution of magnetic moments subject to* $\vec{B}$.

The number of atoms (dN) with a moment within the solid angle ($d\Omega$) shown in Figure 4.9 is given by $dN = A' \, d\Omega$. Given Boltzmann's distribution, we can write that

$$dN = A \exp(+ \frac{\mu_T B \cos \theta}{kT})d\Omega .$$

Taking into account the symmetry of the calculation around the axis Ox, the resultant of the magnetic moments is with respect to the axis. The contribution of dN atoms is therefore

$$dM = \mu_T \cos \theta dN .$$

The resultant of the dipolar moment from all the atoms together with respect to Ox is thus

$$M = \int dM = \int_{\theta=0}^{\theta=\pi} \mu_T \cos\theta \, dN .$$

As each "average" magnetic dipole makes a contribution, it is possible to state:

$$\overline{\mu_T} = \frac{M}{N} = \mu_T \, \overline{\cos\theta} = \mu_T \, L(\beta) \text{ where } \beta = \frac{\mu_T B}{kT} \text{ and } L(\beta) = \coth \beta - \frac{1}{\beta^2} . \text{ When } \beta \text{ is}$$

small (or B is not too intense) we find that $L(\beta) \approx \dfrac{\beta}{3}$. If we suppose that

$B = 1$ Wb m^{-2}, $T = 300$ K, and $\mu_T \approx 10^{-22}$ MKS, we find $\beta \approx 1/400$.
We thus arrive at the definitive equation:

$$\bar{\mu}_T = \mu_T \ L(\beta) \approx \frac{\mu_T \beta}{3} = \frac{\mu_T^2 B}{3kT}.$$

For a given number (n) of atoms per unit volume, the magnetic moment per unit volume is $n \ \overline{\vec{\mu}_T} = n \ \mu_T \ L(\beta) \ \vec{e}_x$, where $\vec{e}_x$ is the unit vector in the direction Ox through which the field $\vec{B}$ is applied. By definition, the magnetic intensity ($\vec{I}$), which is the magnetic moment per unit volume, is precisely $\vec{I} = n \overline{\vec{\mu}_T}$, where $\vec{I} \approx \dfrac{n\mu_T^2 \vec{B}}{3kT}$. As we have represented our magnetic material as a vacuum in which bath magnetic atoms with magnetic moments equal to μ_T, we can write that $\vec{B} = \mu_0 \ \vec{H}$, and therefore

$$\chi_m = \frac{I}{H} = \frac{\mu_0 I}{B} = \frac{n\mu_0 \ \mu_T^2}{3kT}.$$

The relationship

$$\chi_m = \frac{n\mu_0 \ \mu_T^2}{3kT}$$

is in the form $\chi_m = \dfrac{C}{T}$, which is Curie's law, wherein C is Curie's constant that is defined by $C = \dfrac{n\mu_0 \ \mu_T^2}{3k}$. This law shows χ_m to be positive and to vary inversely with temperature.

4.2.3.2. Correction required by quantum theory

Quantum theory gives the magnetic moment as $\vec{\mu}_T = -\dfrac{e}{2m} g\vec{J}$, so that $\mu_T^2 = \dfrac{e^2}{4m^2} g^2 J^2$. Given the particular values for J^2, $<J^2> = \hbar^2 J(J+1)$, and the definition of Bohr magneton (μ_B), $\mu_B = \dfrac{e\hbar}{2m}$, the expression

$$\left\langle \mu_T^2 \right\rangle = \frac{e^2}{4m^2} g^2 <J^2> = \mu_B^2 \ g^2 J(J+1) \text{ gives}$$

$$\chi_m = \frac{n\mu_0 \left\langle \mu_T^2 \right\rangle}{3kT}, \text{ or rather } \chi_m = \frac{n \mu_0 \, J(J+1) \, g^2 \, \mu_B^2}{3kT}.$$

4.2.3.3. Paramagnetism and molecular fields: the Curie-Weiss theory

In solid materials, molecules are not independent of one another and consequently Boltzmann's law, so well adapted to gases, is no longer directly applicable. Particularly when the applied magnetic field is weak and the temperature is low enough to make thermal vibrations weak, the magnetic interactions of electrons and thus neighboring atoms in condensed systems become nonnegligible. Then, atoms subject to the action of an external field $\vec{B}$ and also subject to an additional effect due to a so-called molecular field that results from neighboring molecules or atoms. Weiss hypothesized that this additional effect can be expressed in the form $\vec{H}_m = \eta \, \vec{I}$ and as such must be added to the external field ($\vec{H}$). This hypothesis, that the molecular field is proportional to each magnetic material, seems reasonable as $\vec{I}$ depends on the magnetic moment (μ_T) of the very molecules or atoms that determine the intensity of the magnetic interactions between neighbors.

Thus with a resultant field of the form $\vec{H} + \eta \vec{I}$, we have $\vec{B} = \mu_0(\vec{H} + \eta\vec{I})$.

Taking the equation given in Section 4.2.3.1, namely, $\vec{I} = \dfrac{n\mu_T^2 \vec{B}}{3kT}$, we now find that

$$\vec{I} = \frac{n\mu_0\mu_T^2(\vec{H} + \eta\vec{I})}{3kT} = \frac{C}{T}(\vec{H} + \mu\vec{I}),$$

from which it can be deduced that $\vec{I}\left(1 - \dfrac{\eta C}{T}\right) = \dfrac{C\vec{H}}{T}$, so that

$$\chi_m = \frac{\vec{I}}{\vec{H}} = \frac{\dfrac{C}{T}}{1 - \dfrac{(\eta C)}{T}}, \text{ hence } \chi_m = \frac{C}{T - \eta C}.$$

Given that the Curie temperature (Θ) is such that $\Theta = \eta C$, we find that χ_m is

$$\chi_m = \frac{C}{T - \Theta}.$$

This last equation, or law, accords well with experimental results and is a notable characteristic of paramagnetism for condensed materials.

It is worth noting that in order that $\chi_m > 0$, the temperature must be greater than Θ; so that the temperature Θ is real, it must be greater than zero, so that with $\Theta = \eta C$, η is also greater than zero. As $\vec{H}_m = \eta \ \vec{I}$, $\vec{H}_m$ has the same sign as $\vec{I}$, and therefore the molecular field is positive.

4.2.3.4. Comments

While the Curie-Weiss law is verifiable for most situations, there is a point at low temperatures, notably for $T \leq \Theta$, where a spontaneous ferromagnetism appears. On this and other related points three remarks can be made apparent:

- First, at low temperatures, the approximation $L(\beta) = \coth \beta - \dfrac{1}{\beta} \approx \dfrac{\beta}{3}$ is no longer valid, as β is no longer small. As detailed in Chapter 2, for higher values of β, $L(\beta) \approx 1$, and $I = n \ \overline{\mu}_T = n \mu_T \ L(\beta) \approx n \mu_T = I_s$, where I_s is termed a saturated magnetization and is independent of the applied field. Qualitatively, this means that at low temperatures, thermal agitation no longer limits dipole orientation or the magnetism. More quantitatively, the approximation $L(\beta) \approx \dfrac{\beta}{3}$ is no longer acceptable when $\mu_T B_\ell > kT$; that is to say at a temperature equal to or less than T_c, where T_c is defined by the relationship $kT_c = \mu_T B_\ell$ and is approximately equal to Θ. B_ℓ is the field local to a molecule or atom. However, when $T > Tc$, $L(\beta) \approx \dfrac{\beta}{3}$, then the paramagnetism described above returns due to the creation of structural disorder by thermal agitation.

- Second, we have assumed that the molecular field $\vec{H}_m = \eta \ \vec{I}$ appeared only in the presence of the magnetism $\vec{I}$ originating itself from the effect of orientation by an applied field $\vec{B}$. Therefore, the above-established Curie law will no longer apply to materials that already have a molecular field in the absence of an external field. This molecular field also can orientate the magnetic moments parallel to one another. There is in effect a premagnetism, or spontaneous magnetism, that corresponds to ferromagnetism.

   The Curie law also is not observed by antiferromagnets, such as MnO and Cr_2O_3. While there is still spontaneous magnetism, it is such that particles compensate one another (compensated premagnetization).

- Third, paramagnetism also may be caused by electron spin and is in this case called spin paramagnetism, or Pauli's paramagnetism. Free unpaired electrons, by way of their spin and the resulting spin magnetic moment, can couple with a magnetic field of intensity B. For a spin $\vec{\mu}_s = -\dfrac{e}{m} \vec{s}$ the coupling energy is

$E_p = -\vec{\mu}_s.\vec{B}$. The orbital movement of the electron is not taken into account, otherwise a factor of ½ would have to be introduced into the general theory developed by Thomas who used a frame of reference appropriately tied to the composite spin and orbital movements.

If Oz is the direction along which B is applied, then

$|Ep| = \dfrac{e}{m} B <s_z>$, which gives $|Ep| = \dfrac{e}{m} B \hbar\, m_s$. With $m_s = \pm\dfrac{1}{2}$, two different

values for energy are obtained, namely, $Ep = \pm\mu_B\, B$, where $\mu_B = \dfrac{e\hbar}{2m}$. Two

calculations then can be carried out:

• One with a Boltzmann distribution of the different energy electrons, so that with

$x = \dfrac{\mu_B B}{kT}$, we find $I = (n_- - n_+)\mu_B = n\mu_B\, \mathrm{thx}$ where n_- and n_+ are the number of

electrons per unit volume with spins parallel or antiparallel to the field B,

respectively. With $\mathrm{thx} \approx x$, we arrive at $\chi = \dfrac{\mu_0 I}{B} \approx \dfrac{n\mu_0\mu_B^2}{kT}$, and this law is of the

same type as Curie's law. It is worth noting though that this law is poorly verified for the Pauli paramagnetism or nonferromagnetic metals due to the small susceptibilities and temperature independence of such systems.

• Two, with a Fermi-Dirac distribution, which is better adapted to electron

distributions. The calculation results in finding $\chi = \dfrac{3}{2}\dfrac{\mu_0\mu_B^2}{kT_F}$, where T_F is the Fermi

temperature defined by $E_F = kT_F$. This result was established by Pauli in 1927 and is applicable to free electrons in metals.

4.2.4. Ferromagnetism
4.2.4.1. The orientation of a ferromagnetic bar in a magnetic field
As we saw in Section 4.2.1.1, when a bar is placed in a magnetic field ($\vec{H}_0$) and is:

• In a longitudinal position, parallel to the field, so that $\vec{H}_0 = \vec{H}_L$, we have $\vec{H}_1 = \vec{H}_0 = \vec{H}_L$. The result of this is that $\vec{I} = \vec{I}_L = \chi_m\vec{H}_L = (\mu_r - 1)\vec{H}_L$, and if the susceptibility (χ_m) is large, then $\vec{I}_L$ is also large.

$$\vec{H}_0 = \vec{H}_L$$
$$\vec{I}_L = \chi_m\vec{H}_L$$
$$\vec{H}_1 = \vec{H}_0 = \vec{H}_L$$
$$\vec{H}_1$$

- In a transverse position perpendicular to the field so that $\vec{H}_0 = \vec{H}_T$, we have $B_1 = B_T = B_0$. From this

$$\vec{H}_1 = \frac{B_1}{\mu_0} - \vec{I} = \frac{B_0}{\mu_0} - \vec{I}, \text{ so } \vec{H}_1 = \vec{H}_0 - \vec{I} = \vec{H}_T - \vec{I}.$$

In addition, $\vec{I} = \vec{I}_T = \dfrac{\mu_r - 1}{\mu_r}\vec{H}_T \approx \vec{H}_T$,

$\vec{I}_T \approx \vec{H}_T$
$\vec{H}_1 = \vec{H}_0 - \vec{I}$

$\vec{H}_0 = \vec{H}_T$ $\vec{H}_1$

if μ_r, and consequently χ_m, are large.

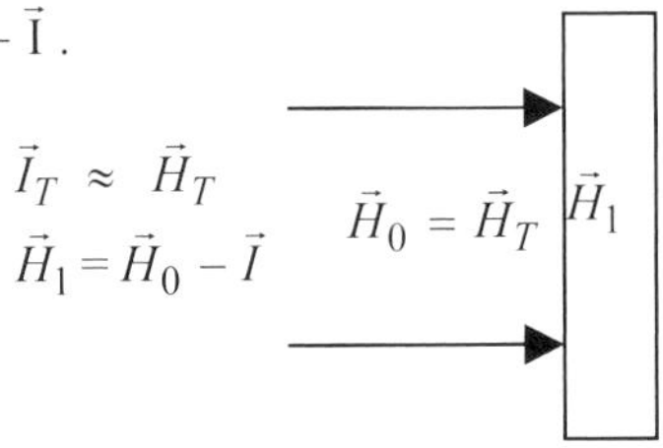

In general terms, a bar of any shape placed in an external field $\vec{H}_0$, within the limiting values of $\vec{H}_1$ detailed above, then

$$\vec{H}_1 = \vec{H}_0 - f\,\vec{I}$$

where f is a form factor that should take on in the above limiting situations the values:
- $f = 0$ (where $\vec{H}_0 = \vec{H}_L$);
- $f = 1$ (where $\vec{H}_0 = \vec{H}_T$).

The value of f decreases as the ellipsoid flattens out.

In addition, if $\vec{H}_1$ is considered in the form $\vec{H}_1 = \vec{H}_0 + \vec{h}_d$, then $\vec{h}_d = -f\,\vec{I}$; in other words the field is demagnetizing.

With respect to the resultant magnetization intensity, we have:

$$\vec{I} = \vec{I}_L + \vec{I}_T = \chi_m\vec{H}_L + \vec{H}_T.$$

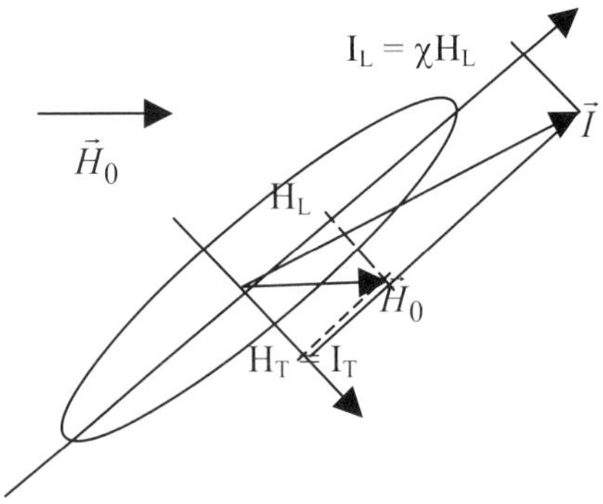

Figure 4.10. *Orientation of* $\vec{I}$ *with respect to the excitation* $\vec{H}_0$.

From this we can go on to draw Figure 4.10, where the resultant magnetization intensity ($\vec{I} = \dfrac{d\vec{\mu}}{d\tau}$, where $d\vec{\mu}$ is the magnetic moment of a bar of any shape such as an iron filing) does not have the same direction as the external field ($\vec{H}_0$). A coupling appears that tends to orientate the filing in such a way that $\vec{I} \, // \, \vec{H}_0$. The potential energy of the system, $E_P \propto -d\vec{\mu}.\vec{B}_0 = -\vec{I}.\vec{B}_0 d\tau$, is at a minimum when $\vec{I} \, // \, \vec{B}_0$ and $\vec{I} \, // \, \vec{H}_0$.

4.2.4.2. Ferromagnets and magnetization plots
4.2.4.2.1. Plot of the primary magnetization

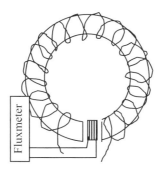

Figure 4.11. *Set up used to plot the first magnetization.*

The device shown in Figure 4.11 shows how the first magnetization can be plotted. There is a large torus that is cut at a cross section so that there is a small gap into which a coil can be placed, which is in turn connected to a fluxmeter. The large iron torus is covered with N turns per unit length, through which flows a current I. For this, $\oint \vec{H}.d\vec{l} = \sum I$, which gives $H = NI$, where H is the field in the torus iron.

Given that the component normal to B does not change throughout, the field inside the torus is the same as that in the air gap, which can be measured using the fluxmeter. Thus knowing both H and B, we can find out the magnetization intensity in the torus from $I = \dfrac{B}{\mu_0}$ - H and then plot $I = f(H)$.

For a sample that has been demagnetized, H is increased from zero and the first magnetization plot shows three main zones, which can be divided as shown in Figure 4.12 a:
• zone 1, which exhibits an essentially linear increase;

- zone 2, in which there is a considerable increase in I with H; and
- zone 3, where the magnetization reaches its saturation point (I_s). The latter is specific to the material under study and is dependent on purity and temperature.

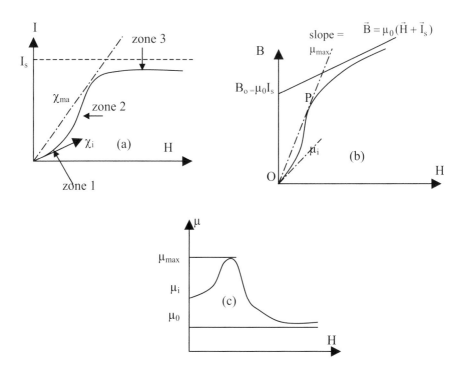

Figure 4.12. *The plots for (a) I(H); (b) B(H); and (c) μ(H).*

As $\vec{B} = \mu_0(\vec{H} + \vec{I})$, the plot showing B(H) in Figure 4.12(b) is the result of plotting $\mu_0 I = f(H)$ [from the plot in (a) with the homothetic ratio μ_0] following an insertion of the linear variation $\mu_0 H$. At higher values of H, the plot tends toward an oblique asymptote such that $\vec{B} = \mu_0(\vec{H} + \vec{I}_s)$ and for which the coordinate of the origin at H = 0 is $B_o = \mu_0 I_s$.

Given the plot B(H) such that $B = \mu_{(H)}.H$, we can plot the line in Figure 4.12(c) which shows $\mu = g(H)$. Geometrically speaking, μ is the slope of the line OP in Figure 4.12(b). For hardened iron, $(\mu_r)_{max} \approx 100$, and this value can reach around 80000 for certain alloys such as Mumetal™.

4.2.4.2.2. Magnetization at the point of saturation

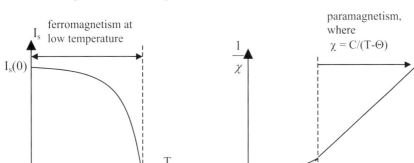

Figure 4.13. *Plots of (a) $I_s = f(T)$ and (b) $1/\chi = f(T)$*

The I_s is temperature (T) dependent. As shown in Figure 4.13(a), as T rises from absolute zero, I_s diminishes quite regularly and then quite quickly before reaching zero at a temperature T_f, which is called Curie's ferromagnetic temperature. Above T_f the material is not ferromagnetic but paramagnetic, as indicated in Figure 4.13(b). At temperatures considerably greater than T_f, the value of χ follows the Curie-Weiss law where $\chi = \dfrac{C}{T - \Theta}$, in which Θ is slightly above T_f. In the case of iron, $T_f = 1043 \text{ K}$ and $\Theta = 1101 \text{ K}$.

4.2.4.2.3. Hysteresis loop and magnetic state

A hysteresis loop can appear on having increased H from 0 to a maximum value (H_{max}) at saturation and then on decreasing H, finding that the current is below that described by the first magnetization plot, shown in Figure 4.14(a). As $\vec{I} = \dfrac{\vec{B}}{\mu_0} - \vec{H}$, this phenomenon results in a delay in B, which is the effect called hysteresis.

By varying H between H_m and $-H_m$, the current follows a closed loop, otherwise known as the hysteresis loop. There are two notable points:
- the remanence magnetization (I_r) remains when H = 0; and
- the coercive field (H_c) is the value of the opposing field H which needs to be applied to remove the magnetization.

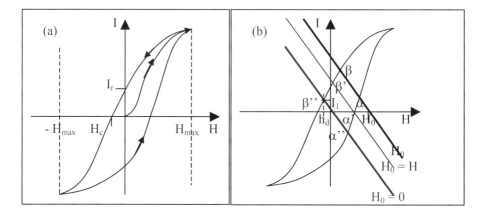

Figure 4.14. *(a) Hysteresis loop and (b) operation.*

For a material with a well-defined hysteresis and a form factor f, its magnetic state can be determined from both:
- its participation in the plot $I = f(H)$, which is characteristic of the hysteresis of a material; and

- its part in the slope $I = -\dfrac{H}{f} + \dfrac{H_0}{f}$ (equation $I = D(H)$ directly deduced from $\vec{H} = \vec{H}_0 - f\,\vec{I}$).

The line has a slope of $-\dfrac{1}{f}$ and is such that $I = 0$ if $H = H_0$ [so that H_0 is at the intersection of the line $I = D(H)$ with the abscissas].

The hysteresis intervenes at two points in Figure 4.14(b) where at the intersection, $I = f(H)$ for a hysteresis along the line $I = D(H)$:
- at α when H is increasing along with I;
- at β when H is decreasing, as is I.

When the external field H_0 changes, the line moves but retains its slope $-\dfrac{1}{f}$.

If $H_0 = 0$, the plot is simplified to $I = -\dfrac{H}{f}$ and goes through the origin. The points of intervention are now α'' and β'' . Given that $\vec{H} = \vec{H}_0 + \vec{h}_d$ and that here $H_0 = 0$, we now find that $\vec{H} = \vec{h}_d$. The intensity of the magnetization, $I_1 = I_{(H = hd)}$, when $H_0 = 0$, is the actual remanence magnetization.

If $H = H_0$, $I = 0$ and the point at which the hysteresis functions is α' (the equation $\vec{H} = \vec{H}_0 + \vec{h}_d$ demonstrates that for this scenario, $\vec{h}_d = 0$).

4.2.4.2.4. Energy loss by hysteresis

For the experimental setup shown in Figure 4.15—which also may be used to study hysteresis loops—Ohm's law also can be written as:

$$u + e = Ri \text{, or rather, } u - \frac{d\phi}{dt} = Ri \text{.}$$

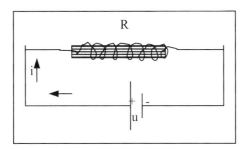

Figure 4.15. *Experimental setup to study hysteresis.*

Multiplying both sides by idt, we have:

$$u \, i \, dt - i \, d\phi = Ri^2 \, dt \text{.}$$

On integrating this differential equation between the points $t = 0$ and $t = T$, $i = 0$ and $i = i$, and $\phi = \phi_0$ and $\phi = \Phi$, we find

$$\int_0^T uidt - \int_0^T id\phi = \int_0^T Ri^2dt \text{, which also gives } \underbrace{\int_0^T uidt}_{(1)} - \underbrace{\int_0^T Ri^2dt}_{(2)} = \underbrace{\int_0^T id\phi}_{(3)} \text{.}$$

Term (1) represents the energy $W_G = \int_0^T uidt$ supplied by the generator of fem u, so

that with $i = \dfrac{dq}{dt}$, we have $W_G = \int_0^T udq = Qu$, if $q = 0$ when $t = 0$ and $q = Q$ when $t = T$.

Term (2) represents the energy lost through the Joule effect between $t = 0$ and $t = T$ which is associated with the resistance of the solenoid.

Term (3), which should be written in the form $\int_{\phi_0}^{\Phi} i d\phi$, represents the energy lost through hysteresis and it is this term that is detailed below.

If we suppose that the solenoid is infinitely long and has N turns each with a surface area S over the solenoid length l, then $H = N\,i$ and $\phi = B\,N\,l\,S$. In turn, excusing the pun, we have $d\phi = N\,l\,S\,dB$, so that $\int_{\phi_0}^{\Phi} i d\phi = \int_{\phi_0}^{\Phi} NlSidB$.

With $V = l\,S$, where V is the volume of the solenoid, and $H = N\,i$, we have:

$$\int_{\phi_0}^{\Phi} i d\phi = V \int_{B'}^{B} HdB,$$

where B' is the field at the initial instant t = 0 and while B is the field at the last instant t = T.

In terms of the coordinates (B,H), the hysteresis takes on the form described in Figure 4.16; H dB is indicated by the hatched surface dS where H is increasing in the first quadrant. When H decreases, H dB is given by an area dS', associated with a value H' < H and such that dS' < dS.

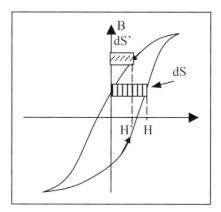

Figure 4.16. *B(H) coordinates of the hysteresis loop.*

Schematized in Figures 4.17(a) and 4.17(b), respectively, are the hatched areas swept in the first quadrant when *H* increases and decrease. The difference between $\int dS$ and $\int dS'$ corresponds to the area (*A*) of the hysteresis loop. As *H* increases, the system gains energy (if $H\uparrow$, dB > 0 and HdB > 0), and when *H* decreases, the system releases energy (if $H\downarrow$, dB < 0 and HdB < 0).

As $\int dS' < \int dS$, the released energy is less than that received, and the energy absorbed through one complete cycle can be written as:

$$\int_{\phi_0}^{\Phi} i \, d\phi = V \int_{B'}^{B} H \, dB = V\,[S - S'] = V.A\,.$$

The higher the value of A, the greater the energy absorbed.

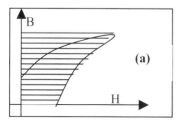 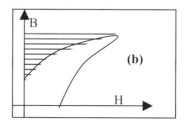

Figure 4.17.   *Areas covered in the first quadrant when (a) H increases;
and (b) H decreases.*

4.2.4.3. Soft and hard ferromagnets
4.2.4.3.1. Soft ferromagnets
Soft ferromagnets are characterized by their weak coercive field, where $H_c < 100$ A.m^{-1}. They magnetization therefore is relatively easy to change. Given that with a low value of H_c hysteresis is small, and energy losses are also small, these materials often are used in transformers, electromagnets, relays, and telephone loud speakers. Examples include Permalloy™ (Fe = 21.5 %, Ni = 78.5 %) and Mumetal™ (Fe = 16%, Ni = 77%, Cu = 5%, Cr = 2%).

4.2.4.3.2. Hard ferromagnets
These magnets exhibit values for $H_c > 10^3$ A.m^{-1} and their remanence is relatively difficult to remove. They are generally used as permanent magnets. Examples include steels with around 1% carbon, or even with Co, Mn, or W. More recently, alloys have been prepared such as the Alnico™ series, which includes Alnico 5 based on Fe 51.5%, Al = 8%, Ni = 13.5%, Co = 24%, and Cu = 3%, or alloys with titanium such as Ticonal™.

4.2.4.4. Aspects of the theory of ferromagnetism
4.2.4.4.1. Theoretical conditions required for spontaneous magnetisation to appear: influence of temperature

It is worth trying to find that conditions required at which a paramagnetic substance remains magnetic $(I \neq 0)$ simply by its molecular field (H_m) and without any external field.

The relationship that defines H_m : $\vec{H}_m = \eta \ \vec{I}$, or rather $\vec{I} = \dfrac{\vec{H}_m}{\eta}$, can be multiplied by T above and below the line to give $\vec{I} = \dfrac{T \ \vec{H}_m}{\eta \ T}$. In the absence of an applied field, H is due only to the molecular field. In the representation $I = f\left(\dfrac{H}{T}\right)$, the magnetization state therefore is shown as a straight line (D) with slope $\dfrac{T}{\eta}$. As detailed in Section 4.2.3.1, the magnetization in the presence of magnetic moments (μ_T) can be given as:

$$I = n \ \bar{\mu}_T = n \ \mu_T \ L(\beta) \text{ where } \beta = \dfrac{\mu_T B}{kT} = \dfrac{\mu_0 \mu_T H}{kT}. \text{ Therefore}$$

$$I = n \ \mu_T \ L\left(\dfrac{\mu_0 \mu_T \ H}{kT}\right).$$

For a material to be spontaneously magnetized, simply by its own molecular field, its magnetization state should be represented by the point A that is both on the line D and on the curve $I = n \ \mu_T \ L(\beta)$, as presented in Figure 4.18.

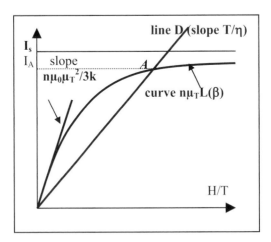

Figure 4.18. *Plots of $I = f(H/T)$ (labeled line D) and $I = n\mu_T L(\beta)$.*
Note: plots are for a given temperature.

Given that at the origin $[\beta \to 0]$ $L(\beta)$ tends toward a straight line with an equation $\dfrac{\beta}{3}$, the function I tends [also when $\beta \to 0$] toward a straight line written by

$$I = n\mu_T \frac{\beta}{3} = \frac{n\mu_0\mu_T^2}{3kT}\frac{H}{} = \frac{n\mu_0\mu_T^2}{3k}\frac{H}{T} \text{, for which } I = f\left(\frac{H}{T}\right) \text{ has a slope equal to}$$

$\dfrac{n\mu_0\mu_T^2}{3k}$; for a point A to exist, the slope of D must be less than the slope of the

tangent to the origin of I, which is represented by a line of slope $\dfrac{n\mu_0\mu_T^2}{3k}$. This

condition is more concisely given by $\dfrac{T}{\eta} < \dfrac{n\mu_0\mu_T^2}{3k}$, or rather

$$T < \Theta = \eta \frac{n\mu_0\mu_T^2}{3k} \ .$$

In order to have T positive, the Curie temperature (Θ) must also be positive. This in turn requires that the molecular field must be positive ($\eta > 0$).

The conclusion therefore is that a paramagnetic material with a positive molecular field is susceptible to being spontaneously magnetized—in the absence of an external field—at a temperature below Θ.

Comment 1. It should be noted that low temperatures enable spontaneous order, and therefore $\beta = \dfrac{\mu_T B}{kT}$ must be relatively high, even if the approximation made by

$L(\beta) \approx \dfrac{\beta}{3}$ is not accurate under such conditions. In reality, it is sufficient that the

slope $\left(\dfrac{T}{\eta}\right)$ of the line D is less than $P < \dfrac{n\mu_0\mu_T^2}{3k}$, that is to say that

$$T < \eta P = T_f < \eta \frac{n\mu_0\mu_T^2}{3k} = \Theta \text{, where } T_f \text{ is the ferromagnetic temperature. As}$$

noted elsewhere in this chapter, representative values given for iron are T_f and Θ equal, respectively, 1043 and 1101 K.

Comment 2. T_f separates the two temperature domains above which a disordered phase reigns resulting in paramagnetism and below which the temperature is sufficiently low for an ordered regime to result in ferromagnetism (see also Comment 1 of Section 4.2.3.4).

4.2.4.4.2. Weiss domains and the Barkhausen effect

Associated with the point A, which represented the ferromagnetic state, is a magnetization intensity (I_A). In the absence of an external field, the intensity is no more than that which is found in small domains, called Weiss domains, which are around a micron cubed in size (Figure 4.19). With respect to the larger macroscopic volumes, the spontaneous magnetization is zero in the absence of an applied field and because of the random domain orientation.

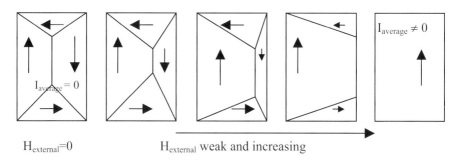

$H_{external}=0$ $H_{external}$ weak and increasing

Figure 4.19. *Evolution of Weiss domains with a growing magnetic field.*

Once an external field (H) is applied, and while H is relatively weak, the walls between the Weiss domains deform to the point where domains facing in the same or nearest direction as H become greater in number than their neighbors. While the displacements (distortions) are relatively small in the weak field, they are also reversible and give rise to the smooth change observed for I(H) in Figure 4.20.

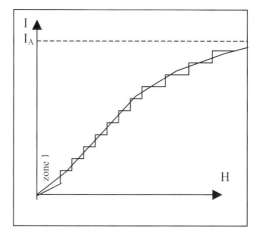

Figure 4.20. *The Barkhausen effect.*

Under a strong field the changes in domain orientation become abrupt so that the magnetization curve becomes discontinuous and resembles a stairway, each step corresponding to the orientation of a single domain in what is known as the Barkhausen effect (Figure 4.20).

The origin of hysteresis can be identified in the movement of these domains. As the displacements required are nonnegligible, they are susceptible to being stopped by obstacles such as impurities and defaults, and the phenomenon is therefore irreversible and nonlinear. The effect of the field becomes delayed and the domain orientation continues, for example, even when the field has disappeared, hence hysteresis. When all the domains are orientated in the direction of the field H, then the average magnetization for all domains tends toward the individual domain value of spontaneous magnetization (I_A).

What now remains in is an explanation of the physical origin of I_A for each domain in the absence of an external magnetic field, that is to say the reason for the existence of an internal field in the absence of an external field. This is opposed to the supposition expressed earlier on paramagnetism where the molecular field appeared due to the generation of order by an external field.

4.2.4.4.3. Origin of the spontaneous magnetization of domains

Ferromagnetism only occurs with elements that have their internal electronic layers incomplete, as is the case with iron which has an incomplete 3d orbital. The unpaired electrons from these inner layers are coupled through spin, with interaction or so-called "exchange" energies being of the form

$$W_e = -2\, J_e\; \vec{s}_1.\vec{s}_2 .$$

In this equation, Heisenberg's theory, J_e is the exchange integral and varies with the overlapping wavefunctions of the electrons. A positive J_e favors an alignment of same-sense spins and spin magnetic moments and little by little we can see that an order can be imposed upon the material through the spin magnetic moments. While for individual atoms J_e is determined by the way in which the different orbitals are filled which follow Hund's rules whereby spins are organized so that S is at a maximum, in a metal Hund's rule is no longer applicable as the atoms contribute to valence bonds. Indeed, the interactions of the 3d orbitals from the atoms placed together result in the formation of the half bands $3d^+$ and $3d^-$, which correspond, respectively, to the parallel or antiparallel alignment of spins. A small shift in the energies of these two bands results in a considerable separation of the two populations, in turn resulting in a large spontaneous magnetization. If the bands are very different, for example, if $3d^+$ is more populous than $3d^-$, then there can be an intense magnetization caused by a high proportion of parallel spins ordering the magnetic moments.

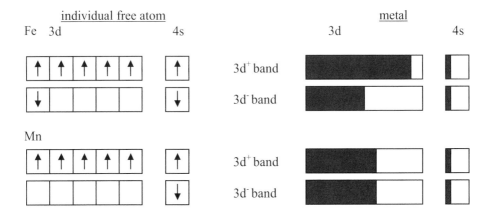

Figure 4.21. *Electronic structure of the transition elements iron (ferromagnetic) and manganese (antiferromagnetic) in both free and metallic states.*

4.2.5. Antiferromagnetism and ferrimagnetism

4.2.5.1. Antiferromagnetism

In the case of chromium or manganese, the $3d^+$ and $3d^-$ bands are pretty well equally populated, so that the average spin magnetic moments are antiparallel and there is no longer any spontaneous magnetization as the spin magnetic moments cancel each other out. Above a certain temperature, the Néel temperature (T_N), this ordered state disappears and χ follows a law of the type $\chi = \dfrac{C}{T + \theta}$. When T increases, χ decreases so that χ goes through a maximum at $T = TN$, in a behavior characteristic of antiferromagnets. For chromium, $T_N = 475\ K$.

(a)   (b)

Figure 4.22. *(a) Organization of spin magnetic moments due to antiferromagnetism; and (b) distribution of spin magnetic moments, of alternating value, in ferrimagnetism.*

4.2.5.2. Ferrimagnetism

For materials based on a mixture of two types of atoms, exchange interactions can orientate all similar atoms in one sense and all the other different atoms in another sense. The overall effect is a nonzero spontaneous magnetization, which can be strong, but not as strong as ferromagnetism. Ferrites make up the main class of ferrimagnets and are based on iron(III) oxides mixed with metals such as Ni, Al, Zn, Mn, and Co in their secondary oxidation states (II). An example is that of Fe_2O_3, MO.

There are soft ferrites, based on mixtures of manganese and zinc, which are of considerable commercial interest. They are insulators ($\rho < 1\ \Omega$ m) and exhibit limited losses through Foucault currents, hence their use in high-frequency transformers. Hard ferrites containing barium, are normally prepared with high temperatures and pressures, and are used as permanent magnets.

4.3. Problems
Dielectrics, electrets, magnets, and the gap in spherical armatures
1. A lhi dielectric of absolute permittivity ε is placed between the electrodes of a spherical capacitor which is defined by spheres of radius a and b centered about O. If $+Q$ and $-Q$ are the charges on the electrodes, where a < b, then:

(a) Determine the vectors $\vec{E}(M)$, $\vec{D}(M)$, $\vec{P}(M)$ for a point M situated at a distance r from O (a < r < b) as well as the potential difference V_a - V_b between the two electrodes, where $\vec{r} = \overline{OM}$.

(b) Calculate the surface and volume densities equivalent to the polarizations σ_a, σ_b, and ρ_P.

(c) Calculate the total quantity of these charges due to the polarizations in the dielectric. What remarks can be made on the results?

2. After having discharged the capacitor used in the pervious problem, the dielectric is changed for an electret, which has a permanent polarization expressed as $\vec{P} = \dfrac{\alpha \vec{r}}{4\pi r^3}$ (where a < r < b and α is a constant). The spheres, of radius a and b, do not carry real charges.

(a) Calculate the charge densities equivalent to the surface polarizations σ'_a, and σ'_b, and the volume polarization ρ'_P.

(b) Use the preceding result to find the vectors $\vec{E}$ and $\vec{D}$ for when r > a .

3. Now the same capacitor has, in place of an electret, a magnet with a permanent magnetization intensity $\vec{I}$ which is such that $\vec{I} = \dfrac{\beta \vec{r}}{4\pi r^3}$ (where a < r < b , and β is a constant).

(a) Calculate the Ampere surface and volume current densities identical to the magnetization.

(b) (i) Calculate the imaginary surface (σ^*_a and σ^*_b) and volume (ρ^*) magnetic mass densities.

(ii) From the preceding result, calculate the total magnetic mass carried by the material with respect to an armature of radius a, and then of radius b. Conclude.

(c) The notion of magnetic mass is used as an intermediate in calculations of the magnetic field at point M located by the vector $\vec{r}$ (where $a < r < b$).

(i) Recalling that at a distance r from a magnetic mass m*, the field is given by:

$$\vec{H}(r) = \frac{1}{4\pi} m * \frac{\vec{r}}{r^3}.$$ Give the form of Gauss's theory for $\vec{H}$.

(ii) Determine $\vec{H}(r)$ when $a < r < b$, and then find $\vec{B}$.

Answers

1.

(a) When $a < r < b$, Gauss's theory states that:

$$\Phi = 4\pi r^2 E = \frac{+Q}{\varepsilon},$$ so that as the field is radial,

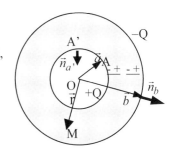

$$\vec{E} = \frac{1}{4\pi\varepsilon} Q \frac{\vec{r}}{r^3}.$$

Therefore,

$$\vec{D} = \varepsilon\vec{E} = \frac{1}{4\pi} Q \frac{\vec{r}}{r^3}$$

and

$$\vec{P} = (\varepsilon - \varepsilon_0)\vec{E} = \frac{(\varepsilon - \varepsilon_0)}{4\pi\varepsilon} Q \frac{\vec{r}}{r^3}.$$

$$V_a - V_b = \int_a^b \vec{E}.\vec{dr} = \frac{Q}{4\pi\varepsilon} \int_a^b \frac{dr}{r^2} = \frac{Q}{4\pi\varepsilon}\left[\frac{1}{a} - \frac{1}{b}\right].$$

(b) As $\rho_\ell = 0$, the localized form of Gauss's theory states that $\rho_P = 0$. Equally valid, a direct calculation can be made when the polarization vector is known:

$$\rho_P(M) = - \mathrm{div}_M \vec{P} = -\frac{(\varepsilon - \varepsilon_0)}{4\pi\varepsilon} Q \, \mathrm{div}_M \frac{\vec{r}}{r^3}$$

$$= \frac{(\varepsilon - \varepsilon_0)}{4\pi\varepsilon} Q \mathrm{div}_M \left(\overrightarrow{\mathrm{grad}}_M \frac{1}{r}\right) = \frac{(\varepsilon - \varepsilon_0)}{4\pi\varepsilon} Q \Delta_M \left(\frac{1}{r}\right) = 0.$$

- When M is at a point A such that $\vec{r} = \vec{a}$, we have

$$\sigma_a = \vec{P}(A).\vec{n}_a = \frac{(\varepsilon - \varepsilon_0)}{4\pi\varepsilon} \frac{Q}{a^3} \vec{a}.\vec{n}_a \ . \ \text{With} \ \vec{a} \ \text{antiparallel at} \ \vec{n}_a \ \text{so that}$$

$\vec{a}.\vec{n}_a = -a$, we find $\sigma_a = -\dfrac{(\varepsilon - \varepsilon_0)}{4\pi\varepsilon}\dfrac{Q}{a^2}$;

σ_a is indeed negative as can be discerned by inspection of the orientation of the dipoles influenced by the polarization (due to the polarization charge $+Q$ on the electrode with radius a) and shown in the figure above.

• In the same way, when M is at the point B, so that $\vec{r} = \vec{b}$, we have:

$$\sigma_b = \vec{P}(B).\vec{n}_b = \frac{(\varepsilon - \varepsilon_0)}{4\pi\varepsilon} \frac{Q}{b^3} \vec{b}.\vec{n}_b \ \text{with} \ \vec{b} \parallel \vec{n}_b , \ \text{so} \ \vec{b}.\vec{n}_b = b, \ \text{and we find that}$$

$\sigma_b = \dfrac{(\varepsilon - \varepsilon_0)}{4\pi\varepsilon}\dfrac{Q}{b^2}$, where $\sigma_b > 0$ (again see the figure above).

(c) If Q_a and Q_b designate the total polarization charges with respect to the armatures with radii a and b, we have $Q_a = 4\pi a^2 \sigma_a$ et $Q_b = 4\pi b^2 \sigma_b$, so that

$$Q_T = Q_a + Q_b = -\frac{(\varepsilon - \varepsilon_0)}{\varepsilon}Q + \frac{(\varepsilon - \varepsilon_0)}{\varepsilon}Q = 0 \ .$$

To conclude, the dielectric material is overall electrically neutral and the polarization has the effect of simply displacing the charges to give localized charge surpluses.

2.

(a) As $\vec{P} = \dfrac{\alpha \vec{r}}{4\pi r^3}$:

- With A such that $\vec{r} = \vec{a}$ and $\vec{a}$ antiparallel at $\vec{n}_a$, then

$$\sigma'_a = \vec{P}(A).\vec{n}_a = \frac{\alpha}{4\pi a^3}\vec{a}.\vec{n}_a = -\frac{\alpha}{4\pi a^2} \ .$$

- With B such that $\vec{r} = \vec{b}$, we have $\vec{b} \parallel \vec{n}_b$, then

$$\sigma'_b = \vec{P}(B).\vec{n}_b = \frac{\alpha}{4\pi b^3}\vec{b}.\vec{n}_b = \frac{\alpha}{4\pi b^2} \ .$$

In addition, $\rho'_P(M) = - \text{div}_M \vec{P} \propto \text{div}_M \dfrac{\vec{r}}{r^3} = 0$ where $\propto$ means proportional to, and as elsewhere we can also state that $\rho_\ell = 0$, so that in the localized form of Gauss's theory, $\rho'_P = 0$.

(b) Writing Gauss's theory for charges sitting in a vacuum thus gives:

- when $a < r < b$, $\Phi(M) = \iint \vec{E}.\overrightarrow{dS} = 4\pi r^2 E = \dfrac{\overset{\text{all charges}}{\sum Q_{int}}}{\varepsilon_0} = \dfrac{Q'_a}{\varepsilon_0}$.

With $Q'_a = 4\pi a^2 \sigma' a = -\alpha$, we have in terms of vectors $\vec{E} = -\dfrac{\alpha}{4\pi\varepsilon_0}\dfrac{\vec{r}}{r^3}$.

- when $r > b$, we have $\sum Q_{int} = 4\pi a^2 \sigma'_a + 4\pi b^2 \sigma'_b = -\alpha + \alpha = 0$, in which $\vec{E} = 0$.

With respect to induction,

- when $a < r < b$: $\vec{D} = \varepsilon_0 \vec{E} + \vec{P} = -\dfrac{\alpha}{4\pi}\dfrac{\vec{r}}{r^3} + \dfrac{\alpha}{4\pi}\dfrac{\vec{r}}{r^3} = 0$; and

- for $r > a$, as $\vec{E}$ and $\vec{P}$ are zero, $\vec{D}$ is also zero.

3.

(a) We have $\vec{I} = \dfrac{\beta\vec{r}}{4\pi r^3}$ when $a < r < b$ so we can state that the Ampere currents are such that:

- volume current: $\vec{J}_a(M) = \overrightarrow{rot}\vec{I} = \dfrac{\beta}{4\pi}\overrightarrow{rot}_M \dfrac{\vec{r}}{r^3} = -\dfrac{\beta}{4\pi}\overrightarrow{rot}_M \overrightarrow{grad}_M \dfrac{1}{r} = 0$; and

- surface current: on an electrode of radius a we have:

$$\vec{j}_a)_{r=a} = \vec{I} \times \vec{n}_a = \dfrac{\beta\vec{a}}{4\pi a^3} \times \vec{n}_a = 0 \text{ as } \sin(\vec{a}, \vec{n}_a) = \sin\pi = 0 .$$

Similarly, $\vec{j}_a)_{r=b} = \vec{I} \times \vec{n}_b = \dfrac{\beta\vec{b}}{4\pi a^3} \times \vec{n}_b = 0$ as $\sin(\vec{a}, \vec{n}_b) = \sin 2\pi = 0$.

(b)(i) For the imaginary magnetic masses equivalent to the magnetization, we have:

- for the volume densities: $\rho^*(M) = -\,\text{div}\vec{I}(M) = \dfrac{\beta}{4\pi}\text{div}_M \overrightarrow{grad}_M \dfrac{1}{r} = \dfrac{\beta}{4\pi}\Delta\left(\dfrac{1}{r}\right) = 0$

- for the surface densities:

    - for A: $\sigma^*(A) = \vec{I}(A).\vec{n}_a = \dfrac{\beta\vec{a}}{4\pi a^3}.\vec{n}_a = -\dfrac{\beta}{4\pi a^2}$ as $\vec{a}$ antiparallel at $\vec{n}_a$,

    - for B : $\sigma^*(B) = \vec{I}(B).\vec{n}_b = \dfrac{\beta\vec{b}}{4\pi b^3}.\vec{n}_b = \dfrac{\beta}{4\pi b^2}$ as $\vec{b} \parallel \vec{n}_a$.

The upshot is that the total magnetic masses with respect to the armatures of radii a and b are, respectively, $m^*_a = 4\pi a^2 \sigma^*_a = -\beta$ and $m^*_b = 4\pi a^2 \sigma^*_b = \beta$.

    We can conclude by verifying that the resulting magnetic mass from the two faces of the magnet is $m^*_R = m^*_a + m^*_b = 0$. This result is analogous to that found in answer 1(b), in that localized excesses in the positive (m^*_a) and negative (m^*_b) magnetic masses appear in the calculation so as to take into account the magnetization.

(b)(ii)

With $\vec{H}(r) = \dfrac{1}{4\pi} m^* \dfrac{\vec{r}}{r^3} = \dfrac{1}{4\pi} m^* \dfrac{\vec{u}}{r^2}$, we can write that

$$\Phi_H = \oiint \vec{H}.\overrightarrow{dS} = \dfrac{m^*}{4\pi} \oiint \dfrac{\vec{u}.\overrightarrow{dS}}{r^2} ,$$

so that with $d\Omega = \dfrac{\vec{u}.\overrightarrow{dS}}{r^2}$ and $\oiint d\Omega = 4\pi$ we arrive at $\Phi_H = m^*$, where m* is such

that $m^* = \oiiint \rho^* d\tau$ and represents the sum of magnetic masses inside the Gaussian surface in the form of Gauss's magnetic theory. If we then consider a point M such that $a < r < b$, with a Gaussian surface being a sphere of radius a, then

$$\Phi_H = \oiint \vec{H}.\overrightarrow{dS} = 4\pi r^2 H = m^* = -\beta , \text{ where } \vec{H} = -\dfrac{1}{4\pi}\beta\dfrac{\vec{r}}{r^3} .$$

From this can be deduced $\vec{B} = \mu_0(\vec{H} + \vec{I}) = \mu_0 \left(-\dfrac{\beta}{4\pi}\dfrac{\vec{r}}{r^3} + \dfrac{\beta}{4\pi}\dfrac{\vec{r}}{r^3} \right) = 0$.

Comment: In nonlinear magnetic materials, the relationship between $\vec{B}$ and $\vec{H}$ is nonlinear and $\vec{B} = \mu(H) \vec{H}$ so that the only usable relationship is $\vec{B} = \mu_0(\vec{H} + \vec{I})$. To be more explicit, it is worth remembering to not write $\vec{B} = \mu \vec{H}$, from which can arise $\vec{B} = 0$ and $\vec{H} = 0$, which would be completely wrong, as we can see in this example!

Chapter 5

Time-Varying Electromagnetic Fields and Maxwell's Equations

5.1. Variable Slow Rates and the Rate Approximation of Quasistatic States (RAQSS)

5.1.1. Definition

In this region of frequencies, the applied field varies sufficiently slowly with time so that it is possible to state that at a given instant the current intensity is the same throughout all parts of a closed circuit. Given then that for any value of S, $I = \iint_S \vec{j}.\overrightarrow{dS} = $ constant implies that the current density flux ($\vec{j}$) across a current "tube" (which delimits a closed surface) is zero, as in $\oiint \vec{j}.\overrightarrow{dS} = 0$ (see also Chapter 1). Following on from Ostrogradsky's theory, it would indicate that $\text{div}\,\vec{j} = 0$, and the conservation of charge therefore would give $\dfrac{\partial \rho}{\partial t} = 0$. This hypothesis indeed could be used as a starting point in defining quasistatic states.

5.1.2. Propagation

Concentrating on systems where the intensity is the same in all parts at a given time means neglecting propagation phenomena that would appear if the intensity were to vary rapidly with time. Once the intensity of the electrical configuration starts to vary, its effects felt at a distance will be delayed related to the velocity of signal propagation (which is the speed of light in a vacuum). When the phenomenon varies periodically, at a frequency ν, a pair of closest neighboring points undergoing the same vibrational state are at a distance of one spatial period, that is to say one wavelength, which is defined by $\lambda = \dfrac{c}{\nu}$. If the length (L) of the circuit is very small with respect to λ, i.e., $L \ll \lambda$ that happens for large λ (that is to say for low ν: variable slow rates), it is possible to make the first approximation that all points in the circuit are in the same vibrational state, as schematized in Figure 5.1.

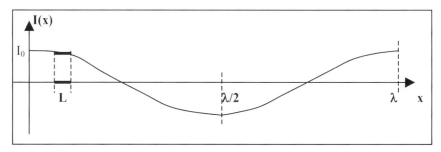

Figure 5.1. *Respective proportions of L and λ, where L << λ and I(x) ≈ constant over L.*

5.1.3. Basics of electromagnetic induction

Mobil charges in a circuit gain energy if subject to a variable magnetic flux, thus giving rise to an electromotive force (emf). If the variation in flux is not due to a displacement by the circuit itself, then magnetostatics cannot explain the problem, as in effect the magnetic force that operates on the charges, given by $\vec{F}_m = q\ \vec{v} \times \vec{B}$, has a direction perpendicular to the velocity of the displacement, and therefore does not participate. An electric induction field (E_i) must be added to the electrostatic field in order to account for the induced emf, and as the former acts as an electromotive field, it is not derived from a potential. The relation $\vec{E} = -\overrightarrow{\text{grad}}\ V$, where $\vec{E}$ is the total field is no longer useful, and by consequence, $\overrightarrow{\text{rot}}\ \vec{E} = 0$ neither is of any value.

5.1.4. Electric circuit subject to a slowly varying rate

5.1.4.1. A conductor without interruption nor capacitance

As detailed in Section 1.3.5, even for very short periods (as little as 10^{-14} sec), $\rho \to 0$. For slowly varied systems, up to around $v \approx 10^{14}$ Hz, the relation $\text{div}\ \vec{j} = 0$ holds true. It is only on reaching frequencies above 10^{14} Hz, nearing the optical region, that the equation $\text{div}\ \vec{j} = 0$ is no longer acceptable.

5.1.4.2. A conductor with a break, and the effect of capacitance

For the current intensity at the level of the surface S at the break (at the capacitor), the superficial charge density (σ) carried by the surface, such that $Q = \sigma\,S$, we have

$$I = \iint \vec{j}.\overrightarrow{dS} = j_n.S$$
$$= \frac{dQ}{dt} = S\frac{d\sigma}{dt} \quad \Rightarrow \quad j_n = \frac{d\sigma}{dt}\,.$$

By imposing the hypothesis of slowly varying frequencies, we suppose that $\dfrac{d\sigma}{dt} \approx 0$, so that $j_n = 0$, which again permits $\operatorname{div} \vec{j} = 0$ (see Section 1.3.3.).

5.1.4.3. Conclusion: electrical characteristics of a circuit subject to low frequencies

In order to calculate the currents subject to RAQSS:

- the duration of the signal is not considered and the intensity of the current is assumed to be the same for all parts of the circuit;
- an electromotive field is added to the applied field when the circuit is placed in a varying flux; and
- it is assumed that the capacitance effects are localized at the surface of the electrodes and that only the capacitance C which introduces the dpp V=Q/C at the terminals of the capacitor is taken into account. Charge variations with time ($d\sigma/dt = 0$) are neglected and as detailed in the following Section 5.2 under a regime of higher frequency fluxes, this effect corresponds to a current (called the "displacement current") which contributes to the magnetic field.

5.1.5. The Maxwell-Faraday relation
5.1.5.1. Lenz's law
For a circuit in a variable flux, either because the circuit is moving or the magnetic field is changing, there is an induction fem due to the electromotive induction field ($\vec{E}_i$) which is such that $\vec{E}_i \neq -\overrightarrow{grad}\ V$. Lenz's law states that in such a case,

$$e = \oint \vec{E}_i.\vec{dl} = -\frac{d\Phi}{dt} = -\iint_S \frac{\partial \vec{B}}{\partial t}.\vec{dS}.$$

5.1.5.2. Form of the resulting electric field
This results in:

$$\left.\begin{array}{l} e = \oint \vec{E}_i.\vec{dl} = \iint_S \overrightarrow{rot}\vec{E}_i\ \vec{dS} \\[2mm] = -\dfrac{d\Phi}{dt} = -\iint_S \dfrac{\partial \vec{B}}{\partial t}\ \vec{dS} \end{array}\right\} \Rightarrow \overrightarrow{rot}\ \vec{E}_i = -\frac{\partial \vec{B}}{\partial t}.$$

By defining the vector potential $\vec{A}$ by the often encountered equation $\vec{B} = \overrightarrow{rot}\ \vec{A}$, it becomes $\overrightarrow{rot}\ \vec{E}_i = -\dfrac{\partial}{\partial t}\overrightarrow{rot}\vec{A} = \overrightarrow{rot}\left(-\dfrac{\partial \vec{A}}{\partial t}\right)$ where $\vec{E}_i = -\dfrac{\partial \vec{A}}{\partial t}$. In a space in which there also is an electrostatic field, the electric field therefore is written as:

$$\vec{E} = -\overrightarrow{grad}\ V - \frac{\partial \vec{A}}{\partial t}$$

5.1.5.3. Maxwell-Faraday's relation

From the preceding equation, we end up with:

$$\overrightarrow{rot}\ \vec{E} = -\overrightarrow{rot}(\overrightarrow{grad}\ \ V) - \frac{\partial}{\partial t}(\overrightarrow{rot}\ \ \vec{A}) = -\frac{\partial \vec{B}}{\partial t} \quad \text{where}$$

$$\boxed{\overrightarrow{rot}\ \ \vec{E} = -\frac{\partial \vec{B}}{\partial t}}$$

and it is this which is the Maxwell-Faraday relation.

5.1.5.4. *Comment: Poisson's equation*

Given the equation for $\vec{E}$, we have:

$$\text{div } \vec{E} = - \text{div } (\overrightarrow{grad}V) - \frac{\partial}{\partial t} \text{div} \vec{A} .$$

Just as for a RAQSS, where $\text{div } \vec{j} = 0$, we have $\text{div } \vec{A} = 0$ (see Section 1.4.5.). The result is that $\text{div } \vec{E} = - \Delta V$, and by using Gauss's theorem, which remains valid (see Section 5.3.1.1 for a more explicit usage), then $\text{div } \vec{E} = \frac{\rho_\ell}{\varepsilon}$, and we finally obtain Poisson's equation:

$$\boxed{\Delta V + \frac{\rho_\ell}{\varepsilon} = 0}$$

5.2. Systems under Frequencies ($\text{div } \vec{j} \neq 0$) and the Maxwell – Ampere Relation

5.2.1. The shortfall of $\overrightarrow{rot}\ \vec{H} = \vec{j}_\ell$ (first form of Ampere's theorem for static regimes)

As opposed to quasistatic regimes, the rapidly varying regimes are such that at a given instant, the intensity differs in different sections of the current cylinder, which also correspond to $\frac{\partial \rho}{\partial t} \neq 0$.

Such regimes therefore are characterized by the relation

$$\text{div } \vec{j} \neq 0 \qquad (1).$$

If Ampere's theorem were to be still valid under its original form, $\overrightarrow{rot}\ \vec{H} = \vec{j}_\ell$, where $\vec{j}_\ell$ is the conduction current density due to a current deliberately applied to the circuit, then by taking the divergence of the two parts the result would be $\text{div } \vec{j}_\ell = 0$. This result is no longer acceptable for rapidly varying regimes, and therefore Ampere's theorem should be modified.

5.2.2. *The Maxwell-Ampere relation*

5.2.2.1. By intervention of vectors $\vec{H}$ and $\vec{D}$

In order to take into account the reality of Eq. (1), Ampere's relation is written in the form:

$$\overrightarrow{\text{rot}}\,\vec{H} \;=\; \vec{j}_\ell + \vec{X}\,, \qquad (2)$$

where $\vec{X}$ is a vector to be determined. Taking the divergence of the two parts in Eq. (2), we have:

$$\text{div}\,\vec{j}_\ell = -\,\text{div}\vec{X}\,. \qquad (3)$$

The introduction of the equation for the conservation of charge, coupled with the localized form of Gauss's theorem $\text{div}\vec{D} = \rho_\ell$ where ρ_ℓ is the volume density of free charges deliberately contributed, gives first $\text{div}\,\vec{j}_\ell + \dfrac{\partial \rho_\ell}{\partial t} = 0$ and then with Gauss's theorem:

$$\text{div}\,\vec{j}_\ell = -\text{div}\,\frac{\partial \vec{D}}{\partial t}\,. \qquad (4)$$

The comparison of Eqs. (3) and (4) would indicate that a vector in X is $\vec{X} = \dfrac{\partial \vec{D}}{\partial t}$.

Equation (2) is thus written:

$$\boxed{\;\overrightarrow{\text{rot}}\,\vec{H} \;=\; \vec{j}_\ell + \frac{\partial \vec{D}}{\partial t}\;}\,. \qquad (5)$$

It is important to note that the vectors $\vec{H}$ and $\vec{D}$ detailed above and also described in Section 3.2.7.2. have the same origin with respect to their sources, respectively, magnetic and electric, but both real.

5.2.2.2. The intervention of vectors $\vec{B}$ and $\vec{E}$

5.2.2.2.1. The intervention of magnetic permeability (μ) and dielectric permittivity (ε) of the material.

On using $\vec{H} = \dfrac{\vec{B}}{\mu}$ and $\vec{D} = \varepsilon\vec{E}$, we obtain directly from Eq. (5):

$$\overrightarrow{\text{rot}}\,\vec{B} \;=\; \mu\left[\vec{j}_\ell + \varepsilon\frac{\partial \vec{E}}{\partial t}\right]\,. \qquad (6)$$

5.2.2.2.2. The intervention of absolute magnetic permeability (μ_0) and absolute dielectric permittivity (ε_0) in a vacuum

With $\vec{H} = \dfrac{\vec{B}}{\mu_0} - \vec{I}$ and $\vec{D} = \varepsilon_0\vec{E} + \vec{P}$ substituted into Eq. (5) we have

$$\overrightarrow{\text{rot}}\left(\dfrac{\vec{B}}{\mu_0} - \vec{I}\right) = \vec{j}_\ell + \varepsilon_0\dfrac{\partial\vec{E}}{\partial t} + \dfrac{\partial\vec{P}}{\partial t}.$$

As $\vec{J}_A = \overrightarrow{\text{rot}}\,\vec{I}$, we finally have:

$$\boxed{\overrightarrow{\text{rot}}\,\vec{B} = \mu_0\left[\vec{j}_\ell + \varepsilon_0\dfrac{\partial\vec{E}}{\partial t} + \dfrac{\partial\vec{P}}{\partial t} + \overrightarrow{\text{rot}}\,\vec{I}\right]}, \qquad (7)$$

and equally, $\overrightarrow{\text{rot}}\,\vec{B} = \mu_0\left[\vec{j}_\ell + \varepsilon_0\dfrac{\partial\vec{E}}{\partial t} + \dfrac{\partial\vec{P}}{\partial t} + \vec{J}_A\right].$ (7')

By making

$$\vec{J}_D = \dfrac{\partial\vec{D}}{\partial t} = \varepsilon_0\dfrac{\partial\vec{E}}{\partial t} + \dfrac{\partial\vec{P}}{\partial t}, \qquad (8)$$

the displacement current in the medium under consideration, then we can also write:

$$\boxed{\overrightarrow{\text{rot}}\,\vec{B} = \mu_0\left[\vec{j}_\ell + \vec{J}_D + \vec{J}_A\right]} \qquad (9)$$

5.2.2.2.3. Conclusion

We have seen that Ampere's theorem for a vacuum, $\overrightarrow{\text{rot}}\,\vec{B} = \mu_0\,\vec{j}_\ell$, must be fulfilled in the case of materials under a regime of high-frequency flux. The current density that intervenes is the density of the total current, $\vec{J}_T = \vec{j}_\ell + \vec{J}_D + \vec{J}_A$, wherein all the currents in a vacuum intervene (with the permeability μ_0 a factor), i.e., the conduction (j_ℓ), displacement (J_D), and Amperian (J_A) currents. We can therefore also write that

$$\boxed{\overrightarrow{\text{rot}}\,\vec{B} = \mu_0\,\vec{J}_T}, \qquad (10)$$

where $\vec{J}_T = \vec{j}_\ell + \vec{J}_D + \vec{J}_A$.

5.2.3. *Physical interpretation of the displacement currents*
5.2.3.1. Recalling the relation for an electric field in a condenser as a function of the superficial charge densities carried by the armatures

Two methodologies with respect to a condenser with electrodes carrying superficial charges $\sigma_T = \sigma_0 + \sigma_P$ were detailed in Chapter 2.

5.2.3.1.1. Dielectric material and its equivalent

A dielectric material can be considered equivalent to a vacuum in which "sit" polarized charges, as shown in Figure 5.2, and from which we can write:

$$E = \frac{\sigma_0}{\varepsilon_0}. \quad (11)$$

This is possible because in the volume, the dipolar charges cancel each other out while the surface charges zero out a certain density (σ_P) of charge on the electrodes such that only the density σ_0 contributes to the generation of an electric field in the vacuum.

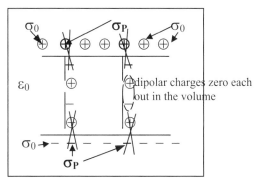

Figure 5.2. *Dielectric material considered as a vacuum in which sit polarization charges.*

5.2.3.1.2. Dielectric material is characterized by a macroscopic absolute dielectric permittivity ε

Permittivity is indeed a macroscopic characteristic and an overall property of a dielectric, as it can be measured through the use of capacitors with dielectrics (C) and with a vacuum (C_0), such that $\varepsilon_r = C/C_0$ where $\varepsilon = \varepsilon_r \varepsilon_0$.

Gauss's theorem can be used here, as indicated in Figure 5.3, as in Eq. (12):

$$E = \frac{\sigma_T}{\varepsilon} = \frac{\sigma_0 + \sigma_P}{\varepsilon}. \quad (12)$$

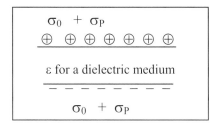

Figure 5.3. *Dielectric material characterized by its permittivity.*

5.2.3.2. Current density on a capacitor
At the level of the electrodes of a capacitor, the current intensity can be written as:

$$I = \iint \vec{j}.\vec{dS} = \vec{j}.\vec{S} = J_n.S, \qquad (13)$$

where S is the surface of the electrodes, and with j_n being the current density along the normal $\vec{n}$ to that of the electrode surface $\vec{S}$.

We can equally write $I = \dfrac{dQ}{dt}$, so that with $Q = Q_T = \sigma_T S = (\sigma_0 + \sigma_P).S$ we obtain

$$I = S \frac{d}{dt}(\sigma_0 + \sigma_P), \qquad (14)$$

By identification of Eqs. (13) and (14):

$$J_n = \frac{d\sigma_0}{dt} + \frac{d\sigma_P}{dt}. \qquad (15)$$

From Eq. (11), $\sigma_0 = \varepsilon_0 E$ and $\sigma_P = P_N = P$ (as the polarization vector is parallel to the electric field itself normal to the electrodes), we arrive at:

$$J_n = \frac{d\sigma_0}{dt} + \frac{d\sigma_P}{dt} = \varepsilon_0 \frac{dE}{dt} + \frac{dP}{dt} = J_{D0} + J_P \overset{D=\varepsilon_0 E+P}{=} \frac{\partial D}{\partial t} = J_D, \qquad (15')$$

with:

- $J_{D0} = \dfrac{d\sigma_0}{dt} = \varepsilon_0 \dfrac{dE}{dt}$ as the vacuum displacement current which is a product of

  the evolution with time of the charges of density σ_0 carried by the electrodes and are free charges located on the electrodes with a vacuum on the opposite facing side (Figure 5.2);

- $J_P = \dfrac{d\sigma_P}{dt} = \dfrac{dP}{dt}$ is the displacement current of polarization charges (also more

  succinctly called polarization current) which is associated with the current resulting from the cycling of dipolar charges distributed within the dielectric created by the rapidly varying electric field. The displacement current, associated with changes in time in the charge densities σ_0 and σ_P, is therefore very similar to Eq. (8):

$$J_D = \varepsilon_0 \frac{dE}{dt} + \frac{dP}{dt} \overset{(15')}{=} j_n = \frac{d\sigma_0}{dt} + \frac{d\sigma_P}{dt}. \qquad (16)$$

5.2.3.3. Comment: alternative methods to realizing the expression for displacement currents

By using both Eq. (12), which makes it possible to state that $(\sigma_0 + \sigma_P) = \varepsilon\, E$, and Eq. (15'), $J_D = \dfrac{d}{dt}(\sigma_0 + \sigma_P)$, one can also deduce that:

$$J_D = \varepsilon\, \frac{dE}{dt}. \qquad (17)$$

As $\varepsilon\, E = D = \varepsilon_0\, E + P$, therefore Eq. (8) is rediscovered:

$$J_D = \frac{dD}{dt} = \varepsilon_0\, \frac{dE}{dt} + \frac{dP}{dt}.$$

An alternative route to Eq. (17) also can be find.

Starting with

$$I = j.S \equiv J_n S$$
$$= \frac{dQ}{dt} = C\frac{dV}{dt} = \frac{S\varepsilon}{e}\frac{dV}{dt}$$

(where e and S represent the thickness and the surface of the capacitor, respectively) and with $E = \dfrac{V}{e}$, one directly obtains Eq. (17):

$$j_n = \varepsilon\frac{dE}{dt} = J_D$$

5.2.4. Conclusion

The displacement current is composed of two terms. They are:

- $J_{D0} = \dfrac{d\sigma_0}{dt} = \varepsilon_0\, \dfrac{dE}{dt}$, which relates to empty space and is an imaginary current in the sense that it does not cycle between the electrodes of the condenser; and

- $J_P = \dfrac{d\sigma_P}{dt} = \dfrac{dP}{dt}$, which corresponds to the movement of charges tied to the polarization of the dielectric with time such that $\dfrac{dP}{dt} \neq 0$.

The variable field (E(t)) results in a current of density $\vec{J}_D$ in the dielectric, which itself then forms a magnetic field in what is a phenomenon similar and complementary to induction processes. The use of what is a displacement current permits the application of Ampere's theorem, by which we must bring into play the total current which is the sum of the conduction ($\vec{j}_\ell$) and displacement ($\vec{J}_D$) currents.

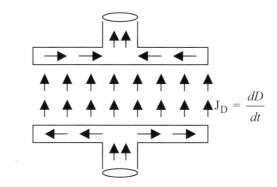

Figure 5.4. *Schematization of displacement current, which appears as an extension of an external current.*

The displacement current, as indicated in Figure 5.4, appears much like an extension of the external current. The imaginary character, that is to say a component that only makes an appearance in mathematical calculations, only comes about in the component $\dfrac{d\sigma_0}{dt} = \varepsilon_0 \dfrac{dE}{dt}$. This corresponds to the displacement current in a vacuum, where there would be no material intervening between the electrodes of the device and is due only to the variation in the surface density of free charges (σ_0) with time to a variation with time which are not dipolar charges at the armatures resulting from capacitance.

The part $\dfrac{dP}{dt}$, for the polarization current, is due to the presence of charges in the dielectric which follow the frequency of E(t), albeit out of phase which can result in leak currents when under alternating systems.

In terms of vectors, the displacement current, being normal to the electrodes [Eq. (13)], as was the case for $\vec{D}$, $\vec{E}$ and $\vec{P}$, permits us to write Eq. (8) as:

$$\vec{J}_D = \frac{d\vec{D}}{dt} = \varepsilon_0 \frac{d\vec{E}}{dt} + \frac{d\vec{P}}{dt}.$$

5.3. Maxwell's Equations

5.3.1. *Forms of* div $\vec{E}$ *and* div $\vec{B}$ *under varying regimes*

5.3.1.1. The Gauss-Maxwell relation

The equation that ties volume density of polarized charges to the polarization vector, $\rho_P = -\,\text{div}\,\vec{P}$, remains valid under varying regimes because it brings to bear with respect to these two terms only spatial considerations, independent of time. It is possible to write a localized form, $\text{div}\,(\varepsilon_0\vec{E}) = \rho_\ell + \rho_P$, which with $\vec{D} = \varepsilon_0\vec{E} + \vec{P}$, we have $\text{div}\vec{D} = \rho_\ell$. By going further and introducing $\vec{D} = \varepsilon\vec{E}$, it is found that:

$$\text{div}\,\vec{E} = \frac{\rho_\ell}{\varepsilon} = \frac{\rho_\ell + \rho_P}{\varepsilon_0}\,.$$

5.3.1.2. Equation for div $\vec{B}$

Just as for the variable regime, we can write $\vec{B} = \overrightarrow{\text{rot}}\vec{A}$, and here again we have $\text{div}\,\vec{B} = 0$.

5.3.2. *Summary of Maxwell's equations.*

Maxwell's equation are brought together below.

$$\text{div}\quad \vec{E} = \frac{\rho_\ell}{\varepsilon} = \frac{\rho_\ell + \rho_P}{\varepsilon_0} \quad (1)$$

$$\text{div}\,\vec{B} = 0 \quad (2) \qquad\qquad \overrightarrow{\text{rot}}\vec{E} + \frac{\partial\vec{B}}{\partial t} = 0 \quad (3)$$

$$\overrightarrow{\text{rot}}\,\vec{B} = \mu\left[\vec{j}_\ell + \varepsilon\frac{\partial\vec{E}}{\partial t}\right] = \mu_0\left[\vec{j}_\ell + \varepsilon_0\frac{\partial\vec{E}}{\partial t} + \frac{\partial\vec{P}}{\partial t} + \overrightarrow{\text{rot}}\,\vec{I}\right] \quad (4)$$

Equations (1) and (4) rewritten for the vectors $\vec{D}$ and $\vec{H}$, respectively, give:

$$\text{div}\,\vec{D} = \rho_\ell \quad (1') \qquad\qquad \overrightarrow{\text{rot}}\,\vec{H} = \vec{j}_\ell + \frac{\partial\vec{D}}{\partial t} \quad (4')$$

Comment: As $\operatorname{div}(\overrightarrow{\mathrm{rot}\,H}) = 0$, Eq. (4') results in

$$\operatorname{div}\left(\vec{j}_{\ell} + \frac{\partial \vec{D}}{\partial t}\right) = 0\,,$$

and this equation follows a conservation in flux of the vector $\vec{J}_{\ell,\mathrm{D}} = \vec{j}_{\ell} + \dfrac{\partial \vec{D}}{\partial t}$.

The same result can be found using $\operatorname{div} \vec{D} = \rho_{\ell}$, from which can be derived

$\operatorname{div} \dfrac{\partial \vec{D}}{\partial t} = \dfrac{\partial \rho_{\ell}}{\partial t}$, which substituted into the equation for the conservation of charge,

$\operatorname{div} \vec{j}_{\ell} + \dfrac{\partial \rho_{\ell}}{\partial t} = 0$, gives $\operatorname{div}\left(\vec{j}_{\ell} + \dfrac{\partial \vec{D}}{\partial t}\right) = 0$.

Comment 2: In the absence of a conduction current we find that $\vec{j}_{\ell} = 0$ and

$\overrightarrow{\mathrm{rot}\,\vec{H}} = \dfrac{\partial \vec{D}}{\partial t}$ which is in a sense an equation for the production of the field $\vec{H}$ in

the absence of a conduction current.

5.3.3. The Maxwell equations and conditions at the interface of two media

5.3.3.1. Continuity of D_n with $\sigma_{\ell} = 0$

Given that Eq. (1') is identical in form for static regimes, it therefore also can give:

$$\boxed{D_{1n} - D_{2n} = \sigma_{\ell}}$$

where σ_{ℓ} is the surface density of real charges, deliberately contributed to the interface. If $\sigma_{\ell} = 0$, there is of course continuity in the normal component of D, so that:

$$\boxed{D_{1n} = D_{2n}}\;.$$

5.3.3.2. Continuity in B_n

As Eq. (2) is identical to the equation used for static regimes, and as we have already shown that

$$\boxed{B_{1n} = B_{2n}}\,,$$

the equation remains valid for a regime of variation.

5.3.3.3. Continuity in E_t

Under static condition, Eq. (3) can be reduced to $\overrightarrow{\text{rot}}\vec{E} = 0$, which brings $\iint \overrightarrow{\text{rot}}\vec{E}\,\overrightarrow{dS} = \oint \vec{E}.\overrightarrow{dl} = 0$ from which can be deduced that $E_{1t} = E_{2t}$. However, under a varying regime, one must write that: $\iint\overrightarrow{\text{rot}}\vec{E}\,\overrightarrow{dS} + \iint\dfrac{\partial\vec{B}}{\partial t}\overrightarrow{dS} = 0$, so that

$\oint\vec{E}.\overrightarrow{dl} = -\iint\dfrac{\partial\vec{B}}{\partial t}\overrightarrow{dS}$. The surface considered in the latter integral is in the form $\iint dS = (A_1 B_1).(B_1 B_2)$ and is such that, as shown in Figure 5.5, in the neighborhood of the interface $(B_1 B_2) \rightarrow 0$. Assuming that this region is finite about the interface $\dfrac{\partial\vec{B}}{\partial t}$, we can consider that $\iint\dfrac{\partial\vec{B}}{\partial t}\overrightarrow{dS} \approx 0$, and from which can be derived $\oint\vec{E}.\overrightarrow{dl} = 0$, so that

$$\boxed{E_{1t} = E_{2t}}\quad.$$

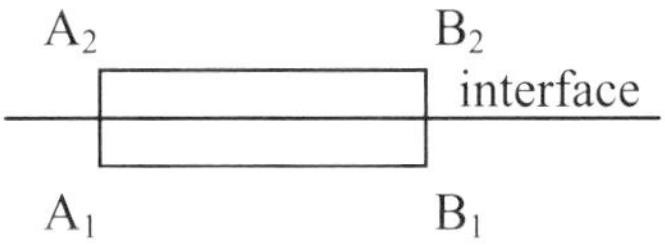

Figure 5.5. *Calculation for the circulation of E near the interface.*

5.3.3.4. Continuity of H_t with $j_\ell = 0$

When $j_\ell = 0$, Eq. (4') can be written as $\overrightarrow{\text{rot}}\,\vec{H} = \dfrac{\partial\vec{D}}{\partial t}$ so that it has the same form as Eq. (3). Permitting $\vec{H}$ to take on the role of $\vec{E}$, in the same way the following equation is derived:

$$\boxed{H_{1t} = H_{2t}\ (\text{where }j_\ell = 0)}\quad.$$

If $j_\ell \neq 0$, the equation for real surface currents intervenes, as is also the case for static regimes.

5.4. Problem

Values for conduction and displacement currents in various media

Copper is a good example of a good conductor. As long as the frequency (v) is such that $v < 100\,\text{GHz}$, it can be assumed that the conductivity (σ) remains constant with respect to its value under continuous current, that is $\sigma_0 \approx 6.10^7\,\Omega^{-1}\text{m}^{-1}$. The dielectric permittivity can be treated as if in a vacuum, i.e., ε_0.

Poor conductors, such as the rare earths or that otherwise known as lanthanides, have conductivities of the order of $\sigma \approx 10^{-4}\,\Omega^{-1}\text{m}^{-1}$ and permittivities close to that of a vacuum.

Nonconductors, or rather good insulators, include the dielectric "plastic" poly(vinyl chloride) (PVC) which is used in electrical goods and has $\sigma \approx 4 \times 10^{-7}\,\Omega^{-1}\text{m}^{-1}$ and $\varepsilon_r \approx 4$. Another very good insulator is Teflon with $\sigma \approx 2 \times 10^{-14}\,\Omega^{-1}\text{m}^{-1}$ and $\varepsilon_r \approx 2$.

(a) Give the expression for conduction (j_ℓ) and displacement (j_D) currents for a material with conductivity (σ) and absolute dielectric permittivity (ε). Also give the value of the ratio R for the moduli of j_D and j_ℓ, i.e., $R = \dfrac{|j_D|}{|j_\ell|}$, using the ratio in the form $R = N\dfrac{v\varepsilon_r}{\sigma}$ where N is a numerical value to be estimated.

(b) Show for copper that while the frequency $v < 100\,\text{GHz}$, the displacement current is negligible with respect to the conduction current.

(c) For a weak conductor, estimate the frequency (v_e) at which $j_D \approx j_\ell$. Write the approximate form for the current density at v_e.

(d) For insulators, estimate for the two examples the value of the frequency (v_i) at which the conduction current can be neglected.

(e)   Summarize the preceding results by writing the approximate form for the Maxwell-Ampere equation for these materials for $v \approx v_e$.

Answers

(a) For an alternating field $E = E_0 e^{i\omega t}$, on one hand we have $j_\ell = \sigma E$ and on the other $j_D = \dfrac{\partial D}{\partial t} = \varepsilon\dfrac{\partial E}{\partial t} = i\omega\varepsilon E$. Thus, $R = \dfrac{|j_D|}{|j_\ell|} = \dfrac{|i\omega\varepsilon E|}{|\sigma E|} = \dfrac{\omega\varepsilon}{\sigma} = \dfrac{2\pi v\varepsilon_r\varepsilon_0}{\sigma}$, and

with $\varepsilon_0 = \dfrac{1}{36\pi 10^9}$ ($\approx 8.85 \times 10^{-12}$ F/m), we obtain $R \approx 5.5\,10^{-11}\left(\dfrac{v\varepsilon_r}{\sigma}\right)$.

(b) For copper, $\varepsilon_r/\sigma \approx 1.7 \times 10^{-8}$, where $R \approx 10^{-18}\nu$ when $\nu < 10^{11}$ Hz so that $R < 10^{-7}$, a value well below unity, so that one can certainly state that for good conductors around this frequency domain, the displacement currents are negligible; $j_T \approx j_\ell$.

(c) For a poor conductor, with $\sigma \approx 10^{-4}\ \Omega^{-1}m^{-1}$ and $\varepsilon_r \approx 1$, we have:
$R \approx 5.5 \times 10^{-7}\nu$ and $j_D \approx j_\ell$ when $R \approx 1$, so that $\nu_e \approx 2$ MHz.
Thus, $j_T \approx j_\ell + j_D \approx 2\, j_\ell \approx 2\, j_D$.
(When $\nu \gg \nu_e$, $R \gg 1$ and $j_D \gg j_\ell$. For $\nu \ll \nu_e$, $j_\ell \gg j_D$.)

(d) With PVC, $\sigma \approx 4\ 10^{-7}\ \Omega^{-1}m^{-1}$ and $\varepsilon_r \approx 4$, from which $R \approx 5.5 \times 10^{-4}\ \nu$. In order to neglect the conduction current, R must be such that $R \gg 1$, so that $\nu > \nu_i \approx 2 \times 10^3 Hz = 2$ kHz.

For Teflon, we have $\sigma \approx 2\ 10^{-14}\ \Omega^{-1}m^{-1}$ and $\varepsilon_r \approx 2$, so that $R \approx 5.5 \times 10^3 \nu$. When $\nu > \nu_i \approx 2 \times 10^{-4}$ Hz, $R \gg 1$. For this material, a high-quality insulator, the conduction currents are *a priori* practically negligible with respect to displacement currents throughout the electromagnetic spectrum.

(e) When $\nu = \nu_e \approx 2$ MHz, we have:
- conducting medium: $j_D \ll j_\ell$ and $j_T \approx j_\ell \Rightarrow \overrightarrow{rotB} \simeq \mu_0 \vec{j}_\ell = \mu_0 \sigma \vec{E}$

- poorly conducting medium: $j_D \approx j_\ell$ and $j_T \approx j_\ell + j_D$
$$\Rightarrow \overrightarrow{rotB} \simeq \mu_0 (\vec{j}_\ell + \vec{j}_D) = \mu_0 (\sigma \vec{E} + \varepsilon \frac{\partial \vec{E}}{\partial t})$$

- insulating medium $j_D \gg j_\ell$ and $j_T \approx j_D \Rightarrow \overrightarrow{rotB} \simeq \mu_0 \vec{j}_D = \mu_0 \varepsilon_0 \varepsilon_r \frac{\partial \vec{E}}{\partial t}$.

The following scheme can be presented as a summary of the results:

Chapter 6

General Properties of Electromagnetic Waves and Their Propagation through Vacuums

6.1. Introduction: Equations for Wave Propagation in Vacuums

6.1.1. Maxwell's equations for vacuums: $\rho_\ell = 0$ and $j_\ell = 0$

From the title, we can state that in this case

$$\text{div } \vec{E} = 0 \quad (1) \qquad\qquad \overrightarrow{\text{rot}}\vec{E} = -\frac{\partial \vec{B}}{\partial t} \quad (3)$$

$$\text{div}\vec{B} = 0 \quad (2) \qquad\qquad \overrightarrow{\text{rot}} \ \vec{B} = \varepsilon_0\mu_0 \frac{\partial \vec{E}}{\partial t} \quad (4)$$

6.1.2. Equations of wave propagation

We can eliminate for example $\vec{B}$ in Eqs. (3) and (4), by taking the rotation of Eq. (3):

$$\overrightarrow{\text{rot}}(\overrightarrow{\text{rot}} \ \vec{E}) = -\frac{\partial}{\partial t}(\overrightarrow{\text{rot}} \ \vec{B}) = -\varepsilon_0\mu_0 \frac{\partial^2 \vec{E}}{\partial t^2} \quad \text{[using Eq. (4)]}$$

$$= \overrightarrow{\text{grad}}(\text{div}\vec{E}) - \Delta\vec{E} = -\Delta\vec{E} \quad \text{[using Eq. (1)]}$$

$$\left. \right\} \Rightarrow \quad \Delta\vec{E} = \varepsilon_0\mu_0 \frac{\partial^2 \vec{E}}{\partial t^2}$$

With $\varepsilon_0\mu_0 = \dfrac{1}{c^2}$, we can write that $\boxed{\Delta\vec{E} - \dfrac{1}{c^2}\dfrac{\partial^2\vec{E}}{\partial t^2} = 0}$. $\quad$ (5)

Similarly, we can eliminate $\vec{E}$ by calculating the rotation of Eq. (4):

$$\Delta\vec{B} = \varepsilon_0\mu_0 \frac{\partial^2 \vec{B}}{\partial t^2}, \text{ from which is deduced that } \boxed{\Delta\vec{B} - \frac{1}{c^2}\frac{\partial^2\vec{B}}{\partial t^2} = 0}. \quad (6)$$

Equation (5) and (6) give rise to six equations based on:

$$\Delta\Phi - \frac{1}{c^2}\frac{\partial^2\Phi}{\partial t^2} = 0 \quad . \qquad (7)$$

where $\Phi = \Phi(x,y,z,t)$ is equal to one of the six components E_x, E_y, E_z, B_x B_y, B_z such that each depends on the same variables (x,y,z,t), so that for example $E_x = E_x$ (x,y,z,t).

6.1.3. Solutions for wave propagation equations such that $\vec{E}$ and $\vec{B}$ depend on one spatial coordinate (z) and time (t)

6.1.3.1. *Forms of the solutions*

In this particular case, we have $\Phi = \Phi(z,t)$, and Eq. (7) results in

$$\frac{\partial^2\Phi}{\partial z^2} - \frac{1}{c^2}\frac{\partial^2\Phi}{\partial t^2} = 0 . \qquad (7')$$

In order in integrate Eq. (7'), the variable should be changed so that:

$$\left.\begin{array}{l} \tau = t + \dfrac{z}{c} \\[2mm] \theta = t - \dfrac{z}{c} \end{array}\right\} \quad \Rightarrow \quad \Phi(z,t) = \Phi(\tau,\theta) .$$

The calculation based on partial derivatives, common enough in the first year but nevertheless extremely tedious, yields:

$$\frac{\partial^2\Phi(\tau,\theta)}{\partial\tau\,\partial\theta} = 0 \quad . \qquad (8)$$

An initial integration with respect to θ gives $\dfrac{\partial\Phi}{\partial\tau} = f(\tau)$, a second with respect to τ gives $\Phi(\tau,\theta) = F(\tau) + G(\theta)$ where $F(\tau)$ is the primitive of $f(\tau)$.

Returning to the initial parameters z and t, we have:

$$\Phi(t,z) = F(t + \frac{z}{c}) + G(t - \frac{z}{c}) .$$

6.1.3.2. *Physical significance of the solutions* $G(t - \frac{z}{c})$ *and* $F(t + \frac{z}{c})$

The function $G(t - \frac{z}{c})$ represents the propagation along points where $z > 0$ for a velocity c. The phenomenon can be examined by studying the position in space of the signal G at times t_1 and t_2, as shown in Figure 6.1.

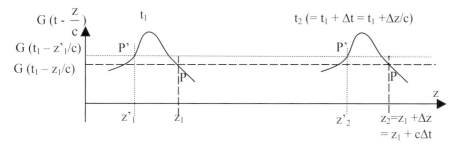

Figure 6.1. *Signal displacement between the times t_1 and t_2.*

The signal represented by the function $G(t_1 - z_1/c)$ at a point (P) on the abscissa (z_1) and at time t_1, will be such that at a following moment in time (t_2) where $t_2 = t_1 + \Delta t$, it can be represented by the function $G(t_2 - \frac{z_2}{c})$ where z_2 is the place on the abscissa where P has reached by the time t_2. The signal is assumed to be the same at t_1 and t_2, and indeed for all the points P, so we have

$$G(t_1 - \frac{z_1}{c}) = G(t_2 - \frac{z_2}{c}), \quad \text{and} \quad \text{if} \quad t_1 - \frac{z_1}{c} = t_2 - \frac{z_2}{c} \quad \text{we} \quad \text{find} \quad \text{that}$$

$z_2 - z_1 = c(t_2 - t_1)$, which can also be written as $\Delta z = c\Delta t$.

As the result above is valid for all the points denoted P, it also can be supposed that it is valid for all the points moving in the signal following the same law, that is:

$$\Delta z = c\Delta t.$$

This indicates that the signal itself, given by the function $G(t - \frac{z}{c})$, undergoes a propagation along $z > 0$ at a velocity c.

For its part, the function $F(t + \frac{z}{c})$ represents the propagation of the signal following a law such that $\Delta z = -c\Delta t$, derived from the condition $t_1 + \frac{z_1}{c} = t_2 + \frac{z_2}{c}$, which indicates that the signal is still moving at a velocity c, but along a point where $z < 0$: the F function represents a retrograde wave.

6.2. Different wave types

6.2.1. Transverse and longitudinal waves

6.2.1.1. Transverse signals

A transverse wave is one that has a movement (of its signal in the most general terms) that is perpendicular to the direction of its propagation. An example is a signal that moves perpendicularly along a stretched rope.

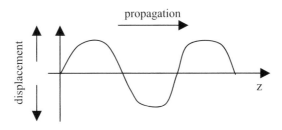

Figure 6.2. *Transverse signal.*

6.2.1.2. Longitudinal signal

A longitudinal wave is one that has a displacement parallel to the direction of the propagation. An example of this is a compression signal applied parallel to and along the length of a spring.

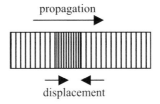

Figure 6.3. *Longitudinal signal.*

6.2.2. Planar waves

6.2.2.1. Definition

A vector ($\vec{A}$) is said to be propagated by a planar wave if at a given moment $\vec{A}$ is the same at all points in a plane π—called the wave plane—parallel to a given plane (π_0) that is perpendicular to the propagation direction (Oz) of the signal. This is schematized in Figure 6.4.

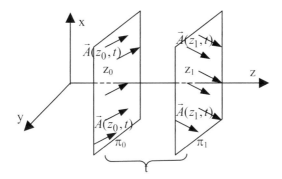

Figure 6.4. *Planar wave at a given point in time (t).*

6.2.2.2. *The implication of a planar wave*

The signal represented by the vector $\vec{A}$ depends only on z and t, so that $\vec{A} = \vec{A}(z, t)$. The solutions looked for in Section 6.1.3 for $\vec{E}$ and $\vec{B}$ also correspond to planar waves, as their solution components were in the form $\Phi(z, t)$. In effect, $\vec{E}$ and $\vec{B}$ propagate as planar waves, so it can be said that we have an electromagnetic planar wave (EMPW).

6.2.2.3. *Planar waves in practice*

In strict terms, in order to have a system of planar waves, all points should be within an infinite plane represented as an example by the vectors $\vec{A}(z_1)$ in an infinite plane denoted π_1. In practical terms that is not possible; nevertheless, at a sufficient distance from a source, identical $\vec{A}$ vectors can be determined within a large enough dimension to be considered a planar wave.

6.2.3. *Spherical waves*

While for a planar wave it is supposed that only one direction of propagation is used (Oz in the above example), for a spherical wave the propagation is isotropic in space. The field components (ξ) depend only on the distance (r) from the observation point of the wave. Thus $\xi = \xi(r, t)$, and the Laplacian is now reduced to

$\Delta\xi = \dfrac{1}{r}\dfrac{\partial^2}{\partial r^2}(r\xi)$ and the propagation equation becomes $\Delta\xi - \dfrac{1}{c^2}\dfrac{\partial^2\xi}{\partial t^2} = 0$, which in

turn gives $\dfrac{\partial^2}{\partial r^2}(r\xi) - \dfrac{1}{c^2}\dfrac{\partial^2(r\xi)}{\partial t^2} = 0$. The solutions for this are in the form

$r\xi = g^+\left(t - \dfrac{r}{c}\right) + f^-\left(t + \dfrac{r}{c}\right)$, so that

$$\xi = \frac{1}{r} g^+ (t - \frac{r}{c}) + \frac{1}{r} f^- (t + \frac{r}{c}).$$

The solution $\frac{1}{r} g^+ (t - \frac{r}{c})$ represents the propagation of a wave of speed c for when

r is positive, with an attenuation of the wave as r increases.

6.2.4. Progressive waves
6.2.4.1. Progressive planar waves (PPW)

As an example, if the function $G(t - \frac{z}{c})$ represents a signal that progresses toward

z > 0 at a speed c, then the sense and direction of the wave are perfectly defined and
we can say that it is a progressive wave (and is also called a "direct" wave given its

propagation toward z > 0). Similarly, $F(t + \frac{z}{c})$ is a progressive wave that goes

toward the points z < 0 also at a speed c. When in addition these progressive waves
are one dimensional, and are only dependent on one component in space, such as the
waves given by $G(z,t)$ or $F(z,t)$, then they are progressive planar waves that have
wave planes defined by z = constant.

6.2.4.2. Monochromatic progressive planar waves (MPPW)

Additionally, if the waves have sinusoidal signals, then the following can be stated:

$$G(t - \frac{z}{c}) = a \cos \left[\omega (t - \frac{z}{c}) \right]$$

$$F(t + \frac{z}{c}) = a \cos \left[\omega (t + \frac{z}{c}) + \varphi \right]$$

G and F waves have been
attributed with the same
amplitude and that F is out of
phase with G by φ.

By introducing the period $T = \frac{2\pi}{\omega}$, it is possible to say that:

$$G(t - \frac{z}{c}) = G(z,t) = a \cos 2\pi \left(\frac{t}{T} - \frac{z}{cT} \right).$$

And then by introducing the wavelength $\lambda = cT$, we have:

$$G(z,t) = a \cos 2\pi \left(\frac{t}{T} - \frac{z}{\lambda} \right).$$

With the introduction of the wavenumber, defined by $k = \frac{\omega}{c}$, the wave can be

written in the form:

$$G(z,t) = a \cos (\omega t - kz).$$

For its part, the wave vector $\vec{k}$ is defined by the relation:

$$\boxed{\vec{k} = k\,\vec{u} = \frac{\omega}{c}\,\vec{u}}\,,$$

where $\vec{u}$ is the unit vector of the direction of the propagation (here $\vec{u} = \vec{e}_z$).

As shown graphically in Figure 6.5, it is possible to see that this function presents both temporal and spatial periodicities.

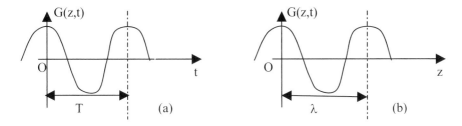

Figure 6.5. *(a) Temporal and (b) spatial periodicity.*

6.2.5. Stationary waves

A stationary wave would require that $S(z,t) = G\,(t - \dfrac{z}{c}\,) + F(t + \dfrac{z}{c}\,)$. Following a simple, direct calculation, we have for a monochromatic wave:

$$S(z,t)= 2\,a\,\cos\!\left(\omega\frac{z}{c} + \frac{\varphi}{2}\right).\cos\!\left(\omega t + \frac{\varphi}{2}\right)=A(z).\cos\!\left(\omega t + \frac{\varphi}{2}\right)=A(z)C(t)\ .$$

in which the term $C(t)$ is the same over all points at any given instant, so that the signal $[S(z,t)]$ is in the same phase at any point. Only the amplitude $[A(z)]$ changes with respect to the abscissa of the point in question, and there is no propagation term. Signals with the form $S(z,t)=A(z)C(t)$ are stationary waves and give rise to nodes at the abscissa where z is such that $\forall t : A(z)=0$ (and also $S(z,t)=0$).

Standing modes

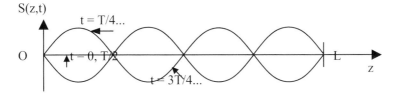

Figure 6.6. *A stationary wave exhibiting nodes at extremities O and L.*

In order to simplify the calculations, additionally we can assume that $\varphi = \pi$

so that: $S(z,t) = 2a \sin(\omega \frac{z}{c}).\sin(\omega t) = 2a (\sin 2\pi \frac{z}{\lambda}) (\sin \omega t)$.

It can be seen immediately that when $z = 0$, $S(z,t) = 0$ and that $S(z,t)$ represents a node at the origin that is repeated at points on the abscissa where $z = m\frac{\lambda}{2}$. For the moment the nodes (and likewise the antinodes) can occur at any value of ω, and it can be stated that the number of normal vibrational modes is infinite, a characteristic of continuous media.

If a second node is now imposed on the other end of the signal at $z = L$, where the signal can be imagined as, for example, a cord now fixed at either extremity or an electric field crossing between two conductors which are such that at their interface $E_{tangential} = 0$, then as shown in Figure 6.6 we have two conditions:
condition (1), $S(z = 0, t) = 0$; and
condition (2), $S(z = L, t) = 0$.

Given the form of $S(z,t) = 2a (\sin 2\pi \frac{z}{\lambda}) (\sin \omega t)$, condition (1) is always fulfilled; however, in order for condition (2) to be fulfilled, it is required that $\sin 2\pi \frac{L}{\lambda} = 0$, so that $2\pi \frac{L}{\lambda} = n\pi$, and that λ must be such that $\lambda = \frac{2L}{n}$. The frequency (v_n) is restricted now to $v_n = \frac{c}{2L}(n)$ and that of the angular frequency (ω_n) is:

$$\omega_n = \frac{\pi c}{L}(n) .$$

These conditions define the standing (stationary) modes and limit their number, as they give rise to discrete values for each of n whole number.

6.3. General Properties of Progressive Planar Electromagnetic Waves (PPEMW) in Vacuums with $\rho_\ell = 0$ and $j_\ell = 0$

With the electromagnetic waves being planar, $\vec{E}$ and $\vec{B}$ only depend on a single spatial coordinate (z) and time (t) in that $\vec{E} = \vec{E}(z,t)$, $\vec{B} = \vec{B}(z,t)$. In this section we consider a direct progressive wave that follows a propagation along $z > 0$ and as solutions for its components has $\vec{E}$ and $\vec{B}$ in the form $G (t - \frac{z}{c})$.

6.3.1. $\vec{E}$ and $\vec{B}$ perpendicular to the propagation: transverse electromagnetic waves

Equation (3), $\overrightarrow{\text{rot}}\vec{E} = -\dfrac{\partial \vec{B}}{\partial t}$, taken along Oz, yields

$$-\frac{\partial B_z(z,t)}{\partial t} = \frac{\partial E_y(z,t)}{\partial x} - \frac{\partial E_x(z,t)}{\partial y} = 0.$$

The result is that B_z must be independent from t.

Equation (2), $\text{div}\vec{B} = 0$, makes it possible to state that:

$$\underbrace{\frac{\partial B_x(z,t)}{\partial x}}_{=0} + \underbrace{\frac{\partial B_y(z,t)}{\partial y}}_{=0} + \frac{\partial B_z(z,t)}{\partial z} = \frac{\partial B_z(z,t)}{\partial z} = 0 \ ,$$

with the result that B_z should be independent of z. Therefore B_z must be independent of z and t and can be only constant. However, this solution does not represent a propagation and cannot be acceptable, and we therefore must simply consider that $B_z = 0$. The same is true for E_z ($E_z = 0$), as Eqs. (4) and (1) have the same structure as Eqs. (3) and (1).

The $\vec{E}$ and $\vec{B}$ therefore have the nonzero components E_x, E_y, B_x, and B_y. The fields $\vec{E}$ and $\vec{B}$ are definitely perpendicular to the direction of the propagation and the electromagnetic wave, by consequence, is termed a transverse electromagnetic (TEM) wave.

6.3.2. The relation between $\vec{E}$ and $\vec{B}$

For example, $\vec{E}$
$$\begin{cases} E_x(z,t) = G_1\left(t - \dfrac{z}{c}\right) \\[2mm] E_y(z,t) = G_2\left(t - \dfrac{z}{c}\right) \qquad (9) \\[2mm] E_z(z,t) = 0 \end{cases}$$

while the following remain as yet unknown $\vec{B}$
$$\begin{cases} B_x(z,t) \\[2mm] B_y(z,t) \\[2mm] B_z(z,t) = 0 \end{cases}$$

Equation (3), $\overrightarrow{\text{rot}}\vec{E} = -\dfrac{\partial \vec{B}}{\partial t}$, gives:

- with respect to Ox, $\dfrac{\partial E_y}{\partial z} = \dfrac{\partial B_x}{\partial t}$ (10); and

- with respect to Oy, $\dfrac{\partial E_x}{\partial z} = -\dfrac{\partial B_y}{\partial t}$.   (11)

Similarly, Eq. (4), $\overrightarrow{\text{rot}}\ \vec{B} = \dfrac{1}{c^2}\dfrac{\partial \vec{E}}{\partial t}$, gives:

- in part $\dfrac{\partial B_y}{\partial z} = -\dfrac{1}{c^2}\dfrac{\partial E_x}{\partial t}$.   (12)

- along with $\dfrac{\partial B_x}{\partial z} = \dfrac{1}{c^2}\dfrac{\partial E_y}{\partial t}$.   (13)

derived with respect to $(t - z/c)$

So from :

Eq. (10), $\dfrac{\partial B_x}{\partial t} = \dfrac{\partial E_y}{\partial z} = -\dfrac{1}{c}G_2'(t-\dfrac{z}{c}) \Rightarrow B_x = -\dfrac{1}{c}G_2(t-\dfrac{z}{c}) + [C^{te}\ \text{wrt t}]$

Eq. (13), $\dfrac{\partial B_x}{\partial z} = \dfrac{1}{c^2}\dfrac{\partial E_y}{\partial t} = \dfrac{1}{c^2}G_2'(t-\dfrac{z}{c}) \Rightarrow B_x = -\dfrac{1}{c}G_2(t-\dfrac{z}{c}) + [C^{te}\text{wrt z}]$

$$\Rightarrow B_x = -\dfrac{1}{c}G_2(t-\dfrac{z}{c}).$$

(Note: wrt means "with respect to".) Similarly, from Eqs. (11) and (12) we can determine that $B_y = \dfrac{1}{c}G_1(t-\dfrac{z}{c})$.

Finally we have $\vec{B}$ such that $\vec{B}$
$$\begin{cases} B_x = -\dfrac{1}{c}G_2(t-\dfrac{z}{c}) = -\dfrac{E_y}{c} \\[3mm] B_y = -\dfrac{1}{c}G_1(t-\dfrac{z}{c}) = \dfrac{E_x}{c} . \end{cases}$$ (14)

The following minor sections give the conclusions to this part.

6.3.2.1. $\vec{E}$ and $\vec{B}$ are perpendicular to one another

If the calculation $\vec{E}\cdot\vec{B}$ is made, we find that $\vec{E}\cdot\vec{B} = -E_x\dfrac{E_y}{c} + E_y\dfrac{E_x}{c} = 0$, and as

both $\vec{E}$ and $\vec{B}$ are perpendicular to the direction of propagation and in the plane Oxy, the configuration shown in Figure 6.7 arises.

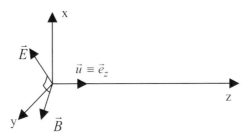

Figure 6.7. *Structure of a progressive planar electromagnetic wave.*

6.3.2.2. Equation for E and B
In fact we have:

$$\frac{E}{B} = c$$

where $E = \sqrt{E_x^2 + E_y^2}$ and $B = \sqrt{\dfrac{E_y^2}{c^2} + \dfrac{E_x^2}{c^2}}$, and hence the result.

6.3.2.3. Involving the unit vector
Bringing in the unit vector $\vec{u} = \vec{e}_z$ for the Oz axis gives rise to two properties that can be condensed into the same equation:

$$\boxed{\vec{E} = c\,\vec{B} \times \vec{u}}\quad . \qquad (15)$$

6.3.3. Breakdown of a planar progressive electromagnetic wave (PPEMW) to a superposition of two planar progressive EM waves polarised rectilinearly

6.3.3.1. Preliminary definition: rectilinear polarized wave
A wave can be termed rectilinearly polarized when the vector $\vec{E}$ stays over all points and instants parallel to a given direction in the plane of the wave. This direction is the direction of polarization.

6.3.3.2 . Breakdown of $\vec{E}$ and $\vec{B}$
Equations (9) and (14) show that $\vec{E}$ and $\vec{B}$ can be considered as vectors in the form $\vec{E} = \vec{E}_1 + \vec{E}_2$ and $\vec{B} = \vec{B}_1 + \vec{B}_2$ in which the vectors $\vec{E}_1$, $\vec{E}_2$, $\vec{B}_1$, and $\vec{B}_2$ are necessarily such that:

$$\vec{E}_1 \begin{cases} E_{1x} = G_1 \\ \\ E_{1y} = 0 \end{cases} \qquad\qquad \vec{B}_1 \begin{cases} B_{1x} = 0 \\ \\ B_{1y} = \dfrac{1}{c}G_1 \end{cases}$$

$$\vec{E}_2 \begin{cases} E_{2x} = 0 \\ \\ E_{2y} = G_2 \end{cases} \qquad\qquad \vec{B}_2 \begin{cases} B_{2x} = -\dfrac{1}{c}G_2 \\ \\ B_{2y} = 0 \end{cases} \qquad (16)$$

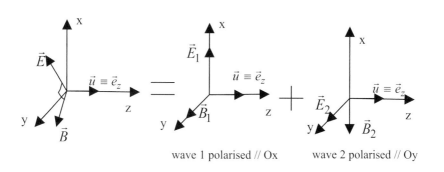

Figure 6.8. *Breakdown of a direct PPEMW in to two rectilinearly polarized PPEMWs.*

As schematized in Figure 6.8, the direct PPEMW ($\vec{E}, \vec{B}$) is such that $\vec{E} = \vec{E}(z,t)$ and $\vec{B} = \vec{B}(z,t)$, can appear thus as a superposition of:

- "wave 1" polarized along Ox, as $\vec{E}_1 = G_1(t - \dfrac{z}{c})\vec{e}_x$; and

- "wave 2" polarized along Oy, as $\vec{E}_2 = G_2(t - \dfrac{z}{c})\vec{e}_y$.

6.3.4. Representation and spectral breakdown of rectilinearly polarized PPEMWs
The first wave polarized along Ox can be studied in three-dimensional space and as a function of time.
It is defined by the function

$$E_{1x} = G_1(t - \dfrac{z}{c}).$$

6.3.4.1. Spatial representation

To have a spatial representation, a given instant, t = 0, is chosen. The signal then progresses at a speed c, so that $E_{1x}(z,t) = E_{1x}(z,0) = G_1(-\frac{z}{c})$. We then have the spatial representation shown in Figure 6.9, where throughout $\vec{E}$ and $\vec{B}$ are normal to each other and $\vec{u}$.

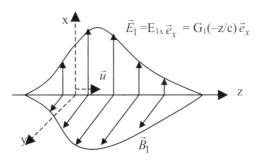

Figure 6.9. *Spatial presence of a rectilinearly polarized PPEMW.*

6.3.4.2. Representation with respect to time

At a given point, z = 0, $E_{1x}(z,t) = E_{1x}(0,t) = G_1(t)$. The same signal can be observed at any point on the abscissa (z_1) shifted in time by $\Delta t = \frac{\Delta z}{c}$. The observed amplitude observed at z = 0 and t = 0 then will be seen after an interval of time at z_1 and t_1 such that $t_1 = \Delta t = \frac{\Delta z}{c} = \frac{z_1}{c}$ (see Figure 6.10 a).

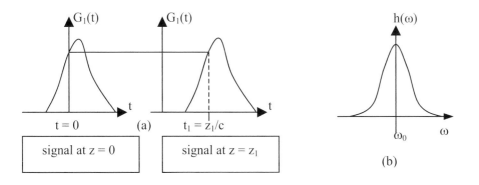

Figure 6.10. *Representations of a rectilinearly polarized PPEMW (a) v.s. time ; and (b) a component centered about ω_0.*

6.3.4.3. Spectral breakdown

If the wave under study is not monochromatic, then it corresponds to a collection of waves that can be seen as the superposition of an infinite number of waves with a distribution of frequencies spread about an average value (ω_0) as shown in Figure 6.10 b. It thus is possible to express the function $G_1(t)$ as a Fourier integral:

$$G_1(t) = \int_{-\infty}^{+\infty} h(\omega) \ e^{i\omega t} d\omega ,$$

where $h(\omega)$ represents a distribution of the amplitude as a function of ω and is such that the reciprocal Fourier transformation gives $h(\omega) = \dfrac{1}{2\pi} \int_{-\infty}^{+\infty} G_1(t) e^{-i\omega t} dt$.

In effect, it is possible that all progressive planar waves can be broken down into an infinite number of monochromatic and rectilinearly polarized progressive planar waves.

6.4. Properties of Monochromatic Planar Progressive Electromagnetic Waves (MPPEMW)

6.4.1. The polarization

Given the above results, a MPPEMW can be seen as a superposition of two monochromatic, rectilinearly polarized, planar progressive waves detailed by the collection of equations denoted (16). The G_1 and G_2 functions can be written in a more general form that takes into account any possible dephasing between G_1 and G_2 in that:

$$\left\{ \begin{array}{l} G_1(t - \dfrac{z}{c}) = a_1 \cos\left[\omega(t - \dfrac{z}{c}) \right], \quad \text{and} \\[2mm] G_2(t - \dfrac{z}{c}) = a_2 \cos\left[\omega(t - \dfrac{z}{c}) - \varphi \right]. \end{array} \right.$$

Here the amplitudes can be denoted as $a_1 = E_{1x} \equiv E_{mx}$ and again, by denotation, $a_2 = E_{2y} \equiv E_{my}$. Additionally, the signals are reproduced at each point in space identical to themselves, with a delay of $\dfrac{\Delta z}{c}$, so that the study can be limited to the point $z = 0$, the origin in space, where the resultant electric field ($\vec{E}$) is such that $\vec{E} = \vec{E}_1 + \vec{E}_2 = \vec{E}_{1x} + \vec{E}_{2y}$. The evolution of the vector $\vec{E}$ as a function of time is given for a point M(x,y,z) at the extremity of the vector $\overrightarrow{OM} = \vec{E}$ such that:

$$\overrightarrow{OM} = \vec{E} \left\{ \begin{array}{l} x = G_1(t) = E_{mx} \cos \omega t \\[1mm] y = G_2(t) = E_{my} \cos(\omega t - \varphi). \qquad (17) \\[1mm] z = 0 \end{array} \right.$$

6.4.1.1. When $\varphi = 0$ (waves "1" and "2" are in phase)

We have $\dfrac{y}{x} = \dfrac{E_{my}}{E_{mx}} = tg\alpha$ and this ratio, equal to that of the amplitudes, is constant

like the angle α. The M denotes a line in the 1st or 3rd quadrants $(tg\alpha = (\dfrac{E_{my}}{E_{mx}}) > 0)$

and the resultant wave ($\vec{E}$) is rectilinearly polarized along a diagonal, schematized in Figure 6.11 a.

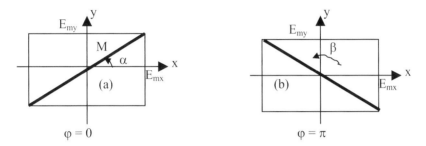

Figure 6. 11. *Rectilinearly polarized wave when (a) $\varphi = 0$; and (b) when $\varphi = \pi$.*

6.4.1.2. When $\varphi = \pi$ (waves "1" and "2" are out of phase)

We have $\dfrac{y}{x} = -\dfrac{E_{my}}{E_{mx}} = tg\beta$ and this ratio is again constant just like the introduced

angle β. The M this time denotes a line through 2nd or 4th quadrants

$(tg\beta = -(\dfrac{E_{my}}{E_{mx}}) < 0)$, and $\vec{E}$ thus is rectilinearly polarized along the other

diagonal, as shown in Figure 6.11 b.

6.4.1.3. When $\varphi = \pm\dfrac{\pi}{2}$ (waves "1" and "2" in quadrature): circular polarization

or elliptic lines (straight ellipse)

If $\varphi = \pm\dfrac{\pi}{2}$, $-\varphi = \mp\dfrac{\pi}{2}$ and the components x, y, z are:

$$\overline{OM} = \vec{E} \begin{cases} x = G_1(t) = E_{mx} \cos\omega t \\ y = G_2(t) = E_{my} \cos(\omega t \mp \dfrac{\pi}{2}) = \pm E_{my} \sin\omega t, \\ z = 0 \end{cases}$$

the calculation directly gives:

$$\frac{x^2}{E_{mx}^2} + \frac{y^2}{E_{my}^2} = 1 .$$

This is the equation for a straight ellipse (elliptical polarization) with a larger axis denoted E_{mx} and a smaller axis denoted E_{my}. The ellipse becomes a circle when the amplitudes are equal, i.e., $E_{mx} = E_{my}$ (see Figure 6.12 a for circular polarization).

We also have $\dfrac{y}{x} = \pm \dfrac{E_{my}}{E_{mx}} \tan \omega t = \pm r \; \tan \omega t = \tan \delta(t)$. The + in front of r is for

when $\varphi = \dfrac{\pi}{2}$ and the – sign is for when $\varphi = -\dfrac{\pi}{2}$ n, while $r = \dfrac{E_{my}}{E_{mx}} > 0$ (ratio of real

amplitudes) is constant.

We thus have:

$$\frac{d}{dt}(\tan \delta) = \frac{1}{\cos^2 \delta}\frac{d\delta}{dt} = \pm r\frac{1}{\cos^2 \omega t}\frac{d(\omega t)}{dt} = \pm r\frac{\omega}{\cos^2 \omega t} .$$

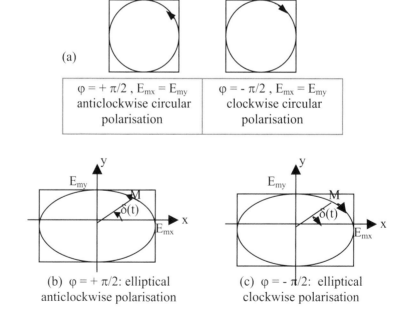

(a)

$\varphi = + \pi/2$, $E_{mx} = E_{my}$ anticlockwise circular polarisation	$\varphi = - \pi/2$, $E_{mx} = E_{my}$ clockwise circular polarisation

(b) $\varphi = + \pi/2$: elliptical anticlockwise polarisation

(c) $\varphi = - \pi/2$: elliptical clockwise polarisation

Figure 6.12. *(a) Circular; (b) anticlockwise elliptical; and (c) clockwise elliptical polarizations.*

For the first quadrant (Figure 6.12 b and c), $\delta(t)$ changes in the same sense as $\delta(t)$, it can be determined that when:

- $\varphi = +\dfrac{\pi}{2}$, we have $\dfrac{d}{dt}(\tan \delta) = +r\dfrac{\omega}{\cos^2\omega t} > 0$, and also $\dfrac{d\delta}{dt} > 0$. The point M describes an ellipse (or circle) in the trigonometric sense and we have an anticlockwise polarization, as in Figure 6.12 b.

- $\varphi = -\dfrac{\pi}{2}$, we have $\dfrac{d}{dt}(\tan \delta) = -r\dfrac{\omega}{\cos^2\omega t} < 0$, and $\dfrac{d\delta}{dt} < 0$. The point M describes a ellipse (or circle) in the opposite direction to the trigonometric sense and we have a clockwise polarization, as in Figure 6.12 c.

6.4.1.4. When φ takes on any value

We have $y = E_{my}\cos(\omega t - \varphi) = E_{my}\left[\cos\omega t\cos\varphi + \sin\omega t\sin\varphi\right]$, so that:

$$\frac{y}{E_{my}} - \cos\omega t\cos\varphi = \sin\omega t\sin\varphi .$$

With

$$\begin{cases} \cos\omega t = \dfrac{x}{E_{mx}} \\[4mm] \sin^2\omega t = 1 - \cos^2\omega t, \text{ from which } \sin\omega t = \left[1 - \dfrac{x^2}{E_{mx}^2}\right]^{1/2} , \end{cases}$$

and then by substitution of the values for $\cos\omega t$ and $\sin\omega t$ into the previous relation :

$$\frac{y}{E_{my}} - \frac{x}{E_{mx}}\cos\varphi = \sin\varphi\left(1 - \frac{x^2}{E_{mx}^2}\right)^{1/2} .$$

Squaring all round, we have:

$$\frac{x^2}{E_{mx}^2} + \frac{y^2}{E_{my}^2} - \frac{2xy}{E_{mx}E_{my}}\cos\varphi = \sin^2\varphi ,$$

which is the equation for an ellipse.

The component of the wave denoted by $x = G_1(t) = E_{mx}\cos\omega t$ is at a maximum when $t = 0$, and is such that $x = E_{mx}$. This is detailed by point A in Figure 6.13 a. At the same time, and according to Eq. (17), the point A has coordinates along Oy that are $y = E_{my}\cos\varphi$.

For this instant when t = 0, we can state that:

$$\begin{cases} \dfrac{dx}{dt} = -\left[\omega E_{mx} \sin \omega t\right]_{t=0} = 0 \\[2ex] \dfrac{dy}{dt} = -\left[\omega E_{my} \sin(\omega t - \phi)\right]_{t=0} = \omega E_{my} \sin \phi \end{cases}$$

$$\Rightarrow \left[\dfrac{d\vec{E}}{dt}\right]_{t=0} = \omega E_{my} \sin \phi \; \vec{e}_y .$$

The sense of direction in which the movement A ≡ M is given by the sign of sin φ.

- If $\phi \in \,]0,\pi[$, $\sin \phi > 0$ and $\dfrac{d\vec{E}_y}{dt} \,/\!/\,\vec{e}_y$, then the sense in which M moves is the trigonometric sense and an anticlockwise elliptical polarization, as in Figure 6.13 a, b, and c; and

- if $\phi \in \,]\pi,2\pi[$, $\sin \phi < 0$ and $\dfrac{d\vec{E}_y}{dt}$ is antiparallel to $\vec{e}_y$, then M moves in a direction opposite to the trigonometric sense and undergoes a clockwise elliptical polarization, as in Figure 6.13 d, e, and f.

It is worth noting that in Figure 6.13 (a), the point A' is such that $y = E_{my}$ is at a maximum. This gives $\omega t = \phi$, while the coordinates for A' in terms of Ox are $x = E_{mx} \cos \omega t = E_{mx} \cos \phi$.

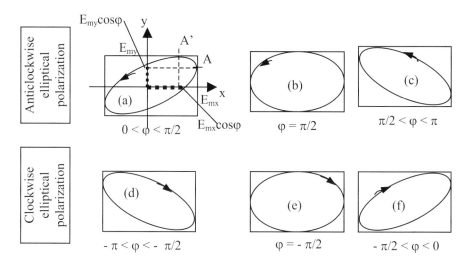

Figure 6.13. *The various polarizations as a function of φ.*

6.4.2. Mathematical expression for a monochromatic planar wave propagating in a direction OH

6.4.2.1. Rectilinearly polarized wave

6.4.2.1.1. Classic system and derivation

If we denote the amplitude of a rectilinearly polarized wave as E_0 and that $\vec{E}_0 \perp \overline{OH}$ so the planar monochromatic wave propagates through OH in the direct sense as indicated in Figure 6.14, then it is possible to write that

$$\vec{E}(h,t) = \vec{E}_0 \cos[\omega(t - \frac{h}{c})], \text{ where } |\overline{OH}| = h .$$

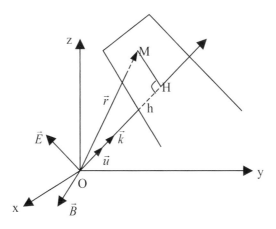

Figure 6.14. *Waves propagating in any direction* $\overline{OH}$.

The field at a point M located in space by the vector $\vec{r} = \overline{OM}$ can be found using the following method. In Figure 6.14 we see that $h = \vec{r}.\vec{u}$ where $\vec{u}$ is the unit vector in the direction of propagation. Given that M is in the same wave plane as H, we thus have:

$$\vec{E}(h,t) = \vec{E}(\vec{r},t) \overset{(\text{notation})}{=} \vec{E} = \vec{E}_0 \cos[\omega(t - \frac{\vec{r}.\vec{u}}{c})].$$

On introducing the wave vector:

$$\boxed{\vec{k} = \frac{\omega}{c}\vec{u}} ,$$

for the fields $\vec{E}$ and $\vec{B}$ we have:

$$\boxed{\vec{E} = \vec{E}_0 \cos[\omega t - \vec{k}\,\vec{r}]}$$

$$\boxed{\vec{B} = \vec{B}_0 \cos[\omega t - \vec{k}\,\vec{r}]} . \qquad (18)$$

6.4.2.1.2. Complex system and derivation

• *Using the normal notation for electrokinetics and dielectrics, where ω is the unit of reference for pulsation*

The goal of this method is to arrive at an expression for $\vec{E}$ that contains the classic electrokinetic term $\exp(+i\omega t)$. As is the tradition, we will omit the symbol R for the "real part" of the waves. A complex system can be given to the waves under consideration, so that for example the direct wave can be written as $\vec{E} = \vec{E}_0 \cos[\omega t - \vec{k}.\vec{r}]$, and in turn:

$$\begin{cases} \vec{E} = \vec{E}_0 \exp i[\omega t - \vec{k}.\vec{r}] = \vec{E}_0 \exp(-i\vec{k}.\vec{r})\exp(i\omega t) = \underline{\vec{E}_0(\vec{r})}\exp(i\omega t), \\ \text{where } \underline{\vec{E}_0(\vec{r})} = \vec{E}_0 \exp(-i\vec{k}.\vec{r}) \text{ for the complex amplitude of the direct wave.} \end{cases} \quad (19)$$

Really it should be written that $\underline{\vec{E}} = \vec{E}_0 \exp(i[\omega t - \vec{k}.\vec{r}]) = \underline{\vec{E}_0(\vec{r})}\exp(i\omega t)$, so that a complex number is in both the left- and right-hand sides of the equation; however, as only the real part ($\vec{E}$) of $\underline{\vec{E}}$ has a physical presence, by notational simplification the waves are often written as presented.

For the retrograde wave, $\vec{E} = \vec{E}_0 \cos[\omega t + \vec{k}.\vec{r}]$ and the corresponding complex form is: $\vec{E} = \vec{E}_0 \exp i[\omega t + \vec{k}.\vec{r}] = \vec{E}_0 \exp(i\vec{k}.\vec{r})\exp(i\omega t) = \underline{\vec{E}_0(\vec{r})}\exp(i\omega t)$ where $\underline{\vec{E}_0(\vec{r})} = \vec{E}_0 \exp(+i\vec{k}.\vec{r})$ is the complex amplitude of a retrograde wave.

• *Using the normal notation for optics, where the unit of reference is the wave vector ($\vec{k}$)*

The goal now in the equations is to bring out the term $\exp(+i\vec{k}.\vec{r})$ in the direct wave and the term $\exp(-i\vec{k}.\vec{r})$ in the retrograde wave. As $\cos(\alpha) = \cos(-\alpha)$, we can write $\vec{E} = \vec{E}_0 \cos[\omega t - \vec{k}.\vec{r}] = \vec{E}_0 \cos[\vec{k}.\vec{r} - \omega t]$ for the direct wave, so that the complex form also can be written:

$$\vec{E} = \vec{E}_0 \exp i[\vec{k}.\vec{r} - \omega t] = \vec{E}_0 \exp(i\vec{k}.\vec{r})\exp(-i\omega t) = \underline{\vec{E}_0(\vec{r})}\exp(-i\omega t), \quad (19')$$
$$\text{where } \underline{\vec{E}_0(\vec{r})} = \vec{E}_0 \exp(+i\vec{k}.\vec{r}) \text{ is the complex amplitude of the direct wave.}$$

For the wave in retrograde, we can write that: $\vec{E} = \vec{E}_0 \cos[\omega t + \vec{k}.\vec{r}] = \vec{E}_0 \cos(-[\omega t + \vec{k}.\vec{r}])$, from which:

$\vec{E} = \vec{E}_0 \exp(-i[\omega t + \vec{k}.\vec{r}]) = \vec{E}_0 \exp(-i\vec{k}.\vec{r})\exp(-i\omega t) = \underline{\vec{E}_0(\vec{r})}\exp(-i\omega t)$, where $\underline{\vec{E}_0(\vec{r})} = \vec{E}_0 \exp(-i\vec{k}.\vec{r})$ is the complex amplitude of the retrograde wave.

It is worth noting that for whatever notation used, in the two exponentials in the equations for $\vec{E}$, the sign that appears must be different for the direct wave, but is the same sign that intervenes in the two exponentials for the retrograde wave.

6.4.2.2. Wave polarized in any sense

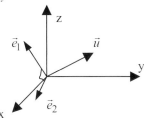

Figure 6.15. *A polarized wave seen as the superposition of two rectilinear waves polarized in $\vec{e}_1$ and $\vec{e}_2$.*

As we have already seen for a more general case, a wave can be considered to be the resultant of two waves rectilinearly polarized in the directions $\vec{e}_1$ *and* $\vec{e}_2$ perpendicular to the plan of the wave. This gives $\vec{E} = \vec{E}_1 + \vec{E}_2$ with (for a direct wave):

$$\vec{E}_1 = E_{m1}\vec{e}_1 \cos(\omega t - \vec{k}.\vec{r}) ; \text{ and}$$

$$\vec{E}_2 = E_{m2}\vec{e}_2 \cos(\omega t - \vec{k}.\vec{r} - \varphi) .$$

The complex form thus can be written:

$$\underline{\vec{E}} = \vec{E}_1 + \vec{E}_2 = \underline{\vec{E}_m} \exp[i(\omega t - \vec{k}.\vec{r})] = \underline{\vec{E}_m} \exp[-i\vec{k}.\vec{r})]\exp[i\omega t]$$

where $\underline{\vec{E}_m} = (E_{m1} \vec{e}_1 + E_{m2}\exp[-i\varphi]\vec{e}_2)$ is a complex vector for an elliptical polarization. By again writing $\vec{E}$ in the form $\underline{\vec{E}} = \underline{\vec{E}_0}(\vec{r})\exp(i\omega t)$, we have:

$$\underline{\vec{E}_0}(\vec{r}) = \underline{\vec{E}_m} \exp[-i\vec{k}.\vec{r})] \qquad (20),$$

where $\underline{\vec{E}_m} = \underline{\vec{E}_0}(0)$.

In this case, $\underline{\vec{E}_m} \neq \vec{E}_0$, while for a rectilinearly polarized wave

$\underline{\vec{E}_m} = \vec{E}_0 = \vec{E}_m$ (actual magnitude).

In this general case, we therefore have in complex notation for the direct wave

$\underline{\vec{E}} = \underline{\vec{E}_m}\exp[-i\vec{k}.\vec{r})]\exp(i\omega t)$,

so that also:

in electrokinetic notation, $\underline{\vec{E}} = \underline{\vec{E}_m} \exp[i(\omega t - \vec{k}.\vec{r})]$, or

in optical notation, $\underline{\vec{E}} = \underline{\vec{E}_m} \exp[i(\vec{k}.\vec{r} - \omega t)]$.

6.4.3. The speed of wave propagation and spatial periodicity

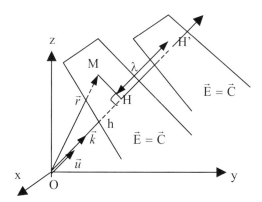

Figure 6.16. *Propagation speed for the wave plane.*

For a monochromatic rectilinear plane polarized electromagnetic wave (MRPPEW) of the form $\vec{E} = \vec{E}_0 \exp i[\omega t - \vec{k}.\vec{r}]$, the phase of the wave is given by $\phi = \omega t - \vec{k}.\vec{r}$. As time varies, the plane of the wave defined by $\vec{E} = \vec{E}_0 \exp i[\omega t - \vec{k}.\vec{r}] = \vec{C}$, where $\vec{C}$ is a constant vector, moves in such a way that $\phi = \omega t - \vec{k}.\vec{r} = \omega t - kh = K$ where K is a constant and as shown in Figure 6.16. For this, $d\phi$ must equal zero so that $(\omega dt - k \, dh) = 0$, from which we find:

$$\boxed{v_\phi = \frac{dh}{dt} = \frac{\omega}{k}}.$$

The point H, and therefore the signal $\vec{E} = \vec{C}$, moves at a speed $v_\phi = \frac{\omega}{k}$, which is the speed of the wave phase, and also therefore the speed of the propagation of the wave plane.

In a vacuum, where $k = \frac{\omega}{c}$, we have $v_\phi = c$.

At a given instant t, $\vec{E} = \vec{C}$ is repeated at intervals in space, equivalent to λ, along the plane of the wave such that $\cos(\omega t - \vec{k}.\vec{r}) = \cos(\omega t - \vec{k}.\vec{r} - k\lambda)$, and so that $k\lambda = 2\pi$, from which we rediscover the physical significance of $\lambda = \frac{2\pi}{k}$ as the spatial period.

6.5. Jones's Representation

6.5.1. Complex expression for a monochromatic planar wave propagating in the direction Oz

A given monochromatic planar wave propagating in a given direction Oz can be seen in the most general of cases, and as detailed in Section 6.3.3.2, as the resultant of two waves rectilinearly polarized in two directions $\vec{e}_x$ and $\vec{e}_y$ (classically taken as Ox and Oy) perpendicular to the plane of the wave, as shown in Figure 6.17.

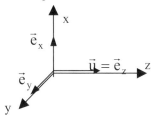

Figure 6.17. *Planar monochromatic EM wave propagating along Oz.*

In general terms, and without considering the origin of the phases on the wave $\vec{E}_x$ otherwise $\varphi_x = 0$ as supposed in Section 6.4.1,

$$\vec{E} = \vec{E}_x + \vec{E}_y, \text{ where}$$

$$\vec{E}_x = E_{mx}\vec{e}_x \cos(\omega t - k.z - \varphi_x)$$
$$\vec{E}_y = E_{my}\vec{e}_y \cos(\omega t - k.z - \varphi_y).$$

The wave can be written under the form:

$$\vec{E}(z) = \underline{\vec{E}}_0(z)\exp(i\omega t),$$

with $\underline{\vec{E}}_0(z) = \underline{\vec{E}}_m \exp[-ik.z)]$, where $\underline{\vec{E}}_m$ is a complex vector such that (relation also noted in Section 6.4.2.2):

$$\vec{E}_0(0) = \underline{\vec{E}}_m \quad \text{Eq. (1)}.$$

In addition:

$$\vec{E}(0) = \underline{\vec{E}}_0(0)\exp(i\omega t) = \vec{E}_x(0) + \vec{E}_y(0)$$

$$= (E_{mx}\exp[-j\varphi_x]\vec{e}_x + E_{my}\exp[-j\varphi_y]\vec{e}_y)\exp(i\omega t),$$

so that:

$$\underline{\vec{E}}_0(0) = (E_{mx}\exp[-j\varphi_x]\vec{e}_x + E_{my}\exp[-j\varphi_y]\vec{e}_y) \quad \text{Eq. (2)}.$$

By comparing Eqs. (1) and (2), we can directly determine that:

$$\underline{\vec{E}}_m = (E_{mx}\exp[-j\varphi_x]\vec{e}_x + E_{my}\exp[-j\varphi_y]\vec{e}_y) = \underline{E}_{mx}\vec{e}_x + \underline{E}_{my}\vec{e}_y \ .$$

By making $\underline{E}_{mx} = E_{mx} \exp\left[-j\varphi_x\right]$ and $\underline{E}_{my} = E_{my} \exp\left[-j\varphi_y\right]$, the complex vector $\vec{\underline{E}}_m$ appears as the vectors for the (complex) components $\underline{E}_{mx}$ and $\underline{E}_{my}$. To sum up then, the wave can be represented in the plane of the wave z = 0 by the vector $\vec{E}(0) \equiv \vec{E}$ (by notation) such that

$$\vec{E} = \vec{\underline{E}}_m \exp(i\omega t)$$

where $\vec{\underline{E}}_m = \vec{\underline{E}}_0(0)$ is a complex vector with components:

$$\vec{\underline{E}}_m \begin{cases} \underline{E}_{mx} = E_{mx} \exp\left[-j\varphi_x\right] \\ \underline{E}_{my} = E_{my} \exp\left[-j\varphi_y\right] . \\ 0 \end{cases}$$

The   ratio   defined   by   the   complex   number,
$\underline{r} = \dfrac{\underline{E}_{my}}{\underline{E}_{mx}} = \dfrac{E_{my}}{E_{mx}} \exp - i[\varphi_y - \varphi_x]$, is such that its modulus is $r = |\underline{r}| = \dfrac{E_{my}}{E_{mx}}$, and its

argument is: $\varphi = \mathrm{Arg}\, \underline{r} = (\varphi_x - \varphi_y)$.

It is worth noting that once again a common abuse of notation has intervened, for example, $\vec{E} = \vec{\underline{E}}_m \exp(i\omega t)$ would be, in a more rigorous notation, $\vec{E} = \vec{\underline{E}}_m \exp(i\omega t)$. The lapse in notation though is because we are simply looking for the physical solution $\vec{E} = R(\vec{\underline{E}})$.

6.5.2. Representation by way of Jones's matrix

The state of a wave at z = 0, given by $\vec{\underline{E}} = \vec{\underline{E}}_m \exp(i\omega t)$, can be simply written using a column matrix (Jones's notation):

$$\begin{pmatrix} \underline{E}_{mx} \\ \underline{E}_{my} \end{pmatrix} = \begin{pmatrix} E_{mx} \exp(-i\varphi_x) \\ E_{my} \exp(-i\varphi_y) \end{pmatrix} = E_{mx} \exp(-i\varphi_x) \begin{pmatrix} 1 \\ r \exp(i\varphi) \end{pmatrix}$$

By choosing the origin of the phases so that $\varphi_x = 0$, one can also write that:

$$\begin{pmatrix} \underline{E}_{mx} \\ \underline{E}_{my} \end{pmatrix} = \begin{pmatrix} E_{mx} \\ E_{my} \exp(-i\varphi_y) \end{pmatrix} = E_{mx} \begin{pmatrix} 1 \\ r \exp(i\varphi) \end{pmatrix}$$

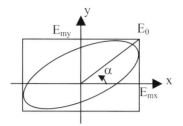

Figure 6.18. *Definition of E_{mx} and E_{my}.*

By making $E_{mx} = E_0 \cos \alpha$ and $E_{my} = E_0 \sin \alpha$, as in Figure 6.18, and still with $\varphi_x = 0$, we have:

$$\underline{\vec{E}}_m \begin{cases} \underline{E}_{mx} = E_{mx} = E_0 \cos \alpha \\ \\ \underline{E}_{my} = E_{my} \exp\left[-j\varphi_y\right] = E_0 \sin \alpha \, \exp\left[-j\varphi_y\right] \end{cases}$$

so that:

$$\begin{pmatrix} \underline{E}_{mx} \\ \underline{E}_{my} \end{pmatrix} = \begin{pmatrix} E_{mx} \\ E_{my} \exp(-i\varphi_y) \end{pmatrix} = \begin{pmatrix} E_0 \cos \alpha \\ E_0 \sin \alpha \; e^{-i\varphi_y} \end{pmatrix} = E_0 \begin{pmatrix} \cos \alpha \\ \sin \alpha \; e^{-i\varphi_y} \end{pmatrix}.$$

For a rectilinear polarization, we can take $\varphi_y = 0$, and the Jones's vector representing the wave is, by notation

$$\left(\underline{\vec{E}}_m\right) = \begin{pmatrix} E_0 \cos \alpha \\ E_0 \sin \alpha \end{pmatrix} = E_0 \begin{pmatrix} \cos \alpha \\ \sin \alpha \end{pmatrix}, \text{ and the normalized Jones's vector is such that:}$$

$$\left(\frac{\underline{\vec{E}}_m}{E_0}\right) = \begin{pmatrix} \cos \alpha \\ \sin \alpha \end{pmatrix}.$$

A rectilinear polarization through Ox corresponds to $\alpha = 0$, so that $\cos \alpha = 1$ and $\sin \alpha = 0$, from which

$$\left(\underline{\vec{E}}_m\right)_x = E_0 \begin{pmatrix} 1 \\ 0 \end{pmatrix} = E_0 (\vec{e}_h)$$

where $\vec{e}_h$ is represented by the column matrix $(\vec{e}_h) = \begin{pmatrix} 1 \\ 0 \end{pmatrix}$.

Similarly, a rectilinear polarization along Oy corresponds to $\alpha = \pi/2$ so that $\cos \alpha = 0$ and $\sin \alpha = 1$, from which

$$\left(\underline{\vec{E}}_m\right)_y = E_0 \begin{pmatrix} 0 \\ 1 \end{pmatrix} = E_0(\vec{e}_v)$$

where $\vec{e}_v$ is represented by the column matrix $(\vec{e}_v) = \begin{pmatrix} 0 \\ 1 \end{pmatrix}$.

For a clockwise polarized wave,

$$\begin{cases} \alpha = \dfrac{\pi}{4} \text{, so that } E_{mx} = E_0 \cos \alpha = E_0 \dfrac{\sqrt{2}}{2} \text{ and } E_{my} = E_0 \sin \alpha = E_0 \dfrac{\sqrt{2}}{2} \\ \varphi_y = -\dfrac{\pi}{2} \text{, from which, with } e^{-i\varphi_y} = e^{i\frac{\pi}{2}} = i \text{, we have:} \end{cases}$$

$$\left(\underline{\vec{E}}_m\right)_d = E_0 \begin{pmatrix} \dfrac{\sqrt{2}}{2} \\ i\dfrac{\sqrt{2}}{2} \end{pmatrix} = \dfrac{E_0\sqrt{2}}{2} \begin{pmatrix} 1 \\ i \end{pmatrix}$$

For an anticlockwise polarized wave, $\alpha = \dfrac{\pi}{4}$ and $\varphi_y = +\dfrac{\pi}{2}$, from which:

$$\left(\underline{\vec{E}}_m\right)_g = E_0 \begin{pmatrix} \dfrac{\sqrt{2}}{2} \\ -i\dfrac{\sqrt{2}}{2} \end{pmatrix} = \dfrac{E_0\sqrt{2}}{2} \begin{pmatrix} 1 \\ -i \end{pmatrix}.$$

These waves can be represented by a linear combination of $(\vec{e}_h)$ and $(\vec{e}_v)$, so for example,

$$\left(\underline{\vec{E}}_m\right)_d = \dfrac{E_0\sqrt{2}}{2} \begin{pmatrix} 1 \\ i \end{pmatrix} = \dfrac{E_0\sqrt{2}}{2} \begin{pmatrix} 1 \\ 0 \end{pmatrix} + \dfrac{E_0\sqrt{2}}{2} \begin{pmatrix} 0 \\ i \end{pmatrix} = \dfrac{E_0\sqrt{2}}{2}(\vec{e}_h) + i\dfrac{E_0\sqrt{2}}{2}(\vec{e}_v).$$

Inversely, the wave obtained by a superposition of the clockwise (negative) and anticlockwise (positive) waves is a rectilinear polarized wave because we have

$$\left(\underline{\vec{E}}_m\right)_d + \left(\underline{\vec{E}}_m\right)_g = \dfrac{E_0\sqrt{2}}{2} \begin{pmatrix} 1 \\ -i \end{pmatrix} + \dfrac{E_0\sqrt{2}}{2} \begin{pmatrix} 1 \\ i \end{pmatrix} = E_0\sqrt{2} \begin{pmatrix} 1 \\ 0 \end{pmatrix} = E_0\sqrt{2}(\vec{e}_h),$$

where the wave is polarized rectilinearly with respect to Ox.

6.6. Problems

6.6.1. Breakdown in real notation of a rectilinear wave into two opposing circular waves

A rectilinearly polarized, monochromatic plane electromagnetic wave propagates along Oz in the form $\vec{E} = \vec{E}_0 \cos(\omega t - kz)$ and is such that $\vec{E} \parallel Oxy$ makes an angle α with the axis Ox.

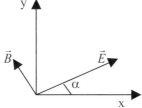

Show that this wave can be seen as a superposition of two planar waves circularly polarized in the opposite sense.

Answers

The components E_x and E_y of $\vec{E}$ on Ox and Oy are, {with

$$\cos a \cos b = \frac{1}{2}\ [\cos(a + b) + \cos(a - b)] \text{ and } \cos a \sin b = \frac{1}{2}\ [\sin(a + b) - \sin(a - b)]\}$$

$$E_x = E_0 \cos(\omega t - kz)\cos\alpha = \frac{E_0}{2}\cos(\omega t - kz + \alpha) + \frac{E_0}{2}\cos(\omega t - kz - \alpha)$$

$$E_y = E_0 \cos(\omega t - kz)\sin\alpha = \frac{E_0}{2}\sin(\omega t - kz + \alpha) - \frac{E_0}{2}\sin(\omega t - kz - \alpha).$$

This wave can be seen as the superposition of two following waves with the components

$$\begin{cases} E_x^{cG} = \dfrac{E_0}{2}\cos(\omega t - kz + \alpha) \\[2mm] E_y^{cG} = \dfrac{E_0}{2}\sin(\omega t - kz + \alpha) \end{cases}$$ anticlockwise circularly polarized wave; and

$$\begin{cases} E_x^{cD} = \dfrac{E_0}{2}\cos(\omega t - kz - \alpha) \\[2mm] E_y^{cD} = -\dfrac{E_0}{2}\sin(\omega t - kz - \alpha) \end{cases}$$ clockwise polarized wave.

Comment: It is worth remembering that a wave with the components

$$\begin{cases} E_x = E_m \cos \omega t \\ E_y = E_m \cos(\omega t - \varphi) \end{cases}$$

is one that has an anticlockwise polarization when $\varphi = +\dfrac{\pi}{2}$, which means that the components of this anticlockwise polarized wave are:

$$\begin{cases} E_x = E_m \cos \omega t \\ E_y = E_m \cos(\omega t - \varphi) = E_m \cos\left(\omega t - \dfrac{\pi}{2}\right) = +E_m \sin \omega t \end{cases}$$

Similarly, a wave which is polarized clockwise corresponds to $\varphi = -\dfrac{\pi}{2}$, and has components

$$\begin{cases} E_x = E_m \cos \omega t \\ E_y = E_m \cos(\omega t - \varphi) = E_m \cos\left(\omega t + \dfrac{\pi}{2}\right) = -E_m \sin \omega t \end{cases}$$

6.6.2. The particular case of an anisotropic medium and the example of a phase-retarding strip

Neutral lines in an anisotropic strip are such that rectilinear polarized waves along the two directions of these neutral lines, which are perpendicular to another, retain their polarization during their propagation. It is supposed that $\vec{e}_x$ and $\vec{e}_y$ are the vectors locating the directions of the neutral lines.

1. In a phase-retarding strip, the rectilinearly polarized waves along $\vec{e}_x$ and $\vec{e}_y$ each propagate as if in an isotropic media but with different indices, which are n_x for $\vec{e}_x$ and n_y for $\vec{e}_y$, and we will suppose that $n_y > n_x$. The components of the incident wave in the plane $z = 0$ are:

$$\vec{E}\begin{cases} E_{mx}\,e^{j\omega t} \\ E_{my}\,e^{j(\omega t - \varphi)} \\ 0 \end{cases}$$

Give the form of the wave emerging from the side plane where $z = e$, and determine the additional difference in phase caused by the wave's propagation between the two components with respect to Ox and Oy through the strip of thickness e. Express the result as a function of the additional differences in optical pathways (δ_L) between the two components of the wave under consideration created by the strip.

2. Under consideration is a half-wave strip such that $\delta_L = \dfrac{\lambda}{2}$. Study the effect of such a strip on an incident rectilinear polarized wave $(\varphi = 0)$.

Answers
1. Phase delay. For the waves polarized rectilinearly with respect to the directions $\vec{e}_x$ and $\vec{e}_y$, if we suppose that $n_y > n_x$, we have $v_x = \dfrac{c}{n_x} > v_y = \dfrac{c}{n_y}$. The axis with respect to $\vec{e}_x$ is termed the fast axis, while the axis along $\vec{e}_y$ is the slow axis.

The two components $\underline{E}_{mx}$ and $\underline{E}_{my}$ of the emerging wave from the strip do not have the same dephasing to that which exists between the two same components of the incident wave.

So the components of the incident wave in the side plane z = 0 are:

$$\vec{E} \begin{cases} E_{mx}e^{j\omega t} \\ E_{my}e^{j(\omega t-\varphi)} \\ 0 \end{cases},$$

and the emerging wave, from the side plane z = e, has the components:

$$\vec{E} \begin{cases} E_{mx}e^{j(\omega t-k_x e)} \\ E_{my}e^{j(\omega t-k_y e-\varphi)} \\ 0 \end{cases}.$$

The additional phase difference, due to the propagation of the wave through a strip of thickness e, between two components with respect to Ox and Oy, is therefore:

$$\Phi = (k_y e - k_x e) = \frac{2\pi}{\lambda}(n_y - n_x)e = \frac{2\pi}{\lambda}\delta_L \text{ where } \delta_L \text{ represents the difference in the}$$

additional optical pathway generated by the strip between the two components of the wave under consideration.

When $\Phi = \dfrac{\pi}{2}$, we have $\delta_L = \dfrac{\lambda}{4}$ and the strip is called a quarter wave strip.

When $\Phi = \pi$, we have $\delta_L = \dfrac{\lambda}{2}$ and the strip is called a half wave strip. In fact, the strip is not a true half wave strip unless $\lambda = 2\,\delta_L = 2\,(n_y - n_x)\,e$.

2.   Effect of a half wave strip on a rectilinearly polarized wave $(\varphi = 0)$. For the incident rectilinearly polarized wave, we can take (at $z = 0$):

$$\vec{E} \begin{cases} E_0 \cos\alpha \quad e^{j\omega t} \\ E_0 \sin\alpha \quad e^{j\omega t} \\ 0 \end{cases}$$

The emergent wave is such that (with $z = e$) : $\vec{E} \begin{cases} E_0 \cos\alpha \quad e^{j(\omega t - k_x e)} \\ E_0 \sin\alpha \quad e^{j(\omega t - k_y e)} \\ 0 \end{cases}$

By changing the origin of times by taking for the temporal origin a value that ensures a zero dephasing of k_x e with respect to the x components, the emerging wave can be written as:

$$\vec{E} \begin{cases} E_0 \cos\alpha \quad e^{j\omega t} \\ E_0 \sin\alpha \quad e^{j(\omega t - k_y e + k_x e)} \\ 0 \end{cases}, \text{ so that } \vec{E} \begin{cases} E_0 \cos\alpha \quad e^{j\omega t} \\ E_0 \sin\alpha \quad e^{j(\omega t - \Phi)} \\ 0 \end{cases}$$

With $\Phi = \pi$ (for the half wave strip) and $e^{-j\pi} = -1$, we reach:

$$\vec{E} \begin{cases} E_0 \cos\alpha \quad e^{j\omega t} \\ -E_0 \sin\alpha \quad e^{j\omega t} \\ 0 \end{cases} .$$

The emergent wave remains rectilinearly polarized and is symmetric to the incident wave with respect to the x axis.

Note: Effect of the quarter wave strip.   Similarly, we can show that a quarter wave strip can transform an elliptical vibration into a rectilinear vibration if the axes of the ellipse coincide with the neutral axes of the strip. All circularly polarized waves are transformed into rectilinearly polarized waves by a quarter wave strip.

6.6.3. Jones's matrix based representation of polarization

In the representation demonstrated by Jones, the effect of a polarizer can be given by a matrix ($\underline{P}$) such that:

$$\left(\vec{E}_m\right) = \begin{pmatrix} E_{mx} \\ E_{my} \end{pmatrix}_{exit} = (\underline{P}) \begin{pmatrix} E_{mx} \\ E_{my} \end{pmatrix}_{entrance}$$

The representation directed along Ox is that of a projection type operator:

$$(\underline{P_X}) = \begin{pmatrix} 1 & 0 \\ 0 & 0 \end{pmatrix}.$$

With respect to the action of a strip that introduces a delay equal to φ_y between the components $\underline{E}_{mx}$ and $\underline{E}_{my}$ of the field, it can be represented by a matrix such as:

$$(\underline{L}) = \begin{pmatrix} 1 & 0 \\ 0 & e^{-i\varphi_y} \end{pmatrix}.$$

1. Indicate the form of the matrix ($\underline{P}_y$) that can be associated with a polarizer directed along Oy so that the matrix is associated with a rotation ψ.
2. Use Jones's formalism to study the effect of a half wave strip on a rectilinearly polarized wave.
3. Give the product matrix that can be associated with an effect of two crossed polarizers between which sits at 45 ° a quartz strip on an incident light.

Answers

1. The action of a polarizer directed along Oy is represented by ($\underline{P}_y$) = $\begin{pmatrix} 0 & 0 \\ 0 & 1 \end{pmatrix}$.

For its part, the rotation (ψ) of an element (strip, polarizer) is represented by the rotational matrix: $R(\psi) = \begin{pmatrix} \cos\psi & -\sin\psi \\ \sin\psi & \cos\psi \end{pmatrix}$.

Finally, the state of polarization on leaving the system can be obtained by bringing together the successive orientational effects subjected on the incident wave. The resultant effect thus is written as a product of the matrices associated with the transformations.

2 The effect of a half wave on a rectilinearly polarized wave.
Using Jones's notation and two dimensions, the incident wave is:

$$\left(\vec{E}_m\right)_{inc} = \left(E_0 \cos\alpha\right)\begin{pmatrix} 1 \\ 0 \end{pmatrix} + \left(E_0 \sin\alpha\right)\begin{pmatrix} 0 \\ 1 \end{pmatrix}.$$

The transformation associated with a half wave strip is given by the matrix:

$$(\underline{L})_{\varphi y=\pi} = \begin{pmatrix} 1 & 0 \\ 0 & e^{-i\varphi_y} \end{pmatrix} = \begin{pmatrix} 1 & 0 \\ 0 & e^{-i\pi} \end{pmatrix} = \begin{pmatrix} 1 & 0 \\ 0 & -1 \end{pmatrix}, \text{ and}$$

$$\left(\vec{E}_m\right)_{exit} = (\underline{L})_{\varphi y=\pi}\left(\vec{E}_m\right)_{inc} = \left(E_0\cos\alpha\right)\begin{pmatrix} 1 & 0 \\ 0 & -1 \end{pmatrix}\begin{pmatrix} 1 \\ 0 \end{pmatrix} + \left(E_0\sin\alpha\right)\begin{pmatrix} 1 & 0 \\ 0 & -1 \end{pmatrix}\begin{pmatrix} 0 \\ 1 \end{pmatrix},$$

so that:

$$\left(\vec{E}_m\right)_{exit} = \left(E_0\cos\alpha\right)\begin{pmatrix} 1 \\ 0 \end{pmatrix} + \left(E_0\sin\alpha\right)\begin{pmatrix} 0 \\ -1 \end{pmatrix} = \left(E_0\cos\alpha\right)\begin{pmatrix} 1 \\ 0 \end{pmatrix} - \left(E_0\sin\alpha\right)\begin{pmatrix} 0 \\ 1 \end{pmatrix}.$$

We rediscover of course (see preceding problem) the conclusion that the emerging wave is symmetrical to the incident wave with respect to the x axis.

3. The effect on a incident wave by two crossed polarizes between which is inserted a strip of quartz at $45°$ is obtained directly from the product of the following matrices:

$$\begin{pmatrix} E_{mx} \\ E_{my} \end{pmatrix}_{exit} = (\underline{P}_y) \, R(45°) \, (\underline{L})_{\varphi=\frac{\pi}{2}} (\underline{P}_x) \begin{pmatrix} E_{mx} \\ E_{my} \end{pmatrix}_{entrance}.$$

Chapter 7

Electromagnetic Waves in Absorbent and Dispersing Infinite Materials and the Poynting Vector

7.1. Propagation of Electromagnetic Waves in an Unlimited and Uncharged Material for Which $\rho_\ell = 0$ and $j_\ell = 0$. Expression for the Dispersion of Electromagnetic Waves

7.1.1. Aide mémoire: the Maxwell equation for a material where $\rho_\ell = 0$ and $j_\ell = 0$

These equations are:

$$\overrightarrow{\text{div}}\vec{E} = 0 \quad (1) \qquad \qquad \overrightarrow{\text{rot}}\vec{E} = -\frac{\partial \vec{B}}{\partial t} \quad (3)$$

$$\text{div}\vec{B} = 0 \quad (2) \qquad \qquad \overrightarrow{\text{rot}}\,\vec{B} = \varepsilon\mu\frac{\partial \vec{E}}{\partial t} \quad (4)$$

7.1.2. General equations for propagation

Just as in a vacuum, we can eliminate $\vec{B}$ between Eqs. (3) and (4) by calculating the rotational of Eq. (3):

$$\overrightarrow{\text{rot}}(\overrightarrow{\text{rot}}\vec{E}) = -\frac{\partial}{\partial t}(\overrightarrow{\text{rot}}\vec{B}) \overset{(4)}{=} -\varepsilon\mu\frac{\partial^2\vec{E}}{\partial t^2} \quad \text{(using Eq. (4))}$$

$$= \overrightarrow{\text{grad}}(\text{div}\vec{E}) - \Delta\vec{E} \overset{(1)}{=} -\Delta\vec{E} \quad \text{(using Eq. (1))}$$

$$\left. \right\} \Rightarrow$$

$$\Delta\vec{E} = \varepsilon\mu\frac{\partial^2\vec{E}}{\partial t^2} = \varepsilon_0\varepsilon_r\mu_0\mu_r\frac{\partial^2\vec{E}}{\partial t^2} \quad (5')$$

With $\varepsilon_0 \mu_0 = \dfrac{1}{c^2}$, we can then go on to write:

$$\overrightarrow{\Delta \vec{E}} - \frac{\varepsilon_r \mu_r}{c^2} \frac{\partial^2 \vec{E}}{\partial t^2} = 0 \qquad (5)$$

And similarly:

$$\overrightarrow{\Delta \vec{B}} - \frac{\varepsilon_r \mu_r}{c^2} \frac{\partial^2 \vec{B}}{\partial t^2} = 0 \qquad (6)$$

7.1.3. A monochromatic electromagnetic wave in a linear, homogeneous, and isotropic material

For a monochromatic electromagnetic (EM) wave in a linear, homogeneous, and isotropic (lhi) material, $\vec{E}$ and $\vec{D}$ are directly related by $\vec{D} = \varepsilon(\omega) \vec{E}$ where $\varepsilon(\omega)$, dependent on ω, has for most materials (i.e., imperfect ones), a complex magnitude given by $\varepsilon(\omega) = \underline{\varepsilon}$.

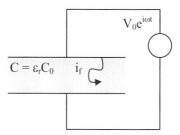

Figure 7.1. A leak current (i_f) in a real dielectric.

The electrical behavior of an imperfect dielectric, otherwise called a real dielectric, does not resemble that of a single capacitor as the presence of leak currents (i_f), caused for example by residual charges, gives rise to a component due to resistance (Figure 7.1). Therefore, $\varepsilon(\omega)$ needs to be written in a complex form in order to give a correct Fresnel diagram for the real dielectric. This means that the current intensity (I) should simultaneously present a component due to a dephasing by $\dfrac{\pi}{2}$ with respect to the applied tension (capacitive component) and a component in phase with the tension due to the resistance associated—in this example—with leak currents.

With $Q = CV = \underline{\varepsilon}_r C_0 V_0 e^{j\omega t}$ where $\underline{\varepsilon}_r$ is the complex relative dielectric permittivity

(given by $\underline{\varepsilon}_r = \varepsilon_r' - j\varepsilon_r''$ and $\underline{\varepsilon}_r = \dfrac{\underline{\varepsilon}}{\varepsilon_0} = \dfrac{\underline{C}}{C_0}$, where C_0 is the capacitance of a

capacitor in a vacuum), it is possible to directly obtain I in terms of two these two components (capacitive and resistive components), as

$$\underline{I} = \frac{dQ}{dt} = \omega \varepsilon_r'' C_0 V + j \, \omega \varepsilon_r' C_0 V = I_R + jI_C \, .$$

It is just worth noting that, as detailed in Chapter 2 of the second volume in this series, C is complex, and to be strictly correct we should write in the preceding equations for ε_r that $\underline{\varepsilon}_r = \underline{C}/C_0$.

Given that the waves are monochromatic, the fields $\vec{E}$, $\vec{D}$, and $\vec{B}$ can be given by:

$$\begin{cases} \underline{\vec{E}}(\vec{r},t) = \underline{\vec{E}}_0(\vec{r})\exp(+j\omega t) \\ \underline{\vec{D}}(\vec{r},t) = \underline{\vec{D}}_0(\vec{r})\exp(+j\omega t) \\ \underline{\vec{B}}(\vec{r},t) = \underline{\vec{B}}_0(\vec{r})\exp(+j\omega t) \end{cases}$$

Following simplification of the two members by $\exp(+j\omega t)$, and given that derivation with respect to time is the same as multiplying by $j\omega$, i.e., $\dfrac{\partial}{\partial t} \Leftrightarrow$ multiplication by $j\omega$, the Maxwell Eqs. (1) to (4) take on the form:

$\text{div}\underline{\vec{E}}_0(\vec{r}) = 0$ (1') $\text{rot}\underline{\vec{E}}_0(\vec{r}) = -j\omega\underline{\vec{B}}_0(\vec{r})$ (3')

$\text{div}\underline{\vec{B}}_0(\vec{r}) = 0$ (2') $\text{rot}\underline{\vec{B}}_0(\vec{r}) = j\omega\underline{\mu}\underline{\vec{E}}_0(\vec{r})$ (4')

Similarly, Eq. (5') for propagation yields:

$$\boxed{\Delta\underline{\vec{E}}_0(\vec{r}) + \omega^2 \, \underline{\varepsilon}\underline{\mu}\underline{\vec{E}}_0(\vec{r}) = 0} \qquad (7)$$

If the medium is simply a nonmagnetic one ($\mu_r = 1$), then

$\underline{\varepsilon}\underline{\mu} = \varepsilon_0\underline{\varepsilon}_r\mu_0 = \dfrac{\underline{\varepsilon}_r}{c^2}$ and Eq. (7) then becomes:

$$\boxed{\Delta\underline{\vec{E}}_0(\vec{r}) + \frac{\omega^2}{c^2}\underline{\varepsilon}_r\underline{\vec{E}}_0(\vec{r}) = 0} \qquad (7')$$

Alternatively, multiplying by $\exp[j\omega t]$ the equation can be given as:

$$\overrightarrow{\Delta \underline{\vec{E}}}(\vec{r}) + \frac{\omega^2}{c^2} \underline{\varepsilon}_r \underline{\vec{E}}(\vec{r}) = 0 \qquad (7'')$$

The relationship $\underline{\varepsilon} = \underline{\varepsilon}(\omega)$ (or $\underline{\varepsilon}_r = \underline{\varepsilon}_r(\omega)$) accounts for the fact that in a vacuum waves of different frequencies do not propagate at the same velocity due to a dispersion phenomenon. Just as was justified above for a real dielectric, $\underline{\varepsilon}(\omega)$ is normally complex, resulting in absorption phenomena associated with its imaginary component.

7.1.4. A case specific to monochromatic planar progressive electromagnetic waves (or MPPEM wave for short)

The complex amplitude of a MPPEM wave is in the form [see Eq. (20) in Chapter 6], that is:

$$\underline{\vec{E}}_0(\vec{r}) = \underline{\vec{E}}_m \exp(-j\underline{\vec{k}}.\vec{r}) \qquad (8)$$

In this case the material can absorb, so in the same way as we made $\varepsilon = \underline{\varepsilon}$, k is used in its complex form: $k = \underline{k}$.

7.1.4.1. Structure of a MPPEM wave

First, Eq. (1') gives us $\mathrm{div}\underline{\vec{E}}_0(\vec{r}) = 0$

$$= \sum_{i=1}^{3} \frac{d}{dxi} \left(\underline{E}_{mxi} \exp[-j(\underline{k}_{x1}x_1 + \underline{k}_{x2}x_2 + \underline{k}_{x3}x_3) \right) = -j \sum_{i=1}^{3} \underline{k}_{xi} \ \underline{E}_{mxi} \ e^{-j\underline{\vec{k}}.\vec{r}} = -j \ \underline{\vec{k}}.\underline{\vec{E}}_0(\vec{r})$$

The result,

$$\boxed{\mathrm{div}\underline{\vec{E}}_0(\vec{r}) = -j \ \underline{\vec{k}}.\underline{\vec{E}}_0(\vec{r})} \qquad (9)$$

is of a general form and concerns the action of the divergence operator on a wave determined by Eq. (8). By using Eq. (1), we can determine that $\underline{\vec{k}}.\underline{\vec{E}}_0(\vec{r}) = 0$, so that simplification with $\exp(-j\underline{\vec{k}}.\vec{r})$ yields:

$$\boxed{\underline{\vec{k}}.\underline{\vec{E}}_m = 0} \qquad (10)$$

Second, Eq. (2') similarly gains $\mathrm{div}\underline{\vec{B}}_0(\vec{r}) = -j \ \underline{\vec{k}}.\underline{\vec{B}}_0(\vec{r}) = 0$, from which we find:

$$\boxed{\underline{\vec{k}}.\underline{\vec{B}}_m = 0} \qquad (11)$$

Third, Eq. (3'), $\overrightarrow{\text{rot}}\ \vec{E}_0(\vec{r}) = -j\omega\ \vec{B}_0(\vec{r})$, demands a calculation around $\overrightarrow{\text{rot}}\ \underline{\vec{E}}_0(\vec{r})$. Therefore,

$$[\overrightarrow{\text{rot}}\ \underline{\vec{E}}_0(\vec{r})]_x = \frac{d}{dy}(\underline{E}_{mz}e^{-j\underline{\vec{k}}.\vec{r}}) - \frac{d}{dz}(\underline{E}_{my}e^{-j\underline{\vec{k}}.\vec{r}}) = -j\underline{k}_y\underline{E}_{mz}\ e^{-j\underline{\vec{k}}.\vec{r}} + jk_z\underline{E}_{my}\ e^{-j\underline{\vec{k}}.\vec{r}}$$

from which yields:

$$[\overrightarrow{\text{rot}}\ \underline{\vec{E}}_0(\vec{r})]_x = -j\left(\underline{\vec{k}}\times\underline{\vec{E}}_0(\vec{r})\right)_x .$$

The general result of the action of the rotational of Eq. (8) finally ends up with

$$\boxed{\left(\overrightarrow{\text{rot}}\ \underline{\vec{E}}_0(\vec{r})\right) = -j\left(\underline{\vec{k}}\times\underline{\vec{E}}_0(\vec{r})\right)} \qquad (12)$$

With the help of Eq. (3'), we find that the relation $\left(\underline{\vec{k}}\times\underline{\vec{E}}_0(\vec{r})\right) = \omega\ \underline{\vec{B}}_0(\vec{r})$, which is simplified on using $\exp(-j\underline{\vec{k}}.\vec{r})$, gives

$$\boxed{\underline{\vec{k}}\times\underline{\vec{E}}_m = \omega\underline{\vec{B}}_m} \qquad (13)$$

Fourth, Eq. (4') is such that we also find:

$$\overrightarrow{\text{rot}}\underline{\vec{B}}_0(\vec{r}) = -j\left(\underline{\vec{k}}\times\underline{\vec{B}}_0(\vec{r})\right)$$

$$= j\omega\ \underline{\varepsilon}\ \mu\ \underline{\vec{E}}_0(r)$$

Therefore, as $\mu_r = 1$, $\varepsilon = \varepsilon_0\ \underline{\varepsilon}_r$ and simplification of the above equation with $\exp(-j\underline{\vec{k}}.\vec{r})$ gives us

$$\boxed{\underline{\vec{k}}\times\underline{\vec{B}}_m = -\omega\ \underline{\varepsilon}\ \mu_0\underline{\vec{E}}_m = -\frac{\omega}{c^2}\ \underline{\varepsilon}_r\ \underline{\vec{E}}_m} \qquad (14)$$

Fifth, the MPPEM in a material has a planar progressive structure that is in direct relationship with the trihedral $(\vec{E}, \vec{B}, \vec{k})$.

7.1.4.2. General equation for dispersion
The calculation of the Laplacian vector for the wave's complex amplitude given in Eq. (8) directly gives:

$$\boxed{\Delta\underline{\vec{E}}_0(\vec{r}) = -\underline{k}^2.\underline{\vec{E}}_0(\vec{r})} \qquad (15)$$

By identification with the value given by $\overrightarrow{\Delta \underline{E}_0(\vec{r})}$ from Eq. (7'), as in

$\overrightarrow{\Delta \underline{E}_0(\vec{r})} = -\dfrac{\omega^2}{c^2}\varepsilon_r \underline{\vec{E}}_0(\vec{r})$ and the additional division by $\exp(-j\underline{k}.\vec{r})$, the result is:

$$\left[\underline{k}^2 - \dfrac{\omega^2}{c^2}\varepsilon_r(\omega) \right]\underline{\vec{E}}_m = 0 .$$

With $\underline{\vec{E}}_m \neq 0$, the general form of the so-called dispersion equation is:

$$\boxed{\underline{k}^2 = \dfrac{\omega^2}{c^2}\varepsilon_r(\omega)} . \qquad (16)$$

7.2. The Different Types of Media
7.2.1. No absorbing media and indices

For nonabsorbent materials with real values for $\varepsilon_r(\omega)$, Eq. (16) indicates that k is also real, within the constraints of $\varepsilon_r(\omega) > 0$, as otherwise $\varepsilon_r(\omega)$ would correspond to an evanescent wave for which $k = \underline{k} = -ik"$ (see Chapter 9). According Eq. (16) can be written more simply as:

$$k = \dfrac{\omega}{c}\sqrt{\varepsilon_r(\omega)} . \qquad (16')$$

The velocity of the phase of a MPPEM wave with frequency ω is now given by:

$$\boxed{v_\varphi = \dfrac{\omega}{k} = \dfrac{c}{\sqrt{\varepsilon_r(\omega)}}} . \qquad (17)$$

For its part, the group velocity of the wave is now:

$$\boxed{v_g = \dfrac{d\omega}{dk}} . \qquad (18)$$

7.2.1.1. In a vacuum

Now $\varepsilon_r = 1$ and therefore from Eq. (17) gives $v_\varphi = c$ (the velocity of the phase is constant with respect to ω and equal to c). As $\omega = ck$, Eq. (18) shows us that $v_g = c$, so that in a vacuum there is $v_\varphi = v_g = c$.

7.2.1.2. In a nondispersing medium

Now ε_r is independent of the frequency and accordingly from Eq. (17):
$v_\varphi = A$, where A is a constant. In addition, as indicated in the plot shown in

Figure 7.2, $\omega = A k$. Indeed, $v_g = \dfrac{d\omega}{dk} = A$, so that for nondispersing media, $v_\varphi = v_g$.

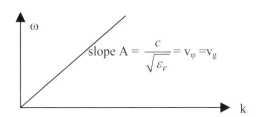

Figure 7.2. *Plot of $\omega = f(k)$ for nondispersing media.*

7.2.1.3. Normal dispersing media

Now we find that if ω increases, then $\varepsilon_r(\omega)$ also increases. However, Eq. (17) indicates that v_φ *decreases*, so that $v_{\varphi 2} < v_{\varphi 1}$ and the tangent $\left(\dfrac{d\omega}{dk}\right)$ to the dispersion curve is above the curve, and therefore $v_g < v_\varphi$, as shown in Figure 7.3a.

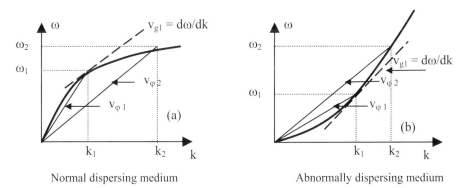

Normal dispersing medium   Abnormally dispersing medium

Figure 7.3. *Dispersion curves showing $\omega = f(k)$ for (a) normal dispersion medium; and (b) abnormal dispersion medium.*

8.2.1.4. Abnormally dispersing media

Now as ω increases, then $\varepsilon_r(\omega)$ *decreases*, and Eq. (17) indicates that v_φ also increases so that $v_{\varphi 2} > v_{\varphi 1}$ and the tangent $\left(\dfrac{d\omega}{dk}\right)$ is below the dispersion curve and $v_g > v_\varphi$, as shown in Figure 7.3b.

7.2.1.5. Indices

An index (n) is defined by: $n = \dfrac{c}{v_\varphi}$ (19)

and as Eq. (17) points out $v_\varphi = v_\varphi(\omega)$, it is also possible to state that $n = n(\omega)$. Placing Eq. (17) into Eq. (19) gives:

$$n(\omega) = \sqrt{\varepsilon_r(\omega)}$$ (19')

from which through Eq. (16') we gain

$$k = \frac{\omega}{c}n = k_0 n .$$ (20)

where $k_0 = \dfrac{\omega}{c}$ and is the wavenumber in a vacuum. As the wavelength in a medium can be determined by $\lambda = v_\varphi T = \dfrac{c}{n}\dfrac{2\pi}{\omega} = \dfrac{2\pi}{k}$, if we introduce $\lambda_0 = \dfrac{2\pi c}{\omega}$ (wavelength *in vacuo*) we arrive at:

$$\lambda = \frac{\lambda_0}{n(\omega)} .$$ (21)

7.2.1.6. Comment

It is possible also to use Eq. (13') by multiplying the two members by $\exp i[\omega t - \vec{k}.\vec{r}]$ so that $\vec{k} \times \underline{\vec{E}} = \omega\underline{\vec{B}}$ which gives in terms of moduli $kE = \omega B$. With $k = \dfrac{\omega}{c}n$, the equation yields:

$$\frac{E}{B} = \frac{c}{n} = v_\varphi .$$ (19'')

7.2.2. Absorbent media and complex indices

7.2.2.1. Absorbent media (dielectric permittivity and wavevector both complex)

7.2.2.1.1. Equation for dielectric permittivity

As above detailed, absorbing materials means that $\varepsilon(\omega)$ is complex, and by notation, $\underline{\varepsilon}(\omega) \equiv \underline{\varepsilon}$. Dielectric losses caused by Joule effects from the resistive component are absorbed by the dielectric—hence the name absorbent. More precisely, and depending on notational convention, complex permittivity can be defined in terms of:

electrokinetics, where $\underline{\varepsilon}(\omega) = \underline{\varepsilon} = \varepsilon' - j\varepsilon''$ (22) or

wave optics, where $\underline{\varepsilon}(\omega) = \underline{\varepsilon} = \varepsilon' + j\varepsilon''$ (22')

   Whether a minus or positive sign is chosen in the above equations (and as detailed below in equations for $\underline{k}$) is relative to the final expression used for the wave, wherein its application to dielectric media signifies an absorption and not a (spontaneous) amplification of the wave, the latter of which would be unrealistic.

   Staying with dielectrics, in Section 7.1.3 the calculation of $\underline{I} = \dfrac{dQ}{dt}$ was performed using an electrokinetic notation, i.e., $V = V_0 \exp(+j\omega t)$ from which with $\underline{\varepsilon} = \varepsilon' - j\varepsilon''$ from Eq. (22') it can be determined that $\underline{I} = \omega\varepsilon''_r C_0 V + j \omega\varepsilon'_r C_0 V$. The term due to resistance (I_R) which is such that $I_R = +\omega\varepsilon''_r C_0 V$ is in phase with the tension; however, if the term $\underline{\varepsilon}$ had been taken into Eq. (22'), then the result would have been $I_R = -\omega\varepsilon''_r C_0 V$, indicating an unphysical dephasing of the intensity and the tension by π at the resistance terminals. Inversely, if the optical notation were used so that $V = V_0 \exp(-j\omega t)$, then Eq. (22') must be used to obtain a physically acceptable result, as in $I_R = +\omega\varepsilon''_r C_0 V$.

7.2.2.1.2. Equation for the wavenumber and the MPPEM wave

In optical terms, according to Eq. (16) for dispersion, the complex form of $\underline{\varepsilon}$ indicates that k also must be complex ($k \to \underline{k}$). As detailed below, the optical wave can be absorbed during its propagation, and therefore k should be written in the form:

$\underline{k} = k' - j\,k''$ (electrokinetic notation)   (23) and

$\underline{k} = k' + j\,k''$ (optical notation)   (23')

   Again following from Eq. (16), with a $\pm$ sign given for both possibilities,

we have: $\underline{k}^2 = k'^2 - k''^2 \pm 2jk'k'' = \dfrac{\omega^2}{c^2}(\varepsilon'_r \pm j\varepsilon''_r)$,

so that by identification of the real and imaginary parts,

$$
\begin{cases}
k'^2 - k''^2 = (\dfrac{\omega^2}{c^2})\,\varepsilon_r' \\[2mm]
2\,k'\,k'' = (\dfrac{\omega^2}{c^2})\,\varepsilon_r'' .
\end{cases}
$$

With $\varepsilon_r'' > 0$ (dielectric absorption) and $k' > 0$ (propagation in the sense $z > 0$), the second relation indicates that $k'' > 0$ and is an optical absorption.

The MPPEM wave along Oz therefore is such that:

$$\vec{E} = \vec{\underline{E}}_m \exp(j[\omega t - \underline{k}z])$$
$$= \vec{\underline{E}}_m \exp(-k''z)\exp(j[\omega t - k'z]) \qquad \text{(using electrokinetic notation)}$$

$$\vec{E} = \vec{\underline{E}}_m \exp(j[\underline{k}z - \omega t]) = \vec{\underline{E}}_m \exp(-k''z)\exp(j[k'z - \omega t]) \quad \text{(by optical notation)}.$$

The term $\exp(-k''z)$ represents the exponential absorption—also called attenuation in optics—of the wave during its propagation. It is worth noting that the wrong choice of sign in Eqs. (23) and (23') would mean instead of representing an attenuation of the signal as it propagates that the signal would be progressively amplified, a physical impossibility.

7.2.2.2. Complex index

In turn, the index also must be complex given that the defining equation, as an extension to Eqs. (19') and (20) where k is complex, is:

$$\underline{n}^2 = \varepsilon_r(\omega) \quad \text{where} \quad \underline{k} = \frac{\omega}{c}\underline{n} \qquad (24)$$

Wherein:

$$\boxed{\begin{array}{l} \underline{n} = n' - jn'' \quad \text{(electrokinetic notation)} \\[2mm] \underline{n} = n' + jn'' \quad \text{(optical notation)} \end{array}}$$
(25)
(25')

where the negative sign is correctly used to represent the absorption in electrokinetic notation and the positive sign is correct in optical terms. Identification of the real and imaginary parts gives:

$$\left\{ \begin{array}{l} n'^2 - n''^2 = \varepsilon_r' \\[3mm] 2\,n'\,n'' = \varepsilon_r'' \end{array} \right. \qquad \left\{ \begin{array}{l} k' = n'\dfrac{\omega}{c} \\[3mm] k'' = n''\dfrac{\omega}{c} \end{array} \right.$$

The progressive wave along Oz therefore can be represented by:

$$\vec{E} = \underline{\vec{E}}_m \exp(j[\omega t - \underline{k}z]) = \underline{\vec{E}}_m \exp(j[\omega t - \frac{\omega}{c}\underline{n}z])$$

(electrokinetic notation)

$$= \vec{E}_m \exp(-n"\frac{\omega}{c}z)\exp(j[\omega t - n'\frac{\omega}{c}z])$$

(26)

$$\vec{E} = \underline{\vec{E}}_m \exp(j[\underline{k}z - \omega t]) = \underline{\vec{E}}_m \exp(j[\frac{\omega}{c}\underline{n}z - \omega t])$$

(optical notation)

$$= \vec{E}_m \exp(-n"\frac{\omega}{c}z)\exp(j[n'\frac{\omega}{c}z - \omega t])$$

The term for absorption and attenuation is thus in the form $\exp(-n"\frac{\omega}{c}z)$ where the component n" for the index is called the extinction index while n' is the refraction index associated with the passage of the wave through an additional medium.

7.3.   The Energy of an Electromagnetic Plane Wave and the Poynting Vector

7.3.1. Definition and physical significance for media of absolute permittivity (ε), magnetic permeability (μ), and subject to a conduction current (j ℓ)

7.3.1.1. Definition
The following two Maxwell equations are used for the titled material:

$$\overrightarrow{rot}\vec{E} = -\frac{\partial \vec{B}}{\partial t} \qquad (3)$$ [Eq. (3) multiplied by $\frac{\vec{B}}{\mu}$]

$$\overrightarrow{rot}\frac{\vec{B}}{\mu} = \vec{j}_\ell + \frac{\partial(\varepsilon\vec{E})}{\partial t} \qquad (4'')$$ [Eq. (4'') multiplied by [-$\vec{E}$]]

Following multiplication of the two members of each equation, as above, the addition of respective members yields:

$$\frac{\vec{B}}{\mu}\overrightarrow{rot}\,\vec{E} - \vec{E}\,\overrightarrow{rot}\frac{\vec{B}}{\mu} = -\frac{\vec{B}}{\mu}\frac{\partial\vec{B}}{\partial t} - \vec{E}\frac{\partial(\varepsilon\vec{E})}{\partial t} - \vec{j}_\ell\,.\vec{E}\;.$$

On using $\operatorname{div}(\vec{a} \times \vec{b}) = \vec{b}\,\overrightarrow{rot}\,\vec{a} - \vec{a}\,\overrightarrow{rot}\,\vec{b}$, the equation is changed to:

$$\operatorname{div}\left(\vec{E} \times \frac{\vec{B}}{\mu}\right) = -\frac{\partial}{\partial t}\left(\varepsilon\frac{E^2}{2} + \frac{B^2}{2\mu}\right) - \vec{j}_\ell.\vec{E}\;. \qquad (27')$$

By definition then, the Poynting vector is:

$$\vec{S} = \vec{E} \times \frac{\vec{B}}{\mu}. \qquad (27)$$

7.3.1.2. Physical significance

By introducing the Poynting vector into Eq. (27'), it is found that:

$$\text{div } \vec{S} = -\frac{\partial}{\partial t}\left(\varepsilon \frac{E^2}{2} + \frac{B^2}{2\mu} \right) - \vec{j_\ell}.\vec{E} \ .$$

For a volume (V) with a surface limited to Σ, we can state that:

$$-\iiint \text{div } \vec{S} \ d\tau = \frac{\partial}{\partial t} \iiint \left(\varepsilon \frac{E^2}{2} + \frac{B^2}{2\mu} \right) d\tau + \iiint \vec{j_\ell}.\vec{E} \ d\tau, \ \text{ which also yields}$$

$$-\oiint_\Sigma \vec{S}.d\vec{\Sigma} = \underbrace{\frac{\partial}{\partial t} \iiint \left(\varepsilon \frac{E^2}{2} + \frac{B^2}{2\mu} \right) d\tau}_{(1)} + \underbrace{\iiint \vec{j_\ell}.\vec{E} \ d\tau}_{(2)} \ .$$

7.3.1.2.1. Physical significance of term (1)
It is well known (at least in vacuum, where we have ε_0 and μ_0 instead of ε and μ) that the volume density of electrical and magnetic energies are, respectively, equal to $\dfrac{dw_e}{d\tau} = \dfrac{\varepsilon E^2}{2}$ and $\dfrac{dw_m}{d\tau} = \dfrac{B^2}{2\mu}$. In all, the energy is equal to $w = \dfrac{dw_e}{d\tau} + \dfrac{dw_m}{d\tau}$.

Term (1) can be written in the form $\iiint \dfrac{dw}{dt} d\tau$ and represents the power (for a total volume of the material) in electrical and magnetic energies created by the electromagnetic field in the

7.3.1.2.2. Physical significance of term (2)
Simplification using $d\tau = d\vec{\Sigma}.d\vec{r}$ permits the equation
$\iiint \vec{j_\ell}.\vec{E} \ d\tau = \iint \vec{j_\ell}.d\vec{\Sigma} \int \vec{E}.d\vec{r} = I.V = P_J$, which represents the electrical power given by the charges in the system, which typically correspond to the power dissipated by the Joule effect caused by j_ℓ.

To conclude, the Poynting vector flux ($\vec{S}$) through a surface (Σ) is equal to the power which gives rise to the electromagnetic field through Σ of the material under consideration.

7.3.2. *Propagation velocity of energy in a vacuum*

For this calculation we suppose that $j_\ell = 0$ and that the energy is being carried by a plane progressive wave, for which the Poynting vector is directed in the same sense as the propagation given by the vector denoted $\vec{u}$ (with the direct trihedral $\vec{E}, \vec{B}, \vec{u}$). In a vacuum, where the components for the vectors $\vec{E}$ and $\vec{B}$ are given by Eqs. (13) and (14) of Chapter 6, we have:

$$|\vec{S}| = S_z = \left(\vec{E} \times \frac{\vec{B}}{\mu_0}\right)_z = \frac{1}{\mu_0}(E_x B_y - E_y B_x) \qquad (28)$$

$$= \frac{1}{\mu_0 c}(G_1^2 + G_2^2) = \varepsilon_0 c\,(G_1^2 + G_2^2).$$

With respect to a unit surface normal to the direction of the EM wave propagation, it therefore can be written with the power (P) transmitted by the EM field through a surface unit ($\Sigma = 1$) that:

$$P = \iint_{\Sigma=1} S_z d\Sigma = S_z = \varepsilon_0 c\,(G_1^2 + G_2^2). \qquad (29)$$

Additionally, with $j_\ell = 0$, the volume density of the energy associated with the EM wave (w) is equal to:

$$w = \frac{\varepsilon_0 E^2}{2} + \frac{B^2}{2\mu_0} = \frac{1}{2}\varepsilon_0(G_1^2 + G_2^2) + \frac{1}{2\mu_0 c^2}\,(G_1^2 + G_1^2) = \varepsilon_0\,(G_1^2 + G_2^2). \qquad (30).$$

From Eqs. (28) and (30), we can determine that $|\vec{S}| = cw$, which in terms of vectors gives

$$\boxed{\vec{S} = c\,w\,\vec{u}} \qquad (31)$$

Given P through $\Sigma = 1$ and w, with the help of Eqs. (29) and (30), we find the relationship:

$$\boxed{P = c\,w} \qquad (32)$$

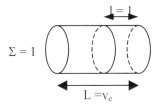

$$\Sigma = 1 \qquad l = 1$$

$$L = v_e$$

Figure 7.4. *Power transmitted through Σ.*

As shown in Figure 7.4., P transmitted as energy per unit time through $\Sigma = 1$ corresponds to the energy contained within the volume (V) given by $V = (\Sigma.L)_{\Sigma=1}$ where $L = v_e\,(\Delta t)_{\Delta t=1} = v_e$ and $V = v_e$ in which v_e is the energy propagation velocity.   As the energy density (w) represents the energy in the unit volume (surface $\Sigma = 1$ and length $l = 1$), we therefore can state that $P = v_e$ w .

As above noted, $P = c\,w$, we can directly find:

$$v_e = c$$

which indicates that the energy propagates at the velocity of light in a vacuum.

7.3.3. Complex notation

Calculations for the Poynting vector or the energy cannot be directly performed using the complex values of the fields $\vec{E}$ or $\vec{B}$ as the real part of the resulting product is not equal to the product of the real parts. Such a problem has been found for $\vec{S}$ or w given by Eqs. (27) and (30) which are not linear with respect to $\vec{E}$ or $\vec{B}$. However, it is possible to obtain the average values of $\vec{S}$ or w.

Given the following expressions for the fields $\vec{E}$ or $\vec{B}$

$$\vec{E} = \vec{E}_0 \exp(i\omega t) ,$$

where a rigorous treatment would mean writing:

$$\vec{E} = R(\vec{E}_0 \exp(i\omega t)) ,$$

where $\vec{E} = R(\underline{\vec{E}})$ and $\underline{\vec{E}} = \vec{E}_0 \exp(i\omega t)$.

As z denotes a complex number, as in $\underline{z} = a + ib$, we find that $R(\underline{z}) = a = \dfrac{\underline{z} + \underline{z}^*}{2}$,

where $\underline{z}^*$ is the conjugated complex of $\underline{z}$.

Given the above, the Poynting vector can be calculated with μ being real by:

$$\vec{S} = \vec{E} \times \frac{\vec{B}}{\mu} = \frac{1}{\mu}\vec{E} \times \vec{B} = \frac{1}{\mu} R(\underline{\vec{E}}) \times R(\underline{\vec{B}})$$

$$= \frac{1}{4\mu}\{\,[\vec{E}_0 \exp(i\omega t) + \vec{E}_0^* \exp(-i\omega t)\,] \times [\vec{B}_0 \exp(i\omega t) + \vec{B}_0^* \exp(-i\omega t)\,]\}$$

$$= \frac{1}{4\mu}\{[\vec{E}_0 \times \vec{B}_0^*] + [\vec{E}_0^* \times \vec{B}_0] + [\vec{E}_0 \times \vec{B}_0 \exp(2i\omega t)] + [\vec{E}_0^* \times \vec{B}_0^* \exp(-2i\omega t)]\} .$$

The first two terms are conjugated complexes and their total value is twice their real parts. The third and fourth terms vary as functions of 2ω so that their average value over a period is equal to zero. This leaves us with

$$\langle \vec{S} \rangle = \frac{1}{2\mu} R(\vec{E}_0 \times \vec{B}_0^*) . \qquad (33)$$

For a planar progressive wave along $\vec{u}$, we have $\vec{E}_0 = \vec{E}_m \exp[-i\vec{k}.\vec{r})]$, which yields:

$$\langle \vec{S} \rangle = \frac{1}{2\mu} R(\vec{E}_m \times \vec{B}_m^*) . \qquad (33')$$

In a vacuum, where $\mu = \mu_0$, $\varepsilon_0 \mu_0 c^2 = 1$, and $B_m = \dfrac{E_m}{c}$:

$$\langle \vec{S} \rangle = \frac{\varepsilon_0 c}{2} \vec{E}_m \times \vec{E}_m^* \, \vec{u} .$$

In the more specific case of a rectilinear plane wave, where $\vec{E}_m = \vec{E}_0$, we have:

$$\langle \vec{S} \rangle = \frac{\varepsilon_0 c}{2} E_0^2 \, \vec{u} .$$

This result can be obtained directly from $\vec{S} = \vec{E} \times \dfrac{\vec{B}}{\mu_0}$ [Eq. (27) written for a vacuum

where $\mu = \mu_0$].

7.3.4. The Poynting vector and the average power for a MPPEM wave in a nonabsorbent (k and n are real) and nonmagnetic ($\mu_r = 1$, so that $\mu = \mu_0$) medium

From Eq. (13), $\vec{B} = \dfrac{1}{\omega} \vec{k} \times \vec{E}$ and $\vec{k} = \dfrac{\omega}{c} n \, \vec{u}$, we have

$$\vec{S} = \vec{E} \times \frac{\vec{B}}{\mu_0} = \frac{n}{c\mu_0} \vec{E} \times (\vec{u} \times \vec{E}) .$$

By using the equation for a paired vector, $\vec{a} \times (\vec{b} \times \vec{c}) = (\vec{a}.\vec{c})\vec{b} - (\vec{a}.\vec{b})\vec{c}$, we obtain

$$\vec{S} = n c \varepsilon_0 [(\vec{E}.\vec{E}).\vec{u} - (\vec{E}.\vec{u})\vec{E}] = n c \varepsilon_0 E^2 \vec{u} .$$

By taking an average value over a given period, for the power transmitted (P) by the EM wave,

$$<P> = <S_z> = \frac{1}{2} n c \varepsilon_0 E_0^2 .$$

7.3.5. Poynting vector for a MPPEM wave in an absorbent dielectric such that μ is real

Using the optical notation given in Section 7.2.2.1,

$$\vec{E} = \vec{E}_m \exp(-k"z)\exp(j[k'z - \omega t])$$

and $\vec{B} = \dfrac{k}{\omega}(\vec{u} \times \vec{E}) = \dfrac{n}{c}(\vec{u} \times \vec{E}).$

According to Eq. (33), established for a real value of μ, we find that

$$<S_z> = \frac{1}{2\mu} R(\vec{E}_0 \times \vec{B}_0^*), \text{ where } \begin{cases} \vec{E}_0 = \vec{E}_m \exp(-k"z)\exp(jk'z) \\ \\ \vec{B}_0 = \dfrac{n}{c}(\vec{u} \times \vec{E}_0) \end{cases}$$

Therefore

$$\vec{E}_0 \times \vec{B}_0^* = \vec{E}_0 \times \left[\frac{n}{c}(\vec{u} \times \vec{E}_0)\right]^* = \vec{E}_0 \times \left[\frac{n^*}{c}(\vec{u} \times \vec{E}_0)^*\right] = \frac{n^*}{c}\vec{E}_0 \times (\vec{u} \times \vec{E}_0)^*$$

$$= \frac{n^*}{c}\vec{E}_0.\vec{E}_0^*\vec{u} = \frac{n^*}{c}|\vec{E}_m|^2\exp(-2k"z) \text{ , from which}$$

$$<S_z> = \frac{1}{2\mu}R(\vec{E}_0 \times \vec{B}_0^*) = \frac{1}{2\mu}\frac{n'}{c}|\vec{E}_m|^2\exp(-2k"z),$$

so that for a nonmagnetic medium ($\mu = \mu_0$)

$$<S_z> = \frac{cn'\varepsilon_0}{2}|\vec{E}_m|^2\exp(-2k"z).$$

The power transmitted by the electromagnetic field exponentially decreases with the distance traversed. The extinction coefficient for this is given by $2k" = 2n" \dfrac{\omega}{c}.$

7.4. Problem

Poynting vector
It is worth recalling that the physical magnitude of flux represents an amount associated with a physical magnitude which traverses a unit surface per unit time.

1. **The analogy between the charge current density and the energy current density vectors**

(a) Here $\vec{j}$ denotes the charge current density vector and ρ the volume charge density. The flux of the vector $\vec{j}$ through a surface $d\Sigma$ is $dI = \vec{j}\,d\Sigma$. What does j alone represent? Give the equation for the conservation of charge for a rapidly varying regime in an isolated system.

(b) By analogy, the energy flux (which thus exhibits a power per unit surface) through the surface $d\vec{\Sigma}$ is defined by a relationship of the type $dP = \vec{S} . d\vec{\Sigma}$ in which $\vec{S}$ represents the energy current density, otherwise termed the Poynting vector. What does S alone represent? By introducing the magnitude w_T, which represents the volume density of the total energy, give an equation for the conservation of energy analogous to that for the conservation of charge ($\vec{S}$ plays the role of $\vec{j}$ and w_T that of ρ).

2. The total volume density of energy w_T can be considered as the sum of two terms, one being due to the density of kinetic energy (w_c) and the other due to the energy density (w_{em}), which will be detailed below.

(a) The theorem for kinetic energy makes it possible to state that the variation in kinetic energy is equal to the work of the applied forces. A variation in the kinetic energy density (for a unit volume containing a charge ρ) is given by $dw_c = \vec{F} . \vec{v}\, dt$

with $\vec{F} = q\left[\vec{E} + \vec{v} \times \vec{B}\right]$ when charges with volume density ρ and velocity $\vec{v}$ are placed in an electromagnetic field characterized by the fields $\vec{E}$ and $\vec{B}$. Determine from the equation for the conservation of energy the value of the expression

$$\text{div}\vec{S} + \vec{E}.\vec{j} + \frac{dw_{em}}{dt}.$$

(b) Give the Maxwell-Faraday (M-F) relationship (M-F which gives $\overrightarrow{\text{rot}}\,\vec{E}$) and the Maxwell-Ampere (M-A) relationship (M-A which gives $\overrightarrow{\text{rot}}\,\dfrac{\vec{B}}{\mu_0}$) for a non-magnetic medium traversed by a real current ($\vec{j}$). By multiplying M-F by $[\dfrac{\vec{B}}{\mu_0}]$ and M-A by $[-\vec{E}]$, calculate $\vec{E}.\vec{j}$.

(c) Determine the expressions for $\vec{S}$ and w_{em} by identification between the results of 2a and 2b. Conclude.

3.
(a) It is worth recalling here that the real part of a product of two complex numbers is different from the product of the real parts of two complex numbers. Show how, in a similar way, the complex amplitude of the product of two complexes is different from the complex amplitude of these two complexes. What is under consideration when dealing with the vectorial products in place of the simple products?
(b) If $\underline{\vec{E}}_0$ and $\underline{\vec{B}}_0$ are the complex amplitudes of a complex field, then the complex electric field and the complex magnetic field:

$$\underline{\vec{E}} = \underline{\vec{E}}_0(\vec{r})\exp(i\omega t) \overset{notation}{=} \underline{\vec{E}}_0\exp(i\omega t) \text{ and } \underline{\vec{B}} = \underline{\vec{B}}_0\exp(i\omega t), \text{ are such that:}$$

$$\vec{E} = Re(\underline{\vec{E}}) \text{ and } \vec{B} = Re(\underline{\vec{B}}).$$

From the calculation for $\vec{S} = \dfrac{1}{\mu_0}[Re(\underline{\vec{E}}) \times Re(\underline{\vec{B}})]$, show that the average value of $\vec{S}$

is of the form:

$$\langle \vec{S} \rangle = \frac{1}{2}Re\left(\frac{\underline{\vec{E}}_0 \times \underline{\vec{B}}^*_0}{\mu_0}\right),$$

where $\underline{\vec{B}}^*_0$ is the complex conjugated with $\underline{\vec{B}}_0$.

(c)   Calculate $\langle \vec{S} \rangle$ for a monochromatic plane progressive wave propagating in the direction of the unit vector $\vec{u}_z$ in a medium where the wave vector $\vec{k}$ is real.

Answers

1.

(a) By definition, j represents the flux of electric charges, also termed the charge current density. This flux therefore represents the amount of electrical charges that traverse a unit surface per unit time. Quantitatively, we can say that this quantity of charges is localized within a cylinder (or parallelepiped) with a cross-sectional unit area $(\Sigma = 1)$ and length (l) given by

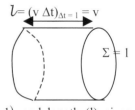

$l = (v \Delta t)_{\Delta t = 1} = v$, where v is the velocity of the charges. If $\rho = n\,q$, is the volume charge density where n is the charge density, i.e., the number of charges per unit volume, then the quantity of charge found in the volume (V) given by $V = (\Sigma)_{\Sigma = 1}\,(v \Delta t)_{\Delta t = 1} = v$ is such that $j = n\,q\,(\Sigma)_{\Sigma = 1}\,(v \Delta t)_{\Delta t = 1} = \rho\,v$.

The vector $\vec{j}$ is defined by $\vec{j} = \rho\vec{v}$. For its part, the equation for the conservation of charge is written

$$div\vec{j} + \frac{\partial \rho}{\partial t} = 0. \quad (1)$$

(b) Similarly, if $\vec{S}$ represents the energy current density vector, S represents the quantity of energy that traverses a unit surface in a unit time. S therefore can be seen as the electrical power through a unit surface.

If w_T represents the total energy, the equation for the conservation of energy can be written by analogy to the equation for the conservation of charge as:

$$\text{div}\vec{S} + \frac{\partial w_T}{\partial t} = 0 . \quad (2)$$

2.

(a) Therefore, $w_T = w_c + w_{em}$, and $\dfrac{dw_T}{dt} = \dfrac{dw_c}{dt} + \dfrac{dw_{em}}{dt}$. (3)

In addition, the theorem for the kinetic energy allows the equation $dw_c = \vec{F}.\vec{v}\, dt$ so that by taking into account the form of $\vec{F}$, we have:

$$\frac{dw_c}{dt} = \vec{F}.\vec{v} = \rho[\vec{E} + \vec{v}\times\vec{B}]\vec{v} = \rho\vec{E}\vec{v} = \vec{j}.\vec{E} . \quad (4)$$

By moving this expression into Eq. (3), we find $\dfrac{dw_T}{dt} = \vec{j}.\vec{E} + \dfrac{dw_{em}}{dt}$. By introducing this equation into Eq. (2) for the conservation of energy, we arrive at

$$\text{div}\vec{S} + \vec{j}.\vec{E} + \frac{dw_{em}}{dt} = 0 . \quad (5)$$

(b) We have:

(M-F) $\overrightarrow{\text{rot}}\ \vec{E} = -\dfrac{\partial\vec{B}}{\partial t}$ (equation multiplied by $\dfrac{\vec{B}}{\mu}$)

(M-A) $\overrightarrow{\text{rot}}\ \dfrac{\vec{B}}{\mu_0} = \vec{j} + \dfrac{\partial(\varepsilon\vec{E})}{\partial t}$ (equation multiplied by $[-\vec{E}\,]$).

Following multiplication of the two members of each equation as indicated, we find by addition that:

$$\frac{\vec{B}}{\mu_0}\overrightarrow{\text{rot}}\vec{E} - \vec{E}\,\overrightarrow{\text{rot}}\,\frac{\vec{B}}{\mu_0} = -\frac{\vec{B}}{\mu_0}\frac{\partial\vec{B}}{\partial t} - \vec{E}\frac{\partial(\varepsilon\vec{E})}{\partial t} - \vec{j}.\vec{E} .$$

Using the relation $\text{div}\left(\vec{a}\times\vec{b}\right) = \vec{b}\,\overrightarrow{\text{rot}}\,\vec{a} - \vec{a}\,\overrightarrow{\text{rot}}\,\vec{b}$, the above equation becomes

$$\text{div}\left(\vec{E}\times\frac{\vec{B}}{\mu_0}\right) = -\frac{\partial}{\partial t}\left(\varepsilon\frac{E^2}{2} + \frac{B^2}{2\mu_0}\right) - \vec{j}.\vec{E}\ \text{ which can also be written as}$$

$$\text{div}\left(\vec{E}\times\frac{\vec{B}}{\mu_0}\right) + \vec{j}.\vec{E} + \frac{\partial}{\partial t}\left(\varepsilon\frac{E^2}{2} + \frac{B^2}{2\mu_0}\right) = 0 . \quad (6)$$

(c) By identification with Eq. (5), it can be determined that $\vec{S} = \vec{E}\times\dfrac{\vec{B}}{\mu_0}$ which is the Poynting vector, and that

$$w_{em} = \varepsilon \frac{E^2}{2} + \frac{B^2}{2\mu_0}$$ is the density of electromagnetic energy in the form:

$$w_{em} = w_e + w_m, \text{ with } w_e = \varepsilon \frac{E^2}{2} \text{ and } w_m = \frac{B^2}{2\mu_0}.$$

These two expressions are for the electrical and magnetic energy densities under a stationary regime and they prove that they are still valid for variable regimes.

3.

(a) The use of complex numbers needs to take into account (with $z_n = a_n + i\,b_n$):

- if the real part of the sum of two complex numbers is actually equal to the sum of the real parts of the two complex numbers, as in $R(z_1 + z_2) = R(z_1) + R(z_2) = a_1 + a_2$; and

- or if the real part of the two complex numbers is not equal to the product of the two real parts of the two complex numbers, as in

$$R(z_1 z_2) = R([a_1 + i\,b_1]\,[a_2 + i\,b_2]) = R\,(a_1 b_1 - a_2 b_2 + i[a_1 b_2 + a_2 b_1])$$
$$= a_1 b_1 - a_2 b_2 \neq a_1 b_1 = R(z_1)R(z_2).$$

Similarly, if the product ($\underline{P}$) of the complexes is $\underline{P} = \underline{A}.\underline{B}$, then its complex amplitude is such that (with $\underline{A} = \underline{A}_0\, e^{i\omega t}$ and $\underline{B} = \underline{B}_0\, e^{i\omega t}$):

$$\underline{P} = \underline{A}_0\, e^{i\omega t}\, \underline{B}_0\, e^{i\omega t} = \underline{A}_0.\underline{B}_0\, e^{i2\omega t} = \underline{P}_0\, e^{i\omega t}, \text{ from which}$$

$$\underline{P}_0 = \underline{A}_0.\underline{B}_0\, e^{iwt} \neq \underline{A}_0.\underline{B}_0 .$$

The complex amplitude of the product of two complexes appears different from the product of the complex amplitude of two complexes. Similarly, the vectorial product, which brings in the components of the products, differs from the simple products, for example:

$$R(\vec{E} \times \vec{B}) \neq R(\vec{E}) \times R(\vec{B}), \text{ so that here } R(\vec{S}) \neq \frac{1}{\mu_0} R(\vec{E} \times \vec{B}) .$$

(b) However, if $\vec{E} = \vec{E}_0 \exp(i\omega t)$ and $\vec{B} = \vec{B}_0 \exp(i\omega t)$, we have:

$$\langle \vec{S} \rangle = \frac{1}{2} \text{Re}\left(\frac{\vec{E}_0 \times \vec{B}^*_0}{\mu_0} \right).$$

In effect, $\vec{S} = \frac{1}{\mu_0}\vec{E} \times \vec{B} = \frac{1}{\mu_0}\left[R(\vec{E}) \times R(\vec{B}) \right] \neq \frac{1}{\mu_0}R(\vec{E} \times \vec{B}) .$

With $\vec{E} = \vec{E}_0 \exp(i\omega t) = A + iB$, we have:

$$R(\underline{\vec{E}}) = A = \frac{\underline{\vec{E}} + \underline{\vec{E}}^*}{2} = \frac{\underline{\vec{E}}_0 e^{i\omega t} + \underline{\vec{E}}_0^* e^{-i\omega t}}{2} \ , \text{ from which}$$

$$\vec{S} = \frac{1}{4\mu_0} \left\{ [\underline{\vec{E}}_0 \exp(i\omega t) + \underline{\vec{E}}_0^* \exp(-i\omega t)] \times [\underline{\vec{B}}_0 \exp(i\omega t) + \underline{\vec{B}}_0^* \exp(-i\omega t)] \right\}$$

$$= \frac{1}{4\mu_0} \left\{ [\underline{\vec{E}}_0 \times \underline{\vec{B}}_0^*] + [\underline{\vec{E}}_0^* \times \underline{\vec{B}}_0] + [\underline{\vec{E}}_0 \times \underline{\vec{B}}_0 \exp(2i\omega t)] + [\underline{\vec{E}}_0^* \times \underline{\vec{B}}_0^* \exp(-2i\omega t)] \right\} .$$

The first two terms are conjugated complexes and their sum is equal to twice their real parts [as $\underline{z}_1 + \underline{z}_1{}^* = (a_1 + i\,b_1) + (a_1 - i\,b_1) = 2\,a_1$)]. The third and fourth terms each vary as a function of 2ω so that they have average values over a half period, and therefore also a full period, equal to zero. Therefore:

$$\langle \vec{S} \rangle = \frac{1}{2\mu_0} R(\underline{\vec{E}}_0 \times \underline{\vec{B}}_0^*) .$$

(c) If the plane wave progresses along Oz, then the electrical field can be supposed, for example, to be moving with respect to Ox. Therefore,

$$\underline{\vec{E}}_0 \overset{\text{notation}}{=} \underline{\vec{E}}_0(r) = \underline{\vec{E}}_m e^{-i\vec{k}.\vec{r}} = \underline{E}_m e^{-i\vec{k}.\vec{r}} \vec{u}_x .$$

Similarly, for a magnetic field moving through Oy, we have

$$\underline{\vec{B}}_0 \overset{\text{notation}}{=} \underline{\vec{B}}_0(r) = \underline{\vec{B}}_m e^{-i\vec{k}.\vec{r}} = \underline{B}_m e^{-i\vec{k}.\vec{r}} \vec{u}_y .$$

Therefore:

$$\langle \vec{S} \rangle = \frac{1}{2\mu_0} R(\underline{\vec{E}}_0 \times \underline{\vec{B}}_0^*) = \frac{1}{2\mu_0} R(\underline{E}_m . \underline{B}_m{}^* \ \vec{u}_x \times \vec{u}_y) = \frac{1}{2\mu_0} R(\underline{E}_m . \underline{B}_m{}^* \ \vec{u}_z) .$$

Given that $\underline{E}_m = v_\varphi \underline{B}_m$ in a material (or $\underline{E}_m = c \underline{B}_m$ if in a vacuum), we have:

$$\langle \vec{S} \rangle = \frac{1}{2\mu_0 v_\varphi} R(\underline{E}_m . \underline{E}_m{}^* \ \vec{u}_z) = \frac{n}{2\mu_0 c} |\underline{E}_m|^2 \vec{u}_z = \frac{\varepsilon_0}{2} nc |\underline{E}_m|^2 \vec{u}_z .$$

If the MPPEM wave is polarized rectilinearly, so that $\underline{E}_m = E_0$ is real, then

$$\langle \vec{S} \rangle = \frac{\varepsilon_0}{2} nc \ E_0{}^2 \ \vec{u}_z .$$

Chapter 8

Waves in Plasmas and Dielectric, Metallic, and Magnetic Materials

8.1. Interactions between Electromagnetic Wave and Materials

8.1.1. Parameters under consideration

So that geometrical considerations do not become a problem, this chapter will look at the interactions of electromagnetic (EM) waves with materials that have infinite dimensions. Materials with limited forms will be looked at, most notably, in Chapters 11 and 12.

The propagation of EM waves can be studied by considering:

- a description of the EM wave in the material with Maxwell's equations; and
- a representation of the material as a collection of electronic and ionic charges that interact with the electromagnetic field through the Lorentzian force, which can be written as:

$$m\frac{d\vec{v}}{dt} = q\vec{E} + q\vec{v} \times \vec{B} .$$

In terms of moduli, the ratio of magnetic and electronic contributions are given by

$$\frac{F_m}{F_e} = \frac{qvB}{qE} = \frac{vB}{E} .$$

For a plane EM wave, the $E = v_\varphi B$ and the ratio $\dfrac{F_m}{F_e} = \dfrac{v}{v_\varphi}$. As $v \ll v_\varphi \approx c$, the magnetic force ($F_m$) is negligible with respect to the electrical force (F_e). With the magnetic force being weak, generally it is stated that the magnetization intensity ($\vec{I}$) is approximately equal to zero, so that $\vec{B} \approx \mu_0\vec{H}$ and $\overrightarrow{rot\vec{I}} = \vec{J}_A \approx 0$.

Different types of forces can be involved, depending on the response of the material under the deformation caused by the incident EM wave.

8.1.2. The various forces involved in conventionally studied materials

8.1.2.1. Dielectric materials

The various movements and polarizations which can come about depends on whether it is electrons, ions, or permanent dipoles that are submitted to F_e.

8.1.2.1.1. Electronic polarization

Much like a spring, valence electrons displaced by F_e are returned to the equilibrium position about their respective atoms by a force (f_r) given by $f_r = -kr = -m\omega_{0e}^2 r$. In addition, if the electrons move within a very dense medium, they can be thought of as being subject to a friction force (f_t) which gives rise to a Joule effect and has an intensity proportional to their velocity, so that $\vec{f}_t = -\dfrac{m}{\tau}\vec{v}$ where τ is the relaxation time of the system and has the dimension of time (so that the equation is dimensionally correct).

8.1.2.1.2. Ionic polarization

The movement of ions under an electric field approximately resembles that of electrons. Their returning force is $f_r = -M_{nucleus}\,\omega_{0i}^2 r$ where, due to the greater mass of ions and their relative inertia, their ionic pulsations (ω_{0i}) are much smaller than the equivalent electronic movements (ω_{0e}). This is detailed further in Chapter 3 of Volume 2.

8.1.2.1.3. Polarization and dipole orientation

It is assumed that dipoles subject to a varying electrical force are predominantly subject to a friction force associated with their rotational movement. Given that they are relatively large due to the chemical association of atoms, they present a high degree of inertia toward excitation by an electric field. The dipoles can only follow—with a dephasing—relatively low frequencies, but tend to contribute considerably to dielectric absorption at such frequencies (see also Chapters 1 and 3 of Volume 2).

8.1.2.2. In plasma

In a plasma, which has a low density and is electrically neutral, it is assumed that electrons make up an electronic gas. Any displacement of the electrons from their equilibrium position by a perturbation resulting in a compression or dilation of the electron gas can be considered the result of the application of a sinusoidal electric field, which has an plasma oscillation pulse (ω_p) due to the returning force of ions in the medium that are assumed to be immobile. Given the assumption that the electrons move freely, we can ignore friction and mechanical returning forces so that ω_p is automatically stabilized by the longitudinal electric field associated with the permanent returning force (thus directly related to this electric field). The corresponding frequency is therefore $v_p = \dfrac{\omega_p}{2\pi}$, and as detailed further in this chapter, can appear as a breaking or "drag" frequency for EM waves.

8.1.2.3. Metals

The electrons that interact most easily with EM waves are those external to atomic or molecular orbitals, in other words conduction electrons. These electrons are in effect free, or semifree, depending on the degree of approximation, and belong to no specific atom, so that any returning mechanical forces are equal to zero. The medium, however, is sufficiently dense to make friction forces nonzero and it is these forces that result in Joule effects in metals, more specifically due to collisions between electrons. The exact nature of these collisions generally is only considered in solid physics along with collisions in a network (phonons) and with impurities.

8.2. Interactions of EM Waves with Linear, Homogeneous and Isotropic (lhi) Dielectric Materials: Electronic Polarization, Dispersion and Resonance Absorption

Electronic polarizations gives rise to resonance frequencies (ω_{0e}, which is more succinctly denoted ω_0 below) in an absorbing and dispersing material, and this section is limited to studying the plots of these interactions. These interactions can be understood using the question-based tutorial below, which has answers detailed in Sections 8.2.1 through to 8.2.8.

   The polarization of a linear, homogeneous, and isotropic (lhi) material is the result of a total number (N) of electrons per unit volume (n_e) (N = n_e) each having a charge (–q) being subject to the following forces:

- a Coulombic force induced by the alternating field which can be described in the complex form by $\underline{\vec{E}} = \underline{\vec{E}}_0 \exp(i\omega t)$;

- a returning force given by $\vec{F}_r = -m\omega_0^2 \vec{r}$, where me is the mass of a electron;

- frictional forces given by: $\vec{F}_t = -m\Gamma \dfrac{d\vec{r}}{dt} = -\dfrac{m}{\tau}\dfrac{d\vec{r}}{dt}$.

1. Under a forced regime, find the expression for the displacement of electrons based on the complex form $\underline{\vec{r}} = \underline{\vec{r}}_0 \exp(i\omega t)$. Determine $\underline{\vec{r}}_0$.

2. Give the complex expression for the electronic polarization. From this determine the real and imaginary parts of the dielectric susceptibility (χ_e' and χ_e'', respectively) given in electrical notation by $\chi_e = \chi_e' - i\chi_e''$. The result should be expressed in terms of $\omega^2_p = (\dfrac{n_e q^2}{m\varepsilon_0})$.

3. Now the study turns to the functions χ_e' and χ_e'' in terms of ω at two points: when $\omega \cong \omega_0$, and then when $\omega \neq \omega_0$. Parts a, b, and c of this question consider the former $\omega \cong \omega_0$.

(a) give the sign associated with χ_e' when $\omega < \omega_0$ or when $\omega > \omega_0$;

(b) for which value of ω, χ_e'' is a maximum;

(c) the width at half-peak height of the function $\chi_e''(\omega)$ corresponds to angular frequencies ω_1 and ω_2, which are such that $\chi_e''(\omega_1) = \chi_e''(\omega_2) = \dfrac{\chi_e''_{max}}{2}$. Determine ω_1 and ω_2 and show that for these values $\chi_e'(\omega)$ is at an extreme point. Determine the values for $\chi_e'(\omega_1)$ and for $\chi_e'(\omega_2)$; and

(d) give the limiting values for χ_e' and χ_e'' ($\omega \to 0, \infty$). Plot $\chi_e'(\omega)$ et $\chi_e''(\omega)$.

4. This question concerns a regime far off the resonance position $\omega \neq \omega_0$; indeed, it is now the hypothesis $\omega^2\Gamma^2 \ll (\omega^2_0 - \omega^2)^2$ that becomes relevant. Give the corresponding representations for $\chi_e'(\omega)$ and $\chi_e''(\omega)$.

5. Staying with the hypothesis presented in question 4 [$\omega \neq \omega_0$ and $\omega^2\Gamma^2 \ll (\omega^2_0 - \omega^2)^2$], show that there is a value ω_ℓ (to be determined) for ω which is such that the real part of the relative dielectric permittivity (ε_r') is zero at this pulsation. Show that for the value of ω given by $\omega = \omega_\ell$ the magnetic field is zero and that the electric field is longitudinal.

6. Plane, progressive, and traversing EM waves that can be written in the form $\vec{E} = \vec{E}_m \exp[i(\omega t - \vec{k}\vec{r})]$ and angular frequencies between but not including ω_0 and ω_ℓ (outside of the resonance and ω_ℓ) are considered in this question. From the general equation for the dispersion of waves, give the form (real, complex, or purely imaginary) of the wave vector in this domain of angular frequencies while also determining the type of wave involved.

7. Outside of the absorption and the interval [ω_0, ω_ℓ]:

(a) give the explicit relation between ω and j using the expression for the relative permittivity (equation involving term expressed as ω^4);

(b) from the real solutions to this equation, give the relations between ω and k; and

(c) With respect to the limiting case, give the general shape of the curves representing $\omega = f(k)$ for dispersion due to electronic polarization.

8. Find for $\omega \ll \omega_0$ the equation (for dispersion) that ties the optical index to the wavelength on a vacuum $\lambda_0 = \dfrac{2\pi c}{\omega}$.

8.2.1. *The Drude Lorentz model: a representation of a dielectric using an electron gas and a study of the electron movement*

8.2.1.1. The Drude-Lorentz model

In the Drude-Lorentz model, a dielectric material is represented by a group of atoms (N_a) per unit volume distributed in a vacuum. The atoms are surrounded by a field of electrons that are themselves assumed to move in a vacuum around the atoms to which they are attached. When subjected to a sinusoidal electric field, the electrons move from their positions and then tend to return to their original position through the influence of two forces :

- a frictional force given by $\vec{F}_t = -m\Gamma\dfrac{d\vec{r}}{dt} = -\dfrac{m}{\tau}\dfrac{d\vec{r}}{dt}$, which takes into account the viscosity of the medium and results in a dephasing between the excitation and the electronic response. In this case, τ is the relaxation time that represents the time required to establish the electronic polarization following application of the field. In the Drude-Lorentz model, which results in an expression for conductivity and Ohm's law, τ represents the average time in between two successive collisions; and

- an elastic force, given by $\vec{F}_r = -m\omega_0^2\vec{r}$, due to the recall of electrons to their particular bonds. Physically, this force resembles that of a returning force exerted by a spring tying together an electron and its nucleus. For each type of electron, be they internal or external layer electrons, the elastic force varies along with the corresponding value for ω_0.

It is worth noting that for metals, which have electronic properties determined by free conduction electrons, the returning or recall force is not brought into account because the energy required (thermal energy) to remove an electron from its original atom is extremely small. As detailed in Section 8.1.2.3, only F_t is assumed to be exerted.

8.2.1.2. Equation for the movement of electrons

Here the charge of an electron is represented by $-e$. With respect to the direction $\vec{r}$, parallel to the applied field ($\vec{E}$), the fundamental equation for the dynamic of the displacement of electrons is:

$$m \frac{d^2\vec{r}}{dt^2} = \sum \vec{F}_{applied} = \vec{F}_r + \vec{F}_t + \vec{F}_{Coulomb} = - m\omega_0^2 \vec{r} - \frac{m}{\tau} \frac{d\vec{r}}{dt} - e\vec{E} \text{ , so that:}$$

$$\frac{d^2\vec{r}}{dt^2} + \frac{1}{\tau} \frac{d\vec{r}}{dt} + \omega_0^2 \vec{r} = - \frac{e}{m} \vec{E} . \quad (1)$$

Given a cosinusoidal field of the form $\vec{E} = \vec{E}_0 e^{i\omega t}$, the required solution for the forced regime subject to an alternating field, particular to the equation with a second term, is in the form $\vec{r} = \vec{r}_0 \exp(i\omega t)$.

With $\overset{o}{\vec{r}} = i\omega \vec{r}$ and $\overset{o\,o}{\vec{r}} = -\omega^2 \vec{r}$, the solution given in Eq. (1) gives, following simplification of the two terms using $e^{i\omega t}$,

$$- \omega^2 \vec{r}_0 + i \frac{\omega}{\tau} \vec{r}_0 + \omega_0^2 \vec{r}_0 = - \frac{e}{m} \vec{E}_0 \quad (1'), \text{ from which}$$

$$\vec{r}_0 = \frac{-e}{m \left[\left(\omega_0^2 - \omega^2 \right) + i \frac{\omega}{\tau} \right]} \vec{E}_0 . \quad (1'')$$

8.2.2. The form of the polarization, the susceptibility, and the dielectric permittivity

8.2.2.1. Expression for electronic polarization

Moving a charge (q) by $\vec{r}$ is the same as applying a moment dipole $\vec{\mu} = q\vec{r}$, and likewise, moving an electronic charge ($q = -e$) is the same as applying a dipole moment $\vec{\mu} = - e\vec{r}$ (see Chapter 2, section 2.2.2). The polarization for $N = n_e$ electrons per unit volume therefore is given by:

$$\vec{P} = n_e \vec{\mu} = -e\, n_e\, \vec{r} = -e\, n_e\, \vec{r}_0 e^{i\omega t} = \frac{n_e e^2}{m \left[\left(\omega_0^2 - \omega^2 \right) + i \frac{\omega}{\tau} \right]} \vec{E}_0 e^{i\omega t} . \quad (2)$$

8.2.2.2. Expressions for susceptibility and dielectric permittivity (electronic component)

Given that $\vec{P} = \varepsilon_0 \underline{\chi}_e \vec{E} = \varepsilon_0 \underline{\chi}_e \vec{E}_0 e^{i\omega t} = \varepsilon_0 (\underline{\varepsilon}_r - 1) \vec{E}_0 e^{i\omega t}$ (3),

from Eqs. (2) and (3) it is possible to directly determine that:

$$\underline{\chi}_e = \frac{n_e q^2}{\varepsilon_0} \frac{1}{m \left[\left(\omega_0^2 - \omega^2 \right) + i \frac{\omega}{\tau} \right]} . \quad (4)$$

Given that $\omega_p^2 = \dfrac{n_e q^2}{\varepsilon_0 m}$, the result is $\underline{\chi}_e = \omega_p^2 \dfrac{1}{\left[\left(\omega_0^2 - \omega^2\right) + i\dfrac{\omega}{\tau}\right]}$.   (4')

From this can be determined that

$$\underline{\chi}_e = \omega_p^2 \frac{\left(\omega_0^2 - \omega^2\right) - i\dfrac{\omega}{\tau}}{\left(\omega_0^2 - \omega^2\right)^2 + \dfrac{\omega^2}{\tau^2}} = \chi_e' - i\,\chi_e''. \qquad (5)$$

By identification of the real and imaginary parts, and with $\underline{\chi}_e = (\underline{\varepsilon}_r - 1) = (\varepsilon_r' - 1 - i\,\varepsilon_r'')$, it is possible to determine that:

$$\left\{ \begin{array}{l} \chi_e' = (\varepsilon_r' - 1) = \omega_p^2 \dfrac{\left(\omega_0^2 - \omega^2\right)}{\left(\omega_0^2 - \omega^2\right)^2 + \dfrac{\omega^2}{\tau^2}} = \omega_p^2 \dfrac{\left(\omega_0^2 - \omega^2\right)}{\left(\omega_0^2 - \omega^2\right)^2 + \omega^2 \Gamma^2} \qquad (6) \\[2em] \chi_e'' = \varepsilon_r'' = \omega_p^2 \dfrac{\dfrac{\omega}{\tau}}{\left(\omega_0^2 - \omega^2\right)^2 + \dfrac{\omega^2}{\tau^2}} = \omega_p^2 \dfrac{\omega \Gamma}{\left(\omega_0^2 - \omega^2\right)^2 + \omega^2 \Gamma^2}. \qquad (7) \end{array} \right.$$

Eq. (6) can be used to define that $\chi_e'(0) = \dfrac{\omega_p^2}{\omega_0^2} = \dfrac{1}{\omega_0^2} \dfrac{n_e q^2}{\varepsilon_0 m}$ (8).

Following this, the introduction of $\chi_e'(0)$ into Eqs. (4), (6), and (7) yields:

$$\underline{\chi}_e = \chi_e'(0) \frac{\omega_0^2}{\left[\left(\omega_0^2 - \omega^2\right) + i\dfrac{\omega}{\tau}\right]}. \qquad (9)$$

In turn, this gives:

$$\chi_e' = \chi_e'(0)\,\omega_0^2 \frac{\left(\omega_0^2 - \omega^2\right)}{\left(\omega_0^2 - \omega^2\right)^2 + \omega^2 \Gamma^2} \qquad (6') \text{ and}$$

$$\chi_e'' = \varepsilon_r'' = \chi_e'(0)\,\omega_0^2 \frac{\omega \Gamma}{\left(\omega_0^2 - \omega^2\right)^2 + \omega^2 \Gamma^2}. \qquad (7')$$

8.2.2.3. Limiting cases
8.2.2.2.1. Low frequencies where $\omega \ll \omega_0$ ($\omega \to 0$)
According to Eqs. (6) and (7),

$$\chi_e' \to \chi_e'(0) = \frac{\omega_p^2}{\omega_0^2} \text{ and } \varepsilon_r' \to \varepsilon_r'(0) = 1 + \frac{\omega_p^2}{\omega_0^2} \qquad (10), \text{ and}$$

$$\chi_e'' \to \chi_e''(0) = 0 \text{ and } \varepsilon_r'' \to \varepsilon_r''(0) = 0 .$$

The upshot is that in this frequency domain, $\underline{\varepsilon}_r \approx \varepsilon_r'(0)$. With $\underline{\chi}_e$ being real and positive, according to Eq. (3), there is no dephasing between $\vec{P}$ and $\vec{E}$.

8.2.2.2.2. At frequencies where $\omega = \omega_0$
Here,

$$\begin{cases} \chi_e'(\omega_0) = 0 , \text{ so that } \varepsilon_r'(\omega_0) = 1 \\ \\ \chi_e''(\omega_0) = \dfrac{\omega_p^2}{\omega_0 \Gamma} \end{cases}$$

$$\Rightarrow \underline{\chi}_e(\omega_0) = -i \, \chi_e''(\omega_0) = -i \, \frac{\omega_p^2}{\omega_0 \Gamma} . \text{ As } -i = e^{-i \frac{\pi}{2}} , \vec{P} \text{ and } \vec{E} \text{ are in quadrature.}$$

8.2.2.2.3. High frequencies for which $\omega \gg \omega_0$ ($\omega \to \infty$)
According to Eq. (6) and (7), $\chi_e'(\infty) \to 0$ and $\chi_e''(\infty) \to 0$. More precisely, with the help of Eq. (9), by neglecting the ω term but not the ω^2 term (as $\omega \to \infty$) from the denominator, $\underline{\chi}_e \approx -\chi_e'(0)\frac{\omega_0^2}{\omega^2}$ (11). As $-1 = e^{i\pi}$, $\vec{P}$ and $\vec{E}$ have opposing phases.

8.2.2.4. Comment concerning the Lorentz correction

The Lorentz correction replaces in Eq. (1) the field ($\vec{E}$) by the locally effective field given by $\vec{E}_{al} = \vec{E} + \dfrac{\vec{P}}{3\varepsilon_0}$. With $\underline{P} = -e \, n_e \, \vec{r}$ Eq. (1') now becomes

$$-\omega^2 \vec{r}_0 + i \frac{\omega}{\tau} \vec{r}_0 + \omega_0^2 \vec{r}_0 - \frac{n_e q^2}{3m\varepsilon_0} \vec{r}_0 = -\frac{e}{m} \vec{E}_0 . \text{ This results in a change in the expressions}$$

for χ_e and $\underline{\varepsilon}$ of the term ω_0^2 to $\omega_0'^2 = \omega_0^2 - \dfrac{n_e q^2}{3m\varepsilon_0}$.

8.2.3. Study of the curves $\chi_e'(\omega)$ and $\chi_e''(\omega)$ for $\omega \approx \omega_0$ (of the order of the absorption zone)

In this section, the approximations $\omega_0^2 - \omega^2 = (\omega_0 - \omega)(\omega_0 + \omega) \approx 2 \omega_0 (\omega_0 - \omega)$ and $\Gamma\omega \approx \Gamma\omega_0$ are made.

8.2.3.1. Expression for χ_e'

From Eq. (6), for χ_e':

$$\chi_{e\,(\omega \approx \omega 0)}' \approx \omega_p^2 \frac{2\omega_0 \left(\omega_0 - \omega\right)}{4\omega_0^2 \left(\omega_0 - \omega\right)^2 + \omega_0^2 \Gamma^2}, \quad \text{so that dividing above and below by}$$

$4\,\omega_0^2$ gives:

$$\chi_{e\,(\omega \approx \omega 0)}' \approx \frac{\omega_p^2}{2\omega_0} \frac{\left(\omega_0 - \omega\right)}{\left(\omega_0 - \omega\right)^2 + \left(\dfrac{\Gamma}{2}\right)^2} = \frac{\omega_0 \chi_e'(0)}{2} \frac{\left(\omega_0 - \omega\right)}{\left(\omega_0 - \omega\right)^2 + \left(\dfrac{\Gamma}{2}\right)^2}. \quad (12)$$

It is noteworthy that χ_e' changes sign with $(\omega_0 - \omega)$. χ_e' is out of step with respect to $(\omega_0-\omega)$ and $\chi_e'_{(\omega \approx \omega 0)} = 0$.

Also, it is possible to state that the function χ_e' is:

- positive on the left hand side of ω_0. Here $\omega_- = \omega_0 - \varepsilon$ (where $\varepsilon > 0$), from which $\omega_0 - \omega_- = \varepsilon > 0$ and $\chi_e'(\omega_-) > 0$; and

- negative on the right hand side of ω_0 so that $\omega_+ = \omega_0 + \varepsilon$, from which $\omega_0 - \omega_+ = -\varepsilon < 0$, and $\chi_e'(\omega_+) < 0$.

8.2.3.2. Expression for χ_e''

Similarly to that in the previous section, from Eq. (7), χ_e'' can be found in:

$$\chi_{e\,(\omega \approx \omega_0)}'' \approx \omega_p^2 \frac{\omega_0 \Gamma}{4\omega_0^2 \left(\omega_0 - \omega\right)^2 + \omega_0^2 \Gamma^2} = \frac{\omega_p^2}{2\omega_0} \frac{\dfrac{\Gamma}{2}}{\left(\omega_0 - \omega\right)^2 + \left(\dfrac{\Gamma}{2}\right)^2}. \quad (13)$$

For its part, χ_e'' is at a maximum when the dominator is at a minimum, that is to say when $\omega - \omega_0 = 0$, so that $\omega = \omega_0$. Thus:

$$\chi_e''(\omega_0) = \chi_e''_{max} = \frac{\omega_p^2}{\omega_0} \frac{1}{\Gamma}. \quad (14)$$

By substituting this expression in Eq. (13):

$$\chi_{e\,(\omega \approx \omega_0)}'' \approx \chi_e''(\omega_0) \frac{\left(\dfrac{\Gamma}{2}\right)^2}{\left(\omega_0 - \omega\right)^2 + \left(\dfrac{\Gamma}{2}\right)^2} = \chi_e''(\omega_0) \frac{1}{1 + \dfrac{4}{\Gamma^2}\left(\omega_0 - \omega\right)^2}. \quad (13')$$

$\chi_e''_{(\omega \approx \omega_0)}$ is a function paired with $(\omega_0 - \omega)$, and therefore is symmetrical with respect to $\omega = \omega_0$. This type of curve is termed a Lorentzian curve, and as ω is separated from ω_0, $(\omega_0 - \omega)^2$ increases and $\chi_e''(\omega) \to 0$.

8.2.3.3. Evolution of $\chi_e'(\omega)$ and $\chi_e''(\omega)$ when $\omega \approx \omega_0$

The functions $\chi_e'(\omega)$ and $\chi_e''(\omega)$ when $\omega \approx \omega_0$ evolve as shown in Figures 8.1 and 8.2.

The halfpeak width (hpw) of the curve representing $\chi_e''(\omega)$ gives the pulsations ω_1 and ω_2 such that $\chi_e''(\omega_1) = \chi_e''(\omega_2) = \chi_e''(\omega_0)/2$. From Eq. (13'), pulsations ω_i (where i = 1 or 2) should accord to:

$$\chi_e''(\omega_i) = \chi_e''(\omega_0) \frac{1}{1 + \dfrac{4}{\Gamma^2}\left(\omega_0 - \omega_i\right)^2} = \frac{1}{2}\chi_e''(\omega_0).$$

From which the following equation can de determined:

$$(\omega_0 - \omega_i)^2 = \frac{\Gamma^2}{4}, \text{ so that } \omega_i = \omega_0 \pm \frac{\Gamma}{2}.$$

These equations finally give:

$$\begin{cases} \omega_1 = \omega_0 - \dfrac{\Gamma}{2} \\ \omega_2 = \omega_0 + \dfrac{\Gamma}{2} \end{cases}$$

The hpw thus is equal to $\Delta\omega = \omega_2 - \omega_1 = \Gamma = 1/\tau$ (15).

In addition, it is possible to see that the angular frequencies ω_i (ω_1 and ω_2) give rise to the extreme limiting values of $\chi_e'(\omega)$. To verify this, it suffices that $\left[\dfrac{d\chi_e'}{d\omega}\right]_{\omega i} = 0$, a relatively simple calculation.

According to Eq. (12), the corresponding values for $\chi_e'(\omega_i)$ are:

$$\chi_e'(\omega_i) = \frac{\omega_0 \chi_e'(0)}{2} \frac{\pm \dfrac{\Gamma}{2}}{\left(\dfrac{\Gamma}{2}\right)^2 + \left(\dfrac{\Gamma}{2}\right)^2} = \pm \frac{\omega_0 \chi_e'(0)}{2\Gamma} = \pm \frac{\omega_0 \tau}{2}\chi_e'(0) .$$

Therefore:

$$\chi_e'(\omega_1) = \frac{\omega_0}{2\Gamma}\chi_e'(0) = \frac{\omega_0 \tau}{2}\chi_e'(0)$$

$$\chi_e'(\omega_2) = -\frac{\omega_0}{2\Gamma}\chi_e'(0) = -\frac{\omega_0 \tau}{2}\chi_e'(0).$$

(16)

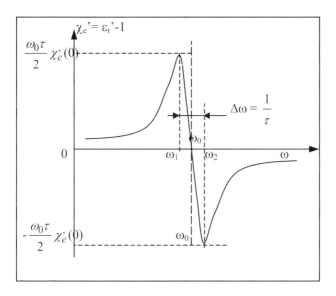

Figure 8.1. *Plot of $\chi_e{}' = f(\omega)$ when $\omega \approx \omega_0$.*

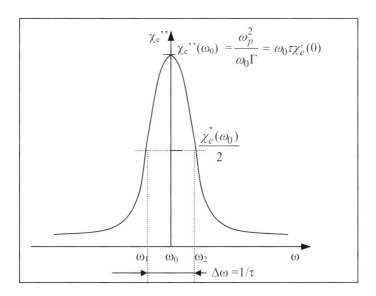

Figure 8.2. *Plot of $\chi_e{}'' = f(\omega)$ when $\omega \approx \omega_0$.*

8.2.4. Study of the curves of $\chi_e'(\omega)$ and $\chi_e''(\omega)$ when $\omega \neq \omega_0$ (well outside of an absorption zone)

Far from an absorption zone, the oscillation can be assumed to undergo only a very weak dragging force. This small force is such that it can be assumed that $\Gamma^2\omega^2 \ll (\omega_0{}^2 - \omega^2)^2$.

8.2.4.1. Form of $\chi_e'(\omega)$

Following Eq. (6'), $\chi_e'(\omega) \approx \chi_e'(0)\dfrac{\omega_0^2}{\left(\omega_0^2 - \omega^2\right)}$.   (17)

The limiting values can be given accordingly:

• when $\omega \ll \omega_0$,

$$\chi_e'(\omega) = \chi_e'(0)\frac{\omega_0^2}{\left(\omega_0 - \omega\right)\left(\omega_0 + \omega\right)} \approx \chi_e'(0)\frac{\omega_0}{\left(\omega_0 - \omega\right)} \underset{\omega\to 0}{\to} \chi_e'(0); \qquad (18)$$

• when $\omega \gg \omega_0$:

$$\chi_e'(\omega) = -\chi_e'(0)\frac{\omega_0^2}{\omega^2} \overset{(8)}{=} -\frac{\omega_p^2}{\omega^2} \underset{\omega\to\infty}{\to} 0^- \qquad (19); \text{ and}$$

$$\chi_e'(\infty) \approx 0^- \quad (\text{when } \omega \to \infty, \chi_e'(\omega) \approx \frac{1}{-\infty} \to 0^-).$$

8.2.4.2. Form of $\chi_e''(\omega)$
According to Eq. (7),

$$\chi_e''(\omega) \approx \omega_p^2 \frac{\omega\Gamma}{\left(\omega_0^2 - \omega^2\right)^2} \approx 0 \approx \varepsilon_r''.$$

A consequence of this equation is that there is no absorption in this zone, and χ and $\underline{\varepsilon}$ are real such that, for example, $\vec{\underline{P}} = \varepsilon_0\chi\,\vec{\underline{E}}$ or $\vec{\underline{D}} = \varepsilon\,\vec{\underline{E}}$. There is no dephasing between the response and the excitation, and it is only the real parts, $\varepsilon(\omega)$ or $\chi(\omega)$, which vary with ω. The medium gives rise to dispersion and is nonabsorbent, which in effect means that it is also transparent.

8.2.4.3. Graphical representation
Figure 8.3 gives a graphical representation of $\chi_e'(\omega)$ far from the resonance frequency (ω_0). This representation gives the electrical component, as at low frequencies it is an ionic or dipolar relaxation contribution that appears.

In general terms, $\varepsilon'_r(\omega) = 1 + \chi_e'(\omega)$. As $\chi_e'(\omega \to \infty) \approx 0$, then $\varepsilon_r'(\omega \to \infty) \approx 1$, and the curve representing $\varepsilon_r'(\omega)$ takes on the shape shown in Figure 8.4. The analytical form for $\varepsilon'_r(\omega)$ at points far from the resonance value is directly obtained from Eq. (17):

$$\varepsilon'_r(\omega) = 1 + \frac{\omega_p^2}{\omega_0^2 - \omega^2}. \qquad (17')$$

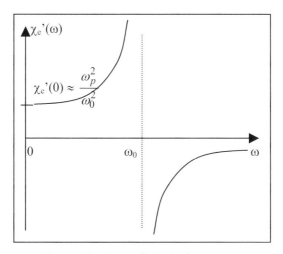

Figure 8.3. *Curve of $\chi'_e(\omega)$ when $\omega \neq \omega_0$.*

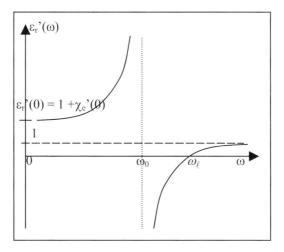

Figure 8.4. *Curve of $\varepsilon'_r(\omega)$ when $\omega \neq \omega_0$.*

8.2.5. *The (zero) pole of the dielectric function, and longitudinal electric waves*

An angular frequency (ω_ℓ), which corresponds to the (zero) pole of a dielectric function, is such that ω_ℓ is different to ω_0, must be such that the following equation is true:

$$\varepsilon'_r(\omega_\ell) = 1 + \frac{\omega_p^2}{\omega_0^2 - \omega_\ell^2} = 0 \; ,$$

the solution of which gives rise to:

$$\omega_\ell^2 = \omega_p^2 + \omega_0^2 \; .$$

This also shows that $\omega_\ell > \omega_0$ (see Figure 8.4), which by consequences means that $\omega_\ell \neq \omega_0$, and justifies the use of Eq. (17') to determine ω_ℓ.

In order to show that the angular frequency ω_ℓ is due to a wave that has a longitudinal structure, the following may be considered. As $\omega \neq \omega_0$, then $\varepsilon_r" \approx 0$, and $\underline{\varepsilon}$ is real ($\underline{\varepsilon} = \varepsilon$). Gauss's equation, which is written div $\underline{\vec{D}}_0 = \varepsilon$ div $\underline{\vec{E}}_0 = 0$, has in reality two solutions, which are either:

- div $\underline{\vec{E}}_0 = 0$ giving for a progressive planar wave, $i\underline{\vec{k}}.\underline{\vec{E}}_0 = 0$, in other terms $\underline{\vec{k}} \perp \underline{\vec{E}}_0$ so that the wave presents a transversal structure; or

- $\varepsilon(\omega) = 0$ which corresponds to an angular frequency value such that $\varepsilon(\omega_\ell) = 0$ and the solution corresponds to the (zero) pole of the dielectric function, which is thus such that div $\underline{\vec{E}} \neq 0$.

  In this case, it remains true that:

  div $\underline{\vec{B}}_0 = 0$

  $\overrightarrow{\text{rot}}\underline{\vec{B}} = \mu_0 \dfrac{\partial \underline{\vec{D}}}{\partial t} = i\omega\mu_0\varepsilon_0\varepsilon_r\underline{\vec{E}}$, so that $\overrightarrow{\text{rot}}\underline{\vec{B}}_0 = i\dfrac{\omega}{c^2}\varepsilon_r\underline{\vec{E}}_0 = 0$ when $\omega = \omega_\ell$ $\left. \right\} \Rightarrow$

with $\underline{\vec{B}}_0 = \underline{\vec{B}}_m e^{-i\vec{k}.\vec{r}}$ (form given for a progressive sinusoidal wave), these two equations give rise to, respectively:

$\left. \begin{array}{l} i\,\vec{k}.\underline{\vec{B}}_0 = 0 \\ i\,\vec{k} \times \underline{\vec{B}}_0 = 0 \end{array} \right\} \quad \Rightarrow \quad \underline{\vec{B}}_0 = 0$ and therefore the wave is completely electric.

The Maxwell-Faraday equation makes it possible to state that:

$\overrightarrow{\text{rot}}\underline{\vec{E}} + \dfrac{\partial \underline{\vec{B}}}{\partial t} = 0$, so that $\overrightarrow{\text{rot}}\underline{\vec{E}}_0 + j\omega\underline{\vec{B}}_0 = 0$. With $\underline{\vec{B}}_0 = 0$, then $\overrightarrow{\text{rot}}\underline{\vec{E}}_0 = 0$, and

$i\,\vec{k} \times \underline{\vec{E}}_0 = 0$. This in turn means that $\underline{\vec{E}}_0 \,/\!/\, \vec{k}$.

The wave is said to have a longitudinal structure as $\vec{E}$ is directed along the wave vector $\vec{k}$ (while for the transversal wave, $\underline{\vec{E}}_0 \perp \vec{k}$).

8.2.6. Behavior of a transverse plane progressive EM wave which is sinusoidal and has a pulsation between ω_0 and ω_ℓ sufficiently far from ω_0 so that $\varepsilon'' \approx 0$

A monochromatic planar progressive electromagnetic (MPPEM) wave that is sinusoidal can be written as:

$$\underline{\vec{E}} = \underline{\vec{E}}_m \exp[i(\omega t - \vec{k}.\vec{r})].$$

When $\omega_0 < \omega < \omega_\ell$, Figure 8.4 indicates how $\varepsilon_r'(\omega) < 0$, while $\varepsilon_r''(\omega) = 0$ at points far from ω_0. Therefore $\underline{\varepsilon}_r = \varepsilon_r'(\omega) - j \varepsilon_r''(\omega) = \varepsilon_r'(\omega)$ which is a real negative, and thus $\underline{\varepsilon}_r = \varepsilon'_r < 0$.

The equation for the dispersion of transverse MPPEM waves is given by $\underline{k}^2 = \dfrac{\omega^2}{c^2}\underline{\varepsilon}_r = \dfrac{\omega^2}{c^2}\varepsilon'_r < 0$, which in turn imposes that $\underline{k} = -ik'' = -i\dfrac{\omega}{c}n''$ where k is a purely imaginary number.

The MPPEM wave thus takes on the form given by $\underline{\vec{E}} = \underline{\vec{E}}_m \exp[-k''r] \exp[i\omega t]$. In effect, there is no wave propagation; the wave is simply attenuated without undergoing absorption (as $\omega \neq \omega_0$). There is no propagation for the wave in the medium (because $k' = 0$), and the wave is termed evanescent.

When $\omega \in \,]\omega_0,\omega_\ell[$, the material can be considered a perfect reflector.

8.2.7. Study of an MPPEM wave both outside the absorption zone ($\omega \neq \omega_0$) and the range [ω_0, ω_ℓ]

8.2.7.1. Relationship between ω and k

The sections above considered an MPPEM was within the range $\omega \in \,]\omega_0,\omega_\ell[$. This section will look at a wave outside this range with $\omega \ll \omega_0$ and $\omega > \omega_\ell$.

In these two domains, $\varepsilon_r''(\omega) = 0$, and $\underline{\varepsilon}_r = \varepsilon'_r > 0$ (see Figure 8.4, noting that when $\omega < \omega_0 , \varepsilon'_r > 1$ and when $\omega > \omega_\ell , 0 < \varepsilon'_r < 1$). Thus, the dispersion equation takes on the form $\underline{k}^2 = \dfrac{\omega^2}{c^2}\underline{\varepsilon}_r = \dfrac{\omega^2}{c^2}\varepsilon'_r > 0$, from which $\underline{k} = k' = k$.

With $\varepsilon'_r(\omega)$ given when $\omega \neq \omega_0$ by Eq. (17'), then

$$k^2 = \frac{\omega^2}{c^2}\left[1 + \frac{\omega_p^2}{\omega_0^2 - \omega^2}\right]. \qquad (20)$$

By developing this further, it is possible to obtain:

$$\omega^4 - \omega^2(k^2c^2 + \omega_0{}^2 + \omega_p{}^2) + k^2c^2\omega_0^2 = 0 .$$

8.2.7.2. The relation $\omega = f(k)$

The preceding equation is a "bi-squared" equation of the type $x^4 + b\,x^2 + c = 0$, which has solutions in the form $x^2 = -\dfrac{b}{2} \pm \dfrac{\sqrt{b^2 - 4c}}{2}$, so that it can be stated that:

$$\omega_{\pm}^2 = \frac{k^2c^2 + \omega_0^2 + \omega_p^2}{2} \pm \left[\left(\frac{k^2c^2 + \omega_0^2 + \omega_p^2}{2}\right)^2 - k^2c^2\omega_0^2\right]^{1/2} . \qquad (21)$$

8.2.7.3. Shape of the dispersion curve

When $\omega \ll \omega_0$ and $\omega > \omega_\ell$, then

$$k^2 = \frac{\omega^2}{c^2}\varepsilon'_r ,$$

where $\varepsilon'_r(\omega) = 1 + \dfrac{\omega_p^2}{\omega_0^2 - \omega^2}$

It is worth considering the two limiting values of k:

- $k \approx 0$

  according to the equation for dispersion, there are two possible solutions. These are either:

  o $\omega \approx 0$, which corresponds to the solution $\omega = \omega_-$, [which also can be directly checked by introducing $k = 0$ into Eq. (21), giving a solution for ω_- as $\omega_- = 0$]; or

  o $\varepsilon'_r(\omega) \approx 0$, in which case when $\omega \approx \omega_\ell$, such that $\omega_\ell^2 = \omega_p^2 + \omega_0^2$, the solution for ω_+ is $\omega_+ = \omega_\ell$, [which again can be found by direct substitution of $k = 0$ into Eq. (21)].

- $k \to \infty$ where once again two solutions are possible:

  o $\dfrac{\omega}{c} \to \infty$ and $k \approx \dfrac{\omega}{c}$ for a solution to ω_+. The introduction into Eq. (21) of $k \to \infty$ so that $k^2c^2 \gg \omega_0{}^2$ or $k^2c^2 \gg \omega_p{}^2$, gives $\omega_+ = ck$; or

  o $\varepsilon'_r(\omega) \to \infty$, from which $(1 + \dfrac{\omega_p^2}{\omega_0^2 - \omega^2}) \to \infty$, and in other words means that $\omega \to \omega_0$, hence the solution to ω_-.

A representation of $\omega = f(k)$ therefore shows two branched curves associated with the transversal EM waves (T branches) corresponding to the two solutions ω_+ and ω_-. The curves are separated by a zone, or band, into which propagation of the EM waves cannot occur. The value at which $\omega = \omega_\ell$ shown in Figure 8.5 corresponds for its part to the longitudinal electric wave (L).

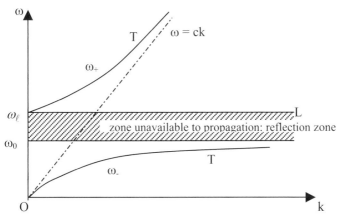

Figure 8.5. *Dispersion curves indicated by T.*

The scheme presented below gives a résumé of the different behaviors of a dielectric undergoing electric polarizations at different frequencies (and pulsations).

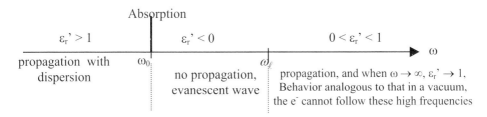

8.2.8. Equation for dispersion $n = f(\lambda_0)$ when $\omega \ll \omega_0$
8.2.8.1. Maxwell's equation for the visible region

When $\omega < \omega_0$, then $\underline{\varepsilon}_r = \varepsilon_r' > 1$, and the relation for dispersion $\underline{k}^2 = \dfrac{\omega^2}{c^2} \varepsilon_r$ gives,

for a real ε_r, $\underline{k} = k = \dfrac{\omega}{c}\sqrt{\varepsilon_r}$. This region is where $\omega < \omega_0$ is in the visible range and

therefore is nonabsorbing and transparent. More precisely, the conditions on ω can be written $\omega_i < \omega < \omega_{0e}$, where ω_i is the resonance pulsation related to the displacement of ions, and in the near infrared, ω_{0e} is the pulsation resonance noted for the most part above as ω_0 and related to the displacement of electrons and therefore also the electric polarization.

Given the equation for the dispersion, the velocity of the y phase can be written $v_\varphi = \dfrac{\omega}{k} = \dfrac{c}{\sqrt{\varepsilon_r}}$. As elsewhere, the index (n) is defined by $n = \dfrac{c}{v_\varphi}$; then it can be written that $n = \sqrt{\varepsilon_r(\omega)}$. This is the Maxwell equation valid for the visible region. It can be simplified further by stating that with a real value for ε_r, the index $\underline{n}$ defined by $\underline{n}^2 = \varepsilon_r$ is also real, so that $\underline{n} = n = \sqrt{\varepsilon_r}$.

As here $\varepsilon_r > 1$ (region $\omega < \omega_{0e}$), $n = \sqrt{\varepsilon_r(\omega)} > 1$, and $v_\varphi = \dfrac{c}{\sqrt{\varepsilon_r}} < c$;

the result is that, with $k_0 = \dfrac{\omega}{c}$:

$$\lambda = \frac{2\pi}{k} = \frac{2\pi}{k_0 n} = \frac{\lambda_0}{n} < \lambda_0,$$

where $\lambda_0 = \dfrac{2\pi c}{\omega}$ and is the wavelength associated with ω in a vacuum.

8.2.8.2. Dispersion equation for $\omega < \omega_{0e}$ and Cauchy's formula

In the region $\omega < \omega_{0e}$, $\varepsilon_r(\omega) = \varepsilon_r'(\omega)$ is given by Eq. (17'), so that with $\omega_0 = \omega_{0e}$ the resulting equation is:

$$\varepsilon_r(\omega) = 1 + \frac{\omega_p^2}{\omega_{0e}^2 - \omega^2} = 1 + \frac{\omega_p^2}{\omega_{0e}^2} \frac{1}{1 - \dfrac{\omega^2}{\omega_{0e}^2}} = 1 + \chi_e(0) \frac{1}{1 - \dfrac{\omega^2}{\omega_{0e}^2}}$$

$$= 1 + \chi_e(0) \left[1 + \frac{\omega^2}{\omega_{0e}^2} + \ldots \right]$$

With $n^2 = \varepsilon_r(\omega)$ and $\omega = \dfrac{2\pi c}{\lambda_0}$ it is possible to say that:

$$n^2 \approx 1 + \chi_e(0) + \frac{\chi_e(0)}{\omega_{0e}^2}\omega^2 = 1 + \chi_e(0) + \frac{\chi_e(0)}{\omega_{0e}^2}(2\pi c)^2 \frac{1}{\lambda_0^2}.$$

By making $A = 1 + \chi_e(0) = \varepsilon_r(0) = n_0^2$, and $B = (2\pi c)^2 \dfrac{\chi_e(0)}{\omega_{0e}^2}$, then $n^2 = A + \dfrac{B}{\lambda_0^2}$.

From this can be determined that $n = A^{1/2}\left(1 + \dfrac{B}{A\lambda_0^2}\right)^{1/2} \approx A^{1/2}\left(1 + \dfrac{B}{2A\lambda_0^2} + ... \right)$,

so that by making $C = \dfrac{B}{2A^{1/2}}$, it can be stated that

$$n \approx n_0 + \frac{C}{\lambda_0^2}. \qquad (22)$$

Equation (22), called Cauchy's equation, shows that the index varies with the wavelength.

8.2.8.3. The Rayleigh relation and group and phase velocities

The group velocity is given by $v_g = \dfrac{d\omega}{dk} = \dfrac{dv}{d(1/\lambda)} = -\lambda^2\dfrac{dv}{d\lambda}$. In addition,

$\lambda_0 = \dfrac{2\pi c}{\omega} = \dfrac{c}{v}$ and $\lambda = \dfrac{\lambda_0}{n} = \dfrac{c}{nv}$, so therefore $v = \dfrac{c}{n\lambda}$. The result is:

$$v_g = -c\lambda^2\frac{d(1/n\lambda)}{d\lambda} = -c\lambda^2\left[\frac{1}{n}\left(-\frac{1}{\lambda^2}\right) + \frac{1}{\lambda}\frac{d\left(\frac{1}{n}\right)}{d\lambda}\right] = \frac{c}{n} - c\lambda\left(-\frac{1}{n^2}\frac{dn}{d\lambda}\right).$$

With $v_\varphi = \dfrac{c}{n}$, it is possible to obtain:

$$v_g = v_\varphi\left(1 + \frac{\lambda}{n}\frac{dn}{d\lambda}\right). \qquad (23)$$

8.2.8.4. Comment: normal and abnormal dispersions

In Sections 7.2.1.3 and 7.2.1.4 normal and abnormal dispersions were defined. In dielectrics, as shown in Figure 8.4 and as a general rule, $\varepsilon_r'(\omega)$ increases with ω. However, as indicated in Figure 8.1, when $\omega \approx \omega_0$, the function $\varepsilon_r'(\omega)$ decreases as ω increases, describing a behavior in the abnormal dispersion zone.

8.3. Propagation of a MPPEM Wave in a Plasma (or the Dielectric Response of an Electronic Gas)

8.3.1. Plasma oscillations and pulsations

8.3.1.1. Definition of a plasma

Overall a plasma is neutral and is made up of ions, assumed to be in fixed positions (due to their high mass which accords them considerable inertia), and by electrons, assumed to be highly mobile. All of this is in a vacuum. At equilibrium (rest state), the ion volume densities (n_0) and the electron volume densities (n_e) are identical. However, following a Coulombic interaction, if the perturbation obliges an electron to move from its equilibrium position, then it will have a tendency to return to original position due to a returning electrical force. Below, electron and ion charges are denoted by –e and +e, respectively.

8.3.1.2. Study of the displacement of electrons in a plasma subject to a mechanical perturbation

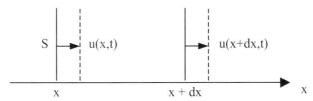

Figure 8.6. *Moving a "slice" of plasma.*

Supposing that under the influence of a mechanical perturbation, for example, an acoustic effect, the electrons shown in Figure 8.6 are moved together along the Ox axis by a small distance $u(x,t)$. The thermal agitation and weights involved are negligible, much as the frictional or mechanical recall forces, as the electrons are not tied to any specific atom as is the case for dielectrics.

To start there is a "slice" of electron "fluid" which has a cross-sectional area S and is placed between x and x + dx at its rest state. The initial corresponding volume of the slice is $V_0 = S\,dx$.

Under the effect of a perturbation, the deformed fluid moves in such a way that its limits move to x + u(x,t) and x + dx + u(x + dx, t) . Therefore its volume also changes to :

$$V_{loss} = S\,[dx + u(x +dx, t) - u(x, t)].$$

With $u(x +dx, t) = u(x) + dx\dfrac{\partial u}{\partial x}$,   then $V_{loss} = S\,[dx + dx\dfrac{\partial u}{\partial x}]$,   and   the   final variation in volume can be considered as :

$$\delta V = V_{pert} - V_0 = S\,dx\,\frac{\partial u}{\partial x}\, .$$

By denoting the concentration of the charge following perturbation by $n_e + \delta n_e$, the conservation of charge in the occupied volume before and after the perturbation can be described by:

$n_e V_0 = (n_e + \delta n_e)\,V_{pert}$, or in other words,

$n_e\,S\,dx = (n_e + \delta n_e)\,S\,[dx + dx\dfrac{\partial u}{\partial x}]$, from which comes

$$n_e \frac{\partial u}{\partial x} + \delta n_e \left(1 + \frac{\partial u}{\partial x}\right) = 0. \text{ With } \frac{\partial u}{\partial x} \ll 1 \text{ , we obtain:}$$

$$\delta ne = -n_e \frac{\partial u}{\partial x}.$$

With the ions assumed to be fixed, their density (n_0) is unchanged and the total charge density therefore is given by

$$\rho = n_0 e - (n_e + \delta n_e)e .$$

With $n_0 = n_e$, then $\rho = -\delta n_e\, e = n_e\, e\, \dfrac{\partial u}{\partial x}.$

Poisson's equation (div $\vec{E} = \dfrac{\rho}{\varepsilon_0}$) makes it possible to write $\dfrac{\partial E}{\partial x} = \dfrac{n_e e}{\varepsilon_0} \dfrac{\partial u}{\partial x}$. With the

limiting condition that $\vec{E} = 0$ when $\bar{u} = 0$, the equation yields

$$\vec{E}(x,t) = \frac{n_e e}{\varepsilon_0} \bar{u}(x,t).$$

This relation ties together the displacement ($\bar{u}(x,t)$) of the electrons to the field ($\vec{E}$) generated by the same displacement. This field is a longitudinal one, as it is parallel to the displacement x.

On applying the dynamic fundamental relation to the electrons (electrons localized in the slice of the fluid) we have:

$$m \frac{d^2\bar{u}}{dt^2} = -e\,\vec{E}(x,t), \text{ so that } \frac{d^2\bar{u}}{dt^2} + \frac{n_e e^2}{m\varepsilon_0}\bar{u} = 0 .$$

The introduction of a pulsation to the plasma, defined by $\omega_p = \sqrt{\dfrac{n_e\, e^2}{m\varepsilon_0}}$, gives

$$u = A\, e^{i\omega_p t} .$$

The latter relationship indicates that the electrons undergo an oscillatory movement at the "plasma pulsation" ω_p.

Numerical application

- $n_e = 10^6 \text{cm}^{-3} = 10^{12} \text{ m}^{-3}$ (ionosphere) $\Rightarrow v_p = 8.97$ MHz (decimeter waves).
- $n_e = 10^{15} \text{ cm}^{-3} = 10^{21} \text{ m}^{-3}$ (dense gas discharge)
  $\Rightarrow v_p = 1.75 \times 10^{12}$ Hz (millimeter waves)
- $n_e = 10^{29} \text{cm}^{-3}$ (metal) $\Rightarrow v_P = 10^{16}$ Hz (UV visible)

To sum, in a plasma the mechanical perturbation that moves electrons forms a longitudinal field (which has a direction depending on the induced displacement). The electric field in turn produces an electronic returning force and therefore a displacement. This can be verified by studying the behavior of a plasma subject to an electric field of a form given by $E = E_0 \, e^{i\omega t}$.

8.3.2. The dielectric response of an electronic gas

As before, the medium is assumed to be neutral overall and consisting of fixed ions and mobile electrons all in a vacuum. In the rest state the ion volume density (n_0) and the electron volume density (n_e) are identical *i.e.* $n_0 = n_e$. As above, electron charges are denoted by –e.

An electric field, acting as the recalling force, ($\vec{E} = \vec{E}_0 e^{i\omega t}$) is applied to the medium along Ox. It is assumed that there are no frictional forces involved.

Now, this study treats the problem by dealing with the following questions:

1.   Looking for the abscissa $\underline{x}$, which defines the position of an electron and is a solution to the permanent regime of form $\underline{x} = \underline{x}_0 e^{i\omega t}$, gives $\underline{x}_0$ with the help of a fundamental dynamics equation.

2. Give the expression for the polarization due to the displacement of the electrons.

3. By determining the expression for the relative dielectric permittivity [$\varepsilon_r(\omega)$], also called the dielectric function, given in the form $\varepsilon_r(\omega) = 1 - \dfrac{\omega_p^2}{\omega^2}$, detail what ω_p is.

4. The electric field ($\vec{E}$) now under consideration is that of a plane sinusoidal EM wave with pulsation denoted by ω and rectilinearly transverse polarized along Ox. It propagates toward increasing values of z and is such that

$\vec{E} = \vec{E}_m e^{i(\omega t - \vec{\underline{k}}.\vec{r})} = E_m e^{i(\omega t - \vec{\underline{k}}.\vec{r})} \vec{u}_x = E_m e^{i(\omega t - \underline{k}.z)} \vec{u}_x = \vec{E}_0 e^{i\omega t}$.

(a)   What is the dispersion equation for transversal plane EM waves? Give this relation with the notations and results of the question.

(b)   When $\omega < \omega_p$, state the exact form and type of the corresponding waves.

(c) When $\omega > \omega_p$, detail the form (progressive or not) of the waves associated with the angular frequency and plot the dispersion curve [$\omega = f(ck)$ where k is the wave vector module and c the speed of light] for these conditions.

(d) In the light of the results conclude about the plasma transparency.

5.   The value of λ_p for alkali metals is around 300 nm ($\lambda_p = \dfrac{2\pi c}{\omega_p}$). If it can be assumed that they can be represented by the electronic gas model, are they transparent to UV light?

6.   Study the (zero) pole of the dielectric function which corresponds to an angular frequency given by ω_L (which is such that $\varepsilon(\omega_L) = 0$). Is it possible to demonstrate that the associated waves are longitudinal?

Answers

1. According to the dynamic fundamental equation, $m\dfrac{d^2x}{dt^2} = -e\underline{E}$, into which

substituting $\underline{x} = \underline{x_0}\,e^{i\omega t}$ gives $i^2\omega^2 m\,\underline{x} = -e\,\underline{E}$, so that $\underline{x} = \dfrac{e\underline{E}}{m\omega^2}$ and $\underline{x_0} = \dfrac{e\underline{E_0}}{m\omega^2}$.

A field moving along Ox given by $\underline{E} = \underline{E_0}e^{i\omega t}$ results in an oscillating movement

given by $\underline{x} = \dfrac{e\underline{E_0}}{m\omega^2}\,e^{i\omega t}$.

2. The movement of a charge $(-e)$ by x is the same as generating a dipole moment by $p = -e\ x$. The corresponding polarization (dipole moment per unit volume) is

therefore $\underline{P} = n_e\,\underline{p} = -n_e\,e\,\underline{x} = -\dfrac{n_e e^2}{m\omega^2}\underline{E}$.

3. The dielectric function $\underline{\varepsilon_r}(\omega)$ is such that $\left.\begin{array}{l}\underline{D}(\omega) = \varepsilon_0\ \underline{\varepsilon_r}(\omega)\ \underline{E}(\omega) \\[4pt] \qquad\qquad = \varepsilon_0\ \underline{E}(\omega) + \underline{P}(\omega)\end{array}\right\} \Rightarrow$

$\underline{\varepsilon_r}(\omega) = 1 + \dfrac{\underline{P}(\omega)}{\varepsilon_0\underline{E}(\omega)}$ so that $\underline{\varepsilon_r}(\omega) = 1 - \dfrac{n_e e^2}{\varepsilon_0 m\omega^2} = \varepsilon_r(\omega)$. With $\omega_p^{\ 2} = \dfrac{n_e e^2}{\varepsilon_0 m}$, then:

$$\varepsilon_r(\omega) = 1 - \dfrac{\omega_p^2}{\omega^2}.$$

4.

(a) The equation for the dispersion of progressive plane EM waves with transverse structures, as here with $\vec{k}$ following Oz, can be written as $\underline{k}^2 c^2 = \omega^2 \varepsilon_r(\omega)$. In this case, ε_r is real. By taking the previous expression for $\varepsilon_r(\omega)$ into the equation for dispersion, we find:

$$\underline{k}^2 = \dfrac{\omega^2}{c^2}\left(1 - \dfrac{\omega_p^2}{\omega^2}\right),$$

which is the equation for the dispersion of EM transversal waves in a plasma.

Comment: The above relationship can be obtained more directly from the Maxwell equations, detailed as Eqs. (10), (11), and (13) in Chapter 7, as in:

$$\vec{k}.\vec{E_0} = 0 \ (1) \qquad \vec{k}.\vec{B_0} = 0 \qquad (2) \quad \vec{k}\times\vec{E_0} = \omega\vec{B_0}. \qquad (3)$$

The equations show by themselves that the EM wave is obligatorily transversal. For its part, the Maxwell-Ampere equation is written in the form:

$\overrightarrow{\text{rot}}\ \vec{B} = \mu_0\left(\vec{j_\ell} + \varepsilon_0\dfrac{\partial\vec{E}}{\partial t}\right)$, with, in this case, $\vec{j_\ell} = \rho\vec{v}$ being the current associated

with the displacement of electrons influenced by the oscillating field. As $\rho = -n_e e$

and $\underline{v} = \dfrac{dx}{dt}$, so with $\underline{x} = \dfrac{eE_0}{m\omega^2}e^{i\omega t}$ it is possible to state that $\underline{v} = \dfrac{ie}{m\omega}E_0e^{i\omega t}$, from

which can be obtained $\vec{\underline{j}}_\ell = -i\dfrac{n_ee^2}{m\omega}\vec{E}$. We can remark also this same current can be

seen   as   the   polarization   current $\vec{\underline{j}}_P = \dfrac{\partial\vec{\underline{P}}}{\partial t}$,   and   thus   with

$\vec{\underline{P}} = -\dfrac{n_ee^2}{m\omega^2}\vec{E} = -\dfrac{n_ee^2}{m\omega^2}\vec{E}_0e^{i\omega t}$, we found again : $\vec{\underline{j}}_P = -i\dfrac{n_ee^2}{m\omega}\vec{E} \equiv \vec{\underline{j}}_\ell$.

Finally the Maxwell-Ampere relation gives $\overrightarrow{\text{rot}}\,\vec{B} = \mu_0\left(-i\dfrac{n_ee^2}{m\omega} + i\varepsilon_0\omega\right)\vec{E}$,

from which can be determined that $-i(\underline{\vec{k}} \times \underline{\vec{B}}_0)=\mu_0\left(-i\dfrac{n_ee^2}{m\omega} + i\varepsilon_0\omega\right)\vec{E}_0$.

By   substituting $\underline{\vec{B}}_0$ from   Eq.   (3)   into   the   last   equation:

$i\dfrac{\underline{k}^2}{\omega}\vec{E}_0 = \mu_0\left(-i\dfrac{n_ee^2}{m\omega} + i\varepsilon_0\omega\right)\vec{E}_0$, from which $\underline{k}^2 = \mu_0\varepsilon_0\omega^2 - \mu_0\varepsilon_0\dfrac{n_ee^2}{m\varepsilon_0}$, and then:

$$\underline{k}^2 = \dfrac{\omega^2}{c^2}\left(1 - \dfrac{\omega_p^2}{\omega^2}\right).$$

(b)   When $\omega < \omega_p$, then in turn $\dfrac{\omega_p^2}{\omega^2} > 1$, and $\varepsilon_r(\omega) < 0$ just as $\underline{k}^2 < 0$. $\underline{k}$ is

therefore purely imaginary. Using the electrokinetic notation where $\underline{k} = k' - i\,k''$,

then $\underline{k} = -ik''$; the wave can be written in the form $\vec{\underline{E}} = \vec{E}_m e^{i(\omega t-\underline{k}z)}=\vec{E}_m e^{-k''z}e^{i\omega t}$.

The upshot is that the wave is no longer propagating but is stationary with an

angular frequency (ω). The wave resembles a evanescent wave that has an amplitude

that decrease exponentially with x.

(c)   When $\omega > \omega_p$, then $1 > \dfrac{\omega_p^2}{\omega^2}$, and $\varepsilon_r(\omega) > 0$ and $\underline{k}^2 > 0$. $\underline{k}$ is therefore real, so

$\underline{k} = k'$, and the wave has the form $\vec{\underline{E}}=\vec{E}_m e^{i(\omega t-\underline{k}z)} = \vec{E}_m e^{i(\omega t-k'z)}$. The wave thus is

progressive and does not exhibit a term for absorption ($k'' = 0$). The dispersion

equation thus can be written, with the more simple $k' = k$, as $\omega = \sqrt{\omega_p^2 + c^2k^2}$.

The curve due to this equation is shown in Figure 8.7. The slope of the dispersion

curve, $v_g = \dfrac{d\omega}{dk}$, gives v_g which is the group velocity. Note that the slope is less

than the slope of the plot $\omega = ck$ and therefore less than the speed of light.

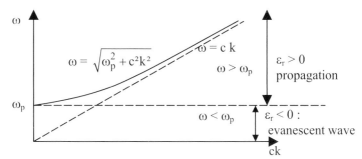

Figure 8.7. *Dispersion curve for a plasma.*

(d) To conclude, the plasma (or electron gas) acts much as a high band filter toward incident waves. As indicated in Figure 8.8, the gas is only transparent when $\varepsilon_r(\omega) > 0$; that is to say when $\omega > \omega_p$. However, when $\omega < \omega_p$, a component due to attenuation by $e^{-k''z}$ without a term for propagation, appears.

5.

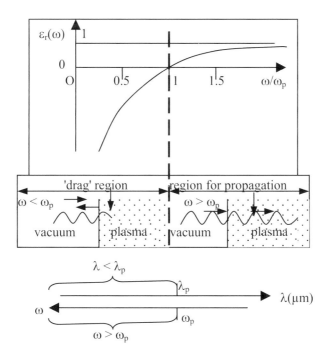

Figure 8.8. *High band filtering characteristics of a plasma.*

Given that $\omega > \omega_p$ corresponds to when $\lambda < \lambda_p$, waves with lengths less than λ_p can propagate with $k'' = 0$, that is without attenuation or in other terms absorption, through the medium under consideration. Thus the alkali metals should be transparent to radiation with wavelengths below 300 nm, and this includes UV radiation.

6. The condition $\varepsilon_{(\omega = \omega_L)} = 0$ makes it possible to determine ω_L, which in turn should be such that $\varepsilon(\omega_L) = 1 - \dfrac{\omega_p^2}{\omega_L^2} = 0$ and that $\omega_L = \omega_p$. The (zeros) poles of the dielectric function therefore are equal to the plasma frequency, that is to say the frequency at which the electron gas undergoes a longitudinal oscillation. The cutoff frequency ($\omega = \omega_p$) of the transversal EM waves corresponds to the longitudinal oscillation mode of the electron gas. The electrons are subject to displacement pulsations, in a direction along that of the associated field ($E = E_0\, e^{i\omega t}$), that is to say in the longitudinal direction of the field and along the line of electron displacement.

8.4. Propagation of an EM Wave in a Metallic Material (Frictional Forces)

A study of the complex conductivity and the dispersion of waves within a metal can be carried out by resolving the following question, which considers the velocity (v) of electrons subject to a Coulombic force generated by an applied alternating field given by $\vec{E} = \vec{E}_0 \exp(i\omega t)$.

The question given in Sections 8.2 and 8.3 is revived here except of course the electrons are now considered to be in a metallic environment so that the volume density for electron charge is given by $\rho_\ell = -n_e\, e$ and the electrons are subject to a frictional force which is in the form $f_t = -\dfrac{m}{\tau} v$, where v and τ are the conduction velocity and relaxation time, respectively. It is assumed that these conduction electrons are not subject to returning forces and that their movement can be studied in one dimension.

1. Give the equation for the movement of each electron and from this determine that the velocity when $\underline{v} = v_0\, e^{i\omega t}$.

2. The material is thought of as a vacuum in which the characteristic conduction electrons are spread. In the following order, calculate:

(a) the conduction current density $j_\ell = \rho_\ell \, \underline{v} = \underline{\sigma}_\ell \, \underline{E}$ and express $\underline{\sigma}_\ell$ as a function

of $\omega_p{}^2 = \dfrac{n_e e^2}{\varepsilon_0 m}$, and also as a function of the conductivity (σ_0) for a stationary

regime, i.e., when $\omega = 0$;
(b) the current density associated with movement through vacuum (j_{D0}); and
(c) from (b) derive the total current density (j_T) and the conductivity ($\underline{\sigma}$), which appears in its complex form.

3.   This   question   concerns $\omega\tau \ll 1$.   In   copper, $\tau \approx 10^{-14}$ s ,   when $v < v_c \approx 100$ GHz , we find that $\omega\tau \leq 10^{-2} \ll 1$.
(a) What form does $\underline{\sigma}_\ell$ take on?

(b) Calculate   the   ratio $\dfrac{\left|j_{D0}\right|}{\left|j_\ell\right|}$ for   copper   where $\sigma_0 \approx 6 \; 10^7 \; \Omega^{-1} m^{-1}$.   What

conclusion can be drawn from the result?

4.   The metallic medium is now considered in its entirety, and therefore the characteristic used is that of its complex relative dielectric permittivity ($\underline{\varepsilon}_r$). Give the form of the displacement current in this medium with $\underline{\varepsilon}_r$. By giving the equality between two forms of complex conductivities, each obtained by the two representations found in questions 2(c) and 4, determine $\underline{\varepsilon}_r$, which is defined within the terms of this question by the relation $\underline{D} = \varepsilon_0 \, \underline{\varepsilon}_r \, \underline{E}$.

5. Here the field under consideration is that of a monochromatic planar progressive electromagnetic (MPPEM) wave which propagates in the same sense as increasing values of z.
(a) Give the relation for the dispersion of these waves.
(b) For $\omega\tau \gg 1$, give the expression for $\underline{\varepsilon}_r$ comparable to that obtained in question 3 above. From the result, draw a conclusion about the nature of the MPPEM wave.
(c) Here, $\omega\tau \ll 1$. Calculate $\underline{\varepsilon}_r$ and indicate the form of the associated wave.

Answers
1. The fundamental dynamic equation along the direction of the velocity is:

$$F = m\frac{dv}{dt} = \Sigma f = -eE - \frac{m}{\tau}v \text{, from which the complex terms derived are:}$$

$$m\frac{d\underline{v}}{dt} + \frac{m}{\tau}\underline{v} = -e\underline{E}$$

With $\underline{E}$ in the form $\underline{E} = \underline{E}_0 e^{i\omega t}$, a solution for v in the form $\underline{v} = \underline{v}_0 e^{i\omega t}$ is required.

By substitution into the differential equation, and with $\dfrac{dv}{dt} = i\omega \underline{v}$, it is possible to

obtain: $im\omega \underline{v} + \dfrac{m}{\tau}\underline{v} = -e\underline{E}$, so that following division by $e^{i\omega t}$, the equation

becomes $im\omega \underline{v}_0 + \dfrac{m}{\tau}\underline{v}_0 = -e\underline{E}_0$.

From this can be derived:

$$\underline{v}_0 = -\frac{e\underline{E}_0 \tau}{m(1 + i\omega\tau)}.$$

2.

(a) Ohm's law gives the conduction current associated with the volume density of electrons (n_e), as in:

$$\underline{j}_\ell = \rho_\ell \ \underline{v} = -n_e \ e \ \underline{v} = \frac{n_e e^2 \tau}{m(1 + i\omega\tau)}\underline{E}_0 e^{i\omega t} = \frac{n_e e^2 \tau}{m(1 + i\omega\tau)}\underline{E} = \sigma_\ell \ \underline{E}, \text{ so that}$$

$$\underline{\sigma}_\ell = \frac{n_e e^2 \tau}{m(1 + i\omega\tau)}.$$

By introducing $\omega_p{}^2 = \dfrac{n_e e^2}{\varepsilon_0 m}$ into $\underline{\sigma}_\ell$, then $\underline{\sigma}_\ell = \dfrac{\varepsilon_0 \omega_p^2 \tau}{1 + i\omega\tau}$.

When $\omega = 0$ (i.e., under a stationary regime), $\sigma_{\ell(\omega=0)} = \sigma_0 = \varepsilon_0 \omega_p^2 \tau$, so that it is possible to also state that:

$$\underline{\sigma}_\ell = \sigma_0 \frac{1}{1 + i\omega\tau}.$$

(b) The displacement current associated with a vacuum, given that in a vacuum $\underline{D} = \varepsilon_0 \ \underline{E}_0 e^{i\omega t}$, is

$$\underline{j}_{D0} = \frac{\partial \underline{D}}{\partial t} = i \omega \varepsilon_0 \ \underline{E}.$$

(c) The total current density is therefore:

$$\underline{j}_T = \underline{j}_\ell + \underline{j}_{D0} = \left(\frac{\varepsilon_0 \omega_p^2 \tau}{1 + i\omega\tau} + i \omega \varepsilon_0\right)\underline{E} = \underline{\sigma} \ \underline{E}, \text{ from which } \underline{\sigma} = \varepsilon_0 \left(\frac{\omega_p^2 \tau}{1 + i\omega\tau} + i\omega\right).$$

3.

(a) With $\underline{\sigma}_\ell = \sigma_0 \dfrac{1}{1 + i\omega\tau}$ and with $\omega\tau \ll 1$, we have $\sigma_\ell \approx \sigma_0$.

(b) We have $\dfrac{\left|j_{D0}\right|}{\left|j_\ell\right|} \approx \dfrac{\omega\varepsilon_0 \left|E\right|}{\sigma_0 \left|E\right|} = \dfrac{\omega\varepsilon_0}{\sigma_0}$, so that for copper and when $\omega\tau \ll 1$ (and for

$v < v_c = 100$ GHz), the calculation gives:

$$\dfrac{\left|j_{D0}\right|}{\left|j_\ell\right|} \approx \dfrac{\omega\varepsilon_0\tau}{\sigma_0\tau} \ll \dfrac{\varepsilon_0}{\sigma_0\tau} \approx \dfrac{10^{-11}}{6.10^7 10^{-14}} \approx 10^{-5}.$$

As a consequence, as long as the frequencies are not too high, i.e., $v < v_c = 100$ GHz , the displacement current in a vacuum is negligible with respect to the internal metal conduction current.

In this frequency range, we therefore find that

$$\underline{\sigma} \approx \underline{\sigma}_\ell = \sigma_0 \dfrac{1}{1 + i\omega\tau} \approx \sigma_0.$$

4.   With $\underline{D} = \varepsilon_0 \underline{\varepsilon}_r\, \underline{E}$, it is possible to directly find $j_D = i\omega\varepsilon_0 \underline{\varepsilon}_r\, \underline{E}$, so here $\underline{\sigma} = i\omega\varepsilon_0 \underline{\varepsilon}_r$.

Equalizing the two expressions obtained for $\underline{\sigma}$ (in 2(c) and just above in this answer) gives:

$$\underline{\varepsilon}_r = 1 + \dfrac{\omega_p^2\tau}{i\omega(1 + i\omega\tau)} = 1 - \dfrac{\omega_p^2\tau}{\omega(\omega\tau - i)} = 1 - \dfrac{\omega_p^2\tau(\omega\tau + i)}{\omega(1 + \omega^2\tau^2)}.$$

5.
(a) For a transverse MPPEM wave, the dispersion equation can be written with permittivity in its complex form ($\underline{\varepsilon}_r$) as

$$k^2 c^2 = \omega^2\, \underline{\varepsilon}_r\,(\omega).$$

(b) When $\omega\tau \gg 1$, the expression for $\underline{\varepsilon}_r$ (obtained in answer 4) becomes

$\underline{\varepsilon}_r = 1 - \dfrac{\omega_p^2}{\omega^2} = \varepsilon_r$ (in its real form). Once again we find the expression previously

given in answer 3 for a problem treated in Section 8.3.2 concerning a plasma wherein the electrons are assumed to undergo negligible frictional forces. The MPPEM waves therefore have the same form as those written above such that when $\omega > \omega_p$ they are progressive and when $\omega < \omega_p$ they are evanescent.

(c)   When $\omega \tau \ll 1$, then from the answer to question 4, $\underline{\varepsilon}_r$ is given by

$$\underline{\varepsilon}_r = 1 - i\frac{\omega_p^2 \tau}{\omega} .$$

The permittivity is complex and the dispersion equation, $\underline{k}^2 c^2 = \omega^2 \underline{\varepsilon}_r (\omega)$, which also

can be written as $\underline{k}^2 = \dfrac{\omega^2}{c^2} \underline{\varepsilon}_r(\omega)$, shows that k is complex in that $\underline{k} = k' - i k''$. The

wave therefore can be given in the form:

$$\psi \propto e^{i(\omega t - \underline{k}z)} = e^{-k''z} e^{i(\omega t - k'z)}$$

term for attenuation   propagation term

8.5. Uncharged Magnetic Media

8.5.1. Dispersion equation in conducting magnetic media

For the medium under consideration, ε represents the dielectric permittivity, its

magnetic permeability $\mu \neq \mu_0$, and its conductivity is such that $\sigma_\ell \overset{\text{notation}}{=} \sigma$. With

the material being electrically neutral in its natural state, then $\rho_\ell = 0$ and Maxwell's

equations from Section 5.3 can be written as:

$$\text{div } \vec{E} = 0 \quad (1) \qquad\qquad \overrightarrow{\text{rot}}\vec{E} + \frac{\partial \vec{B}}{\partial t} = 0 \quad (3)$$

$$\text{div } \vec{B} = 0 \quad (2) \qquad \overrightarrow{\text{rot }} \vec{B} = \mu\left[\vec{j}_\ell + \varepsilon\frac{\partial \vec{E}}{\partial t} \right] = \mu\left[\sigma\vec{E} + \varepsilon\frac{\partial \vec{E}}{\partial t} \right] \quad (4)$$

For MPPEM waves with the form $\underline{\vec{E}} = \underline{\vec{E}}_m \exp[i(\omega t - \vec{k}\vec{r})]$, Eqs. (1), (2), and (3)
respectively give:

$$\underline{\vec{k}}.\underline{\vec{E}} = 0 \ (1') \qquad \underline{\vec{k}}.\underline{\vec{B}} = 0 \ (2') \qquad \underline{\vec{k}} \times \underline{\vec{E}} = \omega\underline{\vec{B}} \quad (3').$$

These relationships were derived from calculations shown in Section 7.1.4 from
which the results were equally applicable to amplitudes (with index m) as complex
vectors, as all that was necessary was to multiply, or divide, the two terms by
$\exp[i(\omega t - \vec{k}\vec{r})]$.

Once again, the structure of the MPPEM wave is the same as if in a vacuum
(transverse wave). Equation (4) details this more specifically, in that

$$\left[-i\,\underline{k}\times\vec{B} \right] = \mu\left[\vec{j}_\ell + \varepsilon\frac{\partial\vec{E}}{\partial t} \right],$$ so that with Eq. (3'), it is possible to state that

$$\left[-i\,\underline{k}\times\frac{\underline{k}}{\omega}\times\vec{E} \right] = \mu\left[\sigma\vec{E} + i\omega\varepsilon\vec{E} \right].$$ Using the formula for the double vectorial

product given by $\vec{a}\times(\vec{b}\times\vec{c}) = (\vec{a}\vec{c})\vec{b} - (\vec{a}\vec{b})\,\vec{c}$, and with $\underline{k} \perp \vec{E}$, from eqn (1'), then:

$$i\frac{\underline{k}^2}{\omega}\vec{E} = i\mu\varepsilon\omega\left(1 + \frac{\sigma}{i\omega\varepsilon} \right)\vec{E},$$ from which comes the dispersion equation, as in:

$$\underline{k}^2 = \mu\varepsilon\omega^2\left(1 - i\frac{\sigma}{\omega\varepsilon} \right). \qquad (5)$$

8.5.2. Impedance characteristics (when k is real)

When the magnetic permeability (μ) is real, $\vec{B}$ and $\vec{H}$ are in phase. Similarly, when

k is real, $\vec{B}$ and $\vec{E}$ are related by $\vec{B} = \dfrac{\vec{k}\times\vec{E}}{\omega}$ and also are in phase. Thus

$$E = \frac{\omega}{k}B = \frac{\omega\,\mu}{k}H,$$ so that $E = Z\,H$ with $Z = \dfrac{\omega\,\mu}{k}$.

Therefore, $Z = \dfrac{E}{H}$, and as E is expressed in V m^{-1} and H in A m^{-1}, Z has the

dimensions V A^{-1}. The Z corresponds to an impedance that is characteristic of the medium under study and is dependent on ω.

In addition, when k, μ, and ε are real, according to the preceding equation, we find that $k = \omega\sqrt{\varepsilon\,\mu}$, and therefore Z also is real and has a value

$$Z = \sqrt{\frac{\mu}{\varepsilon}}. \qquad (6)$$

On setting $Z_0 = \sqrt{\dfrac{\mu_0}{\varepsilon_0}} = 377\,\Omega$ (the characteristic impedance of a vacuum), the

characteristic impedance of a medium is given by $Z = \sqrt{\dfrac{\mu}{\varepsilon}} = \sqrt{\dfrac{\mu_0}{\varepsilon_0}}\sqrt{\dfrac{\mu_r}{\varepsilon_r}}$, so that

$$Z = Z_0\sqrt{\frac{\mu_r}{\varepsilon_r}}. \qquad (7)$$

In general terms, the complex index is defined by $\underline{k} = \underline{n}\dfrac{\omega}{c}$, so $k' - i\,k'' = \dfrac{\omega}{c}(n' - in'')$.

With a medium assumed to be nonabsorbent, then $k'' = 0$. Therefore, $k = k'$ is real and $n'' = 0$, so that $n = n'$. Under these conditions, $k = n\dfrac{\omega}{c}$, and with

$k = \omega\sqrt{\varepsilon\,\mu} = \omega\sqrt{\varepsilon_0\,\mu_0}\sqrt{\varepsilon_r\,\mu_r} = \dfrac{\omega}{c}\sqrt{\varepsilon_r\,\mu_r}$, it is possible to derive: $n = \sqrt{\varepsilon_r\,\mu_r}$.

As $Z = Z_0\sqrt{\dfrac{\mu_r}{\varepsilon_r}} = Z_0\dfrac{\mu_r}{\sqrt{\mu_r\,\varepsilon_r}}$, the final calculation gives:

$$Z = \dfrac{\mu_r}{n}Z_0 . \qquad (8)$$

8.6. Problems

8.6.1. The complex forms for polarization and dielectric permittivity

This question concerns a dielectric that is linear, homogeneous, and isotropic (lhi) and contains N identical molecules per unit volume. Each molecule exhibits in its periphery a free electron with a charge denoted by –e. The dielectric is subject to an electric field applied along the Ox axis, and the polarization of the medium also is studied along that axis while considering that each peripheral electron is subject to a Coulombic force. This force moves the electron from its equilibrium position, to which it tends to return due to a returning force (f_r) given by $f_r = -m\,\omega_0^2\,x$, where ω_0 is homogeneous and of constant angular frequency, and also due to a frictional force (f_t) which tends to brake the movement and is given by $f_t = -m\Gamma\dfrac{dx}{dt}$, where Γ is a homogenous constant that is inverse with respect to time. The movement of the ions is assumed to be negligible given that their mass is considerably greater than that of the electrons.

1. Calculations for real positions:
(a) With the electric field being given by $E(t) = E_0\cos\omega t$, write the fundamental dynamic equation.
(b) From the above equation determine the differential equation that describes the displacement in x of each electron.
(c) Under a forced regime, which corresponds to the particular solution obtained from the second term of the differential equation, the solution is of the form $x(t) = x_0\cos(\omega t - \varphi)$. Determine x_0 and φ. Note that in order to carry this out, it is possible to associate complex numbers to $x(t)$ and $E(t)$ such that $\underline{x} = x_0\exp(-j\varphi)\exp(j\omega t) = \underline{x_0}\exp(j\omega t)$, with $\underline{x_0} = x_0\exp(-j\varphi)$ as the complex amplitude and $\underline{E}(t) = E_0\exp(j\omega t)$.

(d) Give the formula of the dipole moment induced in each individual molecule (which are assumed to not have mutual interactions). From this derive the real polarization given by P(t).

(e) Is this expression an equation of instantaneous proportionality between the response of the system [P(t)] and its excitation [E(t)]? Give a conclusion concerning the dephasing between the establishment of the polarization and the application of the electric field.

2. Calculation of the complex polarization ($\underline{P}$) and the complex induction ($\underline{D}$):

(a) With the help of the results gained from question 1, give $\underline{P}$ along with its complex amplitude P_0.

(b) Recall the defining relationship between the complex susceptibility from which can be given the expression for $\chi(\omega)$.

(c) Give the relation that brings together the complex induction and the complex electric field and which permits a definition of the complex permittivity ($\underline{\varepsilon}_r$). From this, derive the expression for $\underline{\varepsilon}_r(\omega)$.

3. Calculation for the real electric induction (when $\underline{\varepsilon} = \varepsilon_0 \underline{\varepsilon}_r$):

(a) With $\underline{\varepsilon} = \varepsilon' - j\varepsilon''$ give the expression for the real part of the electric induction.

(b) Show how the real electric induction gives rise to a dephasing with respect to the real electric field. Express this dephasing as a function of ε' and ε''.

Answers

1.

(a) The fundamental dynamic equation $\sum \vec{F}_{applied} = m \vec{\gamma}$, makes it possible to state that through Ox, $F_{Coulomb} + F_r + F_f = m \, d^2x/dt^2$ so that

$$- e \, E(t) - m\omega_0^2 x - m\Gamma \frac{dx}{dt} = m \frac{d^2x}{dt^2} .$$

(b) The differential equation that describes the movement, with $E(t) = E_0 \cos \omega t$, therefore is given by $\dfrac{d^2x}{dt^2} + \Gamma \dfrac{dx}{dt} + \omega_0^2 x = -\dfrac{e}{m} E_0 \cos \omega t$.

(c) In order to resolve a differential equation of the second order the solution can be found by adding to a general solution a second term specific to the equation with the second term.

Under a permanent regime, that is when there are forcing oscillations (produced by a Coulombic force given by $\cos \omega t$) applied over a long period of time, the movement is dominated by a specific solution given by:

$$x(t) = x_0 \cos(\omega t - \varphi) .$$

By associating with x(t) and E(t) the complex numbers,

$x = x_0 \exp(-j\varphi) \exp(j\omega t) = \underline{x}_0 \exp(j\omega t)$, where $\underline{x}_0 = x_0 \exp(-j\varphi)$ is the complex

amplitude, so $\underline{E}(t) = E_0 \exp(j\omega t)]$, and hence $\dfrac{d\underline{x}}{dt} = j\omega\underline{x}$, and $\dfrac{d^2\underline{x}}{dt^2} = -\omega^2\underline{x}$.

The differential equation then becomes $-\omega^2\underline{x} + j\omega\Gamma\underline{x} + \omega_0{}^2\underline{x} = -\dfrac{e}{m}\underline{E}$. By dividing

the two terms by $e^{j\omega t}$, the result is:

$-\omega^2\underline{x}_0 + j\omega\Gamma\underline{x}_0 + \omega_0{}^2\underline{x}_0 = -\dfrac{e}{m}E_0$, so that

$$\underline{x}_0 = -\frac{e}{m[(\omega_0^2 - \omega^2) + j\omega\Gamma]}E_0 = x_0\, e^{-j\varphi} .$$

From $\underline{z}_0 = -\underline{x}_0 = \dfrac{c}{a + ib}$, where $c = \dfrac{e}{m}E_0$, $a = (\omega_0{}^2 - \omega^2)$, and $b = \omega\Gamma$, it is

possible to derive $\underline{z}_0 = \dfrac{ac - ibc}{a^2 + b^2} = u - iv = \rho e^{j\theta} = |\underline{z}_0|\, e^{-j\varphi}$. Then

$|\underline{z}_0| = \rho = \sqrt{u^2 + v^2} = \dfrac{c}{\sqrt{a^2 + b^2}} = \dfrac{e}{m[\sqrt{(\omega_0^2 - \omega^2)^2 + \omega^2\Gamma^2}]}E_0$, from which we find

that:

$$x_0 = -\frac{e}{m[\sqrt{(\omega_0^2 - \omega^2)^2 + \omega^2\Gamma^2}]}E_0$$

(in which it can be algebraically verified if E_0 moves according to positive values of x, and thereby if the electrons move toward negative values of x, as detailed in the figure below);

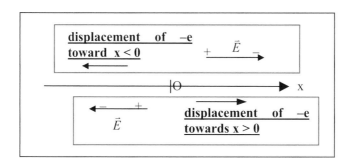

and also that $\tan\theta \ = \ \dfrac{u}{-v} = \tan(-\varphi) = -\tan\varphi$, so that:

$$\varphi = \text{Arc} \tan \frac{u}{v} = \text{Arc} \tan \frac{b}{a} = \text{Arc} \tan \frac{\omega\Gamma}{\left(\omega_0^2 - \omega^2\right)} .$$

To conclude, the movement of an oscillator is cosinusoidal and undergoes the same pulsation (ω) as the excitation force, given by $F_C = -eE\cos\omega t$, the same amplitude (x_0) and the same dephasing (φ) with respect to E. The resonance pulsation is such that x_0 is at a maximum. If $\Gamma = 0$, then the oscillator is unrestricted and x_0 is at a maximum when $\omega = \omega_0$.

(d) Moving a charge (q) by a distance (x) is the same as applying to the system a dipole moment given by $\mu = qx$ (as detailed in Section 2.2.2.1.). Similarly, moving an electronic charge (-e) by x is the same as applying a dipole moment $\mu = -ex$, so that:

$$\mu = -ex_0 \cos(\omega t - \varphi) = \frac{e^2}{m[\sqrt{(\omega_0^2 - \omega^2)^2 + \omega^2\Gamma^2}]} E_0\cos(\omega t - \varphi) .$$

The polarization $P = \sum_i n_i q_i \delta_i$ can be written as $P = N\mu = -eNx$, and hence

$$P = \frac{Ne^2}{m} \frac{1}{[\sqrt{(\omega_0^2 - \omega^2)^2 + \omega^2\Gamma^2}]} E_0 \cos(\omega t - \varphi) .$$

(e) With the applied field being given by $\cos\omega t$, there is no proportionality between the instantaneous polarization, given by $P(t) = P_0 \cos(\omega t - \varphi)$ and the instantaneous field, given by $E(t) = E_0 \cos\omega t$. The polarization is established with a phase delay (φ) with respect to the applied electric field.

2.
(a) The complex polarization is:
$$\underline{P} = -Ne\underline{x} = -Nex_0 \exp(-j\varphi) \exp(j\omega t) = -Ne\underline{x}_0 \exp(j\omega t) = \underline{P}_0 \exp(j\omega t) , \qquad \text{from}$$
which

$$\underline{P}_0 = -Nex_0 \exp(-j\varphi) = -Ne\underline{x}_0 = \frac{Ne^2}{m[(\omega_0^2 - \omega^2) + j\omega\Gamma]} E_0 = P_0\, e^{-j\varphi} \ , \text{ with}$$

$$P_0 = \frac{Ne^2}{m} \frac{1}{[\sqrt{(\omega_0^2 - \omega^2)^2 + \omega^2\Gamma^2}]} E_0 \text{ and } \varphi = \text{Arc} \tan \frac{\omega\Gamma}{\left(\omega_0^2 - \omega^2\right)} .$$

(b) The complex susceptibility is defined by $\underline{P} = \varepsilon_0 \; \underline{\chi} \; \underline{E}$, which means that

$$\underline{P}_0 \, \exp(-j\omega t) = \varepsilon_0 \; \underline{\chi} \; \underline{E}_0 \exp(j\omega t) = \varepsilon_0 \; \underline{\chi} \; E_0 \exp(j\omega t) .$$

From this can be immediately derived that:

$$\left. \begin{array}{l} \underline{P}_0 = \varepsilon_0 \; \underline{\chi} \; \underline{E}_0 \\[2mm] = \dfrac{Ne^2}{m[(\omega_0^2 - \omega^2) + j\omega\Gamma]} E_0 \end{array} \right\} \quad \Rightarrow \quad \underline{\chi} = \dfrac{Ne^2}{m\varepsilon_0[(\omega_0^2 - \omega^2) + j\omega\Gamma]}$$

(c) Here

$$\left. \begin{array}{l} \underline{D} = \varepsilon_0 \; \underline{E} + \underline{P} \\[2mm] = \underline{\varepsilon} \; \underline{E} \end{array} \right\} \quad \Rightarrow \quad \underline{\varepsilon} = \varepsilon_0 + \dfrac{P}{E} = \varepsilon_0 + \varepsilon_0 \; \underline{\chi} = \varepsilon_0 \, (1 + \underline{\chi}) = \varepsilon_0 \; \underline{\varepsilon}_r .$$

So $\underline{\varepsilon}_r = 1 + \underline{\chi} = 1 + \dfrac{Ne^2}{m\varepsilon_0[(\omega_0^2 - \omega^2) + j\omega\Gamma]}$, and by making $\omega_p^{\,2} = \dfrac{Ne^2}{m\varepsilon_0}$, we find

that: $\underline{\varepsilon}_r = 1 + \dfrac{\omega_p^2}{[(\omega_0^2 - \omega^2) + j\omega\Gamma]}$.

3.

(a) With $\underline{\varepsilon} = \varepsilon' - j\,\varepsilon''$, $\underline{D} = \underline{\varepsilon} \; \underline{E} = (\varepsilon' - j\varepsilon'') \; \underline{E}$ and

$\underline{E} = E_0 \, \exp(j\omega t) = E_0 \, \cos\omega t + j \, E_0 \, \sin\omega t$

$\Rightarrow \quad \underline{D} = \varepsilon'(E_0 \, \cos\omega t + j \, E_0 \, \sin\omega t) - j \, \varepsilon'' \, E_0 \, \cos\omega t + \varepsilon'' \, E_0 \, \sin\omega t$, and

$D = R(\underline{D}) = (\varepsilon' \cos \omega t + \varepsilon'' \sin \omega t) \, E_0$, or:

$$D = A\cos \omega t + B\sin \omega t , \text{ with } A = \varepsilon' E_0 \text{ and } B = \varepsilon'' E_0 .$$

(b) D therefore is given by $D = A \cos \omega t + B \sin \omega t$, which can be changed to $D = C \cos (\omega t - \varphi)$. In effect, if we make:

$$\left. \begin{array}{l} A = C \cos \varphi \\[2mm] B = C \sin \varphi \end{array} \right\} \quad \Rightarrow \quad C = \sqrt{A^2 + B^2} \text{ and } \operatorname{tg} \varphi = \dfrac{B}{A} , \text{ then}$$

$D = A \cos \omega t + B \sin \omega t = C \cos \varphi \cos \omega t + C \sin \varphi \sin \omega t = C \cos (\omega t - \varphi) .$

Consequently, the real induction is

$$D = C \cos (\omega t - \varphi), \text{ with } C = \sqrt{A^2 + B^2} = E_0 \sqrt{\varepsilon'^2 + \varepsilon''^2} , \; \operatorname{tg} \varphi = \dfrac{B}{A} = \dfrac{\varepsilon''}{\varepsilon'} .$$

Finally, this can be written as:

$$D = C \cos (\omega t - \varphi) = E_0 \sqrt{\varepsilon'^2 + \varepsilon''^2} \, \cos (\omega t - \varphi) , \text{ with } \varphi = \operatorname{Arc\,tan} \dfrac{\varepsilon''}{\varepsilon'} .$$

8.6.2. A study of the electrical properties of a metal using the displacement law of electrons by x and the form of the induced electronic polarization (based on Section 8.4)

This exercise concerns the polarization, conductivity, and optical properties of a metal assumed to consist of a collection of N fixed ions (per unit volume) with N (N = n_e) free electrons, there being one free electron for each atom, of mass and charge denoted, respectively, by m and –e. All are assumed to be in a vacuum with a permittivity denoted by ε_0. The electronic gas is subjected to the following forces:

• a Coulombic force induced by an alternating field (applied along Ox) that is given in its complex form as $\vec{\underline{E}} = \vec{\underline{E}}_0 \exp(i\omega t)$; and

• a frictional force given by $\vec{F}_t = -\dfrac{m}{\tau}\dfrac{d\vec{\underline{x}}}{dt}$.

The metal of choice for this problem is copper, which has a relaxation time (τ) of 10^{-14} sec. Its conductivity is denoted by $\sigma(0)$ and is approximately equal to 6 x 10^7 Ω^{-1} m^{-1}. The plasma angular frequency (ω_p) is defined by the relationship $\omega_p^2 = \dfrac{Nq^2}{m\varepsilon_0}$ and typically is of the order of 10^{16} rad sec^{-1}.

1. Under a forcing regime, we are looking for the expression for displacement in the form $\vec{\underline{x}} = \vec{\underline{x}}_0 \exp(i\omega t)$ (complex expression). Determine $\vec{\underline{x}}_0$.

2. Give the expression for the complex polarization ($\vec{\underline{P}}$) as a function of $\dfrac{1}{1 + i\omega\tau}$.

3. The complex conductivity ($\underline{\sigma}$) is defined in the general equation: $\vec{\underline{j}} = \underline{\sigma}\vec{\underline{E}} = \dfrac{\partial\vec{\underline{D}}}{\partial t}$.

(a) By giving $\vec{\underline{D}}$ as a function notably of polarization, give the equation for $\underline{\sigma}$ that will then be used to give the static conductivity [$\sigma(0)$] which describes the conductivity when the frequency is zero and can be given as a function of N, q, τ, and m.

(b) Give the physical significance of the two terms that appear in the equation for $\underline{\sigma}$.

(c) Express $\sigma(0)$ and then $\underline{\varepsilon}_r$ as a function of ω_p^2.

Answers

1. The dynamic fundamental solution given for $\sum \vec{F}_{applied} = m\vec{\gamma}$ moving along Ox

gives $-\dfrac{m}{\tau}\dfrac{d\underline{x}}{dt} - e\underline{E} = m\dfrac{d^2\underline{x}}{dt^2}$, so that

$$\frac{d^2\underline{x}}{dt^2} + \frac{1}{\tau}\frac{d\underline{x}}{dt} = -\frac{e}{m}\underline{E}\ .$$

The solution for this differential equation, under a forcing regime where $\underline{E} = \underline{E}_0 \exp(i\omega t)$, is required in the form $\underline{x} = \underline{x}_0 \exp(i\omega t)$. With $\underline{\dot{x}} = i\omega\underline{x}$ and $\underline{\ddot{x}} = -\omega^2\underline{x}$ substituted into the differential equation, it is possible to obtain

$-\omega^2\underline{x}_0 + i\dfrac{\omega}{\tau}\underline{x}_0 = -\dfrac{e}{m}\underline{E}_0$, following division of the two terms by $\exp(i\omega t)$, and

hence :

$$\underline{x}_0 = \frac{e}{m\left[\omega^2 - i\dfrac{\omega}{\tau}\right]}\underline{E}_0 = \frac{-e\tau}{im\omega[1 + i\omega\tau]}\underline{E}_0\ .$$

2. The electronic polarization is such that for each individual electron (of charge $-e$) is displaced by x by the electric field and generates a dipole moment given by $\mu = -e\underline{x}$. Denoting the electron density by n_e, the corresponding polarization is given

by $\underline{P} = -n_e\ e\ \underline{x} = \dfrac{n_e e^2 \tau}{im\omega}\dfrac{1}{[1 + i\omega\tau]}\underline{E}_0 e^{j\omega\tau}$. It is interesting to note that by

introducing $\omega_p{}^2 = \dfrac{n_e e^2}{m\varepsilon_0}$, it is possible to write that:

$$\underline{P} = \frac{\omega_p^2 \varepsilon_0 \tau}{i\omega[1 + i\omega\tau]}\underline{E}_0 e^{j\omega\tau}\ .$$

3.
(a) With $\underline{D} = \varepsilon_0\ \underline{E} + \underline{P}$, it is possible to state that

$\underline{j} = \underline{\sigma}\underline{E} = \dfrac{\partial\underline{D}}{\partial t} = \varepsilon_0\dfrac{\partial\underline{E}}{\partial t} + \dfrac{\partial\underline{P}}{\partial t} = i\omega\left[\varepsilon_0\underline{E} + \dfrac{n_e e^2 \tau}{im\omega}\dfrac{1}{1 + i\omega\tau}\underline{E}\right]$. It is possible to

immediately derive the relationship

$$\underline{\sigma} = i\omega\varepsilon_0 + \frac{n_e e^2 \tau}{m}\frac{1}{[1 + i\omega\tau]} = \underline{\sigma}(\omega)\ .$$

From this, when $\omega = 0$, $\underline{\sigma} = \sigma(0) = \dfrac{n_e e^2 \tau}{m}$, and finally,

$$\underline{\sigma} = i\omega\varepsilon_0 + \frac{\sigma(0)}{[1 + i\omega\tau]}\ .$$

(b) The first term given as $i\omega\varepsilon_0$ corresponds to $\varepsilon_0\dfrac{\partial\underline{E}}{\partial t}$. This term can be considered as being tied to the displacement current through a vacuum, as in $\sigma_{vide} = i\omega\varepsilon_0$. The second term, $\dfrac{\sigma(0)}{[1 + i\omega\tau]}$, corresponds to $\dfrac{\partial P}{\partial t}$, meaning the current due to the volume polarization of the material (that is to say inside the material). This conductivity is an internal conductivity

$$\underline{\sigma}_{\text{internal}} = \frac{\sigma(0)}{[1 + i\omega\tau]} .$$

(c) With $\underline{D} = \varepsilon_0\,\varepsilon_r\,\underline{E}$, we have $j = \underline{\sigma}\underline{E} = \dfrac{\partial \underline{D}}{\partial t} = \varepsilon_0\varepsilon_r\,\dfrac{\partial \underline{E}}{\partial t} = i\omega\varepsilon_0\varepsilon_r\,\underline{E}$, from which

$\underline{\sigma} = i\omega\varepsilon_0\,\varepsilon_r$.

By equalizing the preceding expression for conductivity, $\underline{\sigma} = i\omega\varepsilon_0 + \dfrac{\sigma(0)}{[1 + i\omega\tau]}$, it is

possible to obtain $\underline{\varepsilon}_r = 1 + \dfrac{1}{i\omega\varepsilon_0}\,\dfrac{\sigma(0)}{[1 + i\omega\tau]} = 1 - \dfrac{i}{\omega\varepsilon_0}\dfrac{\sigma(0)}{[1 + i\omega\tau]}$.

(d) With $\sigma(0) = \dfrac{n_e e^2\tau}{m}$

$\left. \begin{array}{l} \\ \omega_p{}^2 = \dfrac{n_e e^2}{m\varepsilon_0} \end{array} \right\} \quad \Rightarrow$

$\sigma(0) = \omega_p{}^2\,\varepsilon_0\,\tau$, and $\underline{\varepsilon}_r = 1 - i\dfrac{\omega_p^2\tau}{\omega[1 + i\omega\tau]}\left(= 1 - \dfrac{\omega_p^2\tau[\omega\tau + i]}{\omega[1 + \omega^2\tau^2]}\right)$.

Chapter 9
Electromagnetic Field Sources, Dipolar Radiation, and Antennae

9.1. Introduction

Until now the properties of electromagnetic (EM) waves have been covered without considering the mechanisms for their production or destruction. In fact, the laws of electromagnetism and classic mechanics applied to quasipoint charges such as electrons gives rise to a theory for wave emission, the principal result of which is that when a particle undergoes an acceleration, an EM wave may be emitted. In this chapter, it will be shown how the sinusoidal oscillation of an electric dipole can yield an EM wave. Electric dipolar radiation can be obtained by either sinusoidally oscillating the distance between dipolar charges or the actual dipolar charges themselves, as long as the dipolar moment is in the form $p = p_0 \, e^{j\omega t}$. This configuration is used in antennae, the principal types of which are detailed below with particular attention being paid to the half-wave antenna.

A preliminary determination of V and $\vec{A}$ potentials, associated with the dipoles, is required to calculate the EM radiation field. In addition, this calculation requires the solutions to Poisson's equations, which guide the propagation of potentials, generally called "retarded potentials".

In this chapter we will limit ourselves to the study of radiation in a vacuum, while in next chapters (Chapter 10 among others) we will look at interactions between EM waves and materials.

Complex notation: It is worth noting that if a signal is in the form $g = G_m \cos(\omega t - \varphi)$, the complex notation is $g = R(G_m \, e^{j\omega t} \, e^{-j\varphi})$, and that by convention we can write more simply $g = G_m \, e^{j\omega t} \, e^{-j\varphi}$. In this case, the complex amplitude is $\underline{G} = G_m \, e^{-j\varphi}$, which makes it possible to state that $g = \underline{G} \, e^{j\omega t}$. Strictly speaking, and as mentioned in Chapter 6, it is best to write g with an underline so that the equation becomes $\underline{g} = \underline{G} \, e^{j\omega t}$; however, the simplified if criticized notation also is used. If g and G_m are the vectors $\vec{g}$ and $\vec{G}_m$, $\underline{G}$ becomes a complex vector denoted by $\vec{G}$, which generally has the components $\underline{G}_x$, $\underline{G}_y$, and $\underline{G}_z$. We thus have for the differentiation or integration operations of $g = \underline{G}\ e^{j\omega t}$ with respect to time, the following:

$$\frac{dg}{dt} = j\omega\ \underline{G}\ e^{j\omega t} = j\omega g,$$

$$\rightleftarrows \frac{dg}{dt} = \frac{d\underline{G}}{dt}\ e^{j\omega t},\ \text{so by identification}\ \frac{d\underline{G}}{dt} = j\omega\underline{G}\ (=\ j\omega\ G_m\ e^{-j\varphi});\ \text{and}$$

$$\int gdt = \frac{1}{j\omega}\ \underline{G}\ e^{j\omega t} = \frac{1}{j\omega}g,$$

$$\rightleftarrows \int gdt = (\int\underline{G}dt)e^{j\omega t},\ \text{and}\ \int\underline{G}dt = \frac{1}{j\omega}\underline{G}.$$

9.2. The Lorentz Gauge and Retarded Potentials

9.2.1. Lorentz's gauge

9.2.1.1. Poisson's equations for potentials within an approximation of quasistationary states

In a homogeneous dielectric media, and under a regime of an approximation of quasistationary states (AQSS), we have:

$$\vec{E} = -\overrightarrow{\text{grad}}\ \ V - \frac{\partial\vec{A}}{\partial t}\quad(1)\qquad\vec{B} = \overrightarrow{\text{rot}}\ \ \vec{A}\quad(2)$$

The V and $\vec{A}$ potentials are expressed as a function of the deliberately applied and acting charge (ρ_ℓ) and current (j_ℓ) densities, as in:

$$V = \frac{1}{4\pi\varepsilon}\iiint\frac{\rho_\ell d\tau}{r}\quad(3)\qquad\vec{A} = \frac{\mu}{4\pi}\iiint\frac{\vec{j}_\ell d\tau}{r}\quad(4)$$

Under an AQSS regime, $\dfrac{\partial\rho_\ell}{\partial t} = 0$, and the equation for the conservation of charge gives $\text{div}\ \vec{j}_\ell = 0$, from which can be deduced that $\text{div}\ \vec{A} = 0$ (see Section 1.4.5). In addition, the potentials follow the Poisson equations, so that:

$$\Delta V + \frac{\rho_\ell}{\varepsilon} = 0\quad(5)\qquad\overrightarrow{\Delta\vec{A}} + \mu\vec{j}_\ell = 0\quad(6)$$

9.2.1.2. Under rapidly changing regimes

For systems subject to rapid variations, the displacement current must be considered

by adding in $\vec{j}_D = \dfrac{\partial \vec{D}}{\partial t}$ to the conduction current $\vec{j}_\ell = \sigma \vec{E}$, so that he current density

which now intervenes is:

$$\vec{j} = \vec{j}_\ell + \vec{j}_D \, .$$

The V and $\vec{A}$ potentials are instantaneous potentials calculated in Eqs. (3) and (4) for a time t at a certain point (r) from charges and currents at the same t. Propagation does not intervene in these equations for the potentials. The instantaneous potentials V and $\vec{A}$ thus appear as intermediates in the calculation of $\vec{E}$ and $\vec{B}$, having brought in a term for the displacement currents rather than the propagation. In effect, in Eq. (4), where $\vec{j} = \vec{j}_\ell + \vec{j}_D$ now takes the place of $\vec{j}_\ell$, $\vec{A}$

is bound to the derivative of $\vec{D}$ (in terms of $\vec{j}_D = \dfrac{\partial \vec{D}}{\partial t}$) and is not an intermediate in

the simple calculation of $\vec{E}$ in Eq. (1) and $\vec{D}$, because knowledge of $\vec{A}$ supposes that the magnitude of $\vec{D}$, or rather the derivative of $\vec{D}$, is known.

In order to resolve the problem more simply, the displacement currents are ignored and other potentials are used, such as those defined and denoted below as V_r and $\vec{A}_r$, to represent the propagation.

9.2.1.3. Lorentz's gauge

In Maxwell's equations, div $\vec{B} = 0$ so that $\vec{B}$ is still derived from a potential vector ($\vec{A}_r$), such that:

$$\vec{B} = \overrightarrow{rot}\vec{A}_r \, . \qquad (7)$$

The equation $\overrightarrow{rot}\vec{E} = -\dfrac{\partial \vec{B}}{\partial t}$ becomes $\overrightarrow{rot}\left(\vec{E} + \dfrac{\partial \vec{A}_r}{\partial t} \right) = 0$, which indicates that

$\vec{E} + \dfrac{\partial \vec{A}_r}{\partial t}$ is derived from a scalar potential (V_r) such that:

$$\vec{E} = -\overrightarrow{grad}V_r - \dfrac{\partial \vec{A}_r}{\partial t} \, . \qquad (8)$$

In a homogeneous media, possibly charged, the Poisson equation, div $\vec{E} = \dfrac{\rho_\ell}{\varepsilon}$, can

be written as:

$$- \Delta V_r - \frac{\partial}{\partial t}(\text{div}\vec{A}_r) = \frac{\rho_\ell}{\varepsilon}. \qquad (9)$$

In addition, Ampere's theorem given by $\overrightarrow{\text{rot}}\vec{B} = \mu\left(\vec{j}_\ell + \varepsilon\frac{\partial\vec{E}}{\partial t}\right)$, brings us to:

$$\overrightarrow{\text{rot}}\vec{B} = \overrightarrow{\text{rot}}\,\overrightarrow{\text{rot}}\vec{A}_r = \mu\left(\vec{j}_\ell + \varepsilon\frac{\partial\vec{E}}{\partial t}\right). \qquad (10)$$

With $\overrightarrow{\text{rot}}\,\overrightarrow{\text{rot}} = \overrightarrow{\text{grad}}\,\text{div} - \vec{\Delta}$, and by bringing Eq. (8) into Eq. (10), we have:

$$\overrightarrow{\text{grad}}\;\text{div}\vec{A}_r - \vec{\Delta}\vec{A}_r = \mu\left(\vec{j}_\ell - \varepsilon\;\overrightarrow{\text{grad}}\frac{\partial V_r}{\partial t} - \varepsilon\frac{\partial^2\vec{A}_r}{\partial t^2}\right). \qquad (11)$$

Equations (9) and (10) can be rewritten in the following form to give Poisson's equations for a rapidly varying regime:

$$\Delta V_r - \varepsilon\mu\frac{\partial^2 V_r}{\partial t^2} + \frac{\rho_\ell}{\varepsilon} = -\frac{\partial}{\partial t}(\text{div}\vec{A}_r + \varepsilon\mu\frac{\partial V_r}{\partial t}) \qquad (9')$$

$$\vec{\Delta}\vec{A}_r - \varepsilon\mu\frac{\partial^2\vec{A}_r}{\partial t^2} + \mu\vec{j}_\ell = \overrightarrow{\text{grad}}(\text{div}\vec{A}_r + \varepsilon\mu\frac{\partial V_r}{\partial t}) \qquad (11')$$

In addition, Eq. (7), $\vec{B} = \overrightarrow{\text{rot}}\vec{A}_r$, can define only $\vec{A}_r$ to the closest gradient, as $\overrightarrow{\text{rot}}\vec{A}_r = \overrightarrow{\text{rot}}\vec{A}_r'$ when $\vec{A}_r' = \vec{A}_r + \overrightarrow{\text{grad}}f$.

For its part, V_r can be determined only to within $\frac{\partial f}{\partial r}$, as $\vec{A}_r = \vec{A}_r' - \overrightarrow{\text{grad}}f$ substituted into Eq. (8) gives:

$$\vec{E} = -\overrightarrow{\text{grad}}\;V_r - \frac{\partial\vec{A}_r'}{\partial t} + \overrightarrow{\text{grad}}\frac{\partial f}{\partial t}$$

$$= -\overrightarrow{\text{grad}}\left(V_r - \frac{\partial f}{\partial t}\right) - \frac{\partial\vec{A}_r'}{\partial t}$$

$$= -\overrightarrow{\text{grad}}V_r' - \frac{\partial\vec{A}_r'}{\partial t}, \text{ with } V_r' = V_r - \frac{\partial f}{\partial t}.$$

The change of (V_r, $\vec{A}_r$) to (V_r', $\vec{A}_r'$) is called a gauge transformation, as in:

$$V_r \rightarrow V_r' = V_r - \frac{\partial f}{\partial t}$$

$$\vec{A}_r \rightarrow \vec{A'}_r = \vec{A}_r + \overrightarrow{\mathrm{grad}f} .$$

The determination of the potentials is carried out by imposing a measuring condition so as to find V_r and $\vec{A}_r$; f is an arbitrary function called a gauge function or simply a gauge. The invariance of $\vec{E}$ and $\vec{B}$ is termed the EM field gauge invariance.

The most convenient method to determine the potentials therefore is to impose on $\vec{A}_r$ and V_r a condition that simplifies Eqs. (9) and (11) by removing in both cases the terms in parentheses. This condition,

$$\mathrm{div}\vec{A}_r + \varepsilon\mu\frac{\partial V_r}{\partial t} = 0 , \qquad (12)$$

is called Lorentz's gauge, noting when in addition $\dfrac{\partial V}{\partial t} = 0$ it takes the place of the condition $\mathrm{div}\ \vec{A} = 0$ for stationary regimes.

Equations (9') and (10') therefore yield:

$$\Delta V_r - \varepsilon\mu\frac{\partial^2 V_r}{\partial t^2} + \frac{\rho_\ell}{\varepsilon} = 0 \quad (13)$$

$$\overrightarrow{\Delta\vec{A}}_r - \varepsilon\ \mu\ \frac{\partial^2 \vec{A}_r}{\partial t^2} + \mu\ \vec{j}_\ell = 0 \qquad (14)$$

In these equations, V_r and $\vec{A}_r$ are decoupled, in contrast to Eqs. (9') and (11'), and follow Eqs. (13) and (14) if f satisfies the wave equations discussed in Comment 3 below. Equations (13) and (14) generalize Eqs. (5) and (6) to rapidly varying regimes.

Comment   1:   We   have $\varepsilon\ \mu = \varepsilon_0\ \mu_0\ \varepsilon_r\ \mu_r = \dfrac{1}{c^2}\varepsilon_r\ \mu_r$,   where $\varepsilon_r\ \mu_r = n^2$

(with $\mu_r = 1$, we again find Maxwell's equation, $\varepsilon_r = n^2$), so that $\varepsilon\ \mu = \dfrac{n^2}{c^2} = \dfrac{1}{v^2}$.

Equations (13) and (14) can also be written as:

$$\Delta V_r - \frac{1}{v^2}\frac{\partial^2 V_r}{\partial t^2} + \frac{\rho_\ell}{\varepsilon} = 0$$

$$\overrightarrow{\Delta\vec{A}}_r - \frac{1}{v^2}\frac{\partial^2 \vec{A}_r}{\partial t^2} + \mu\ \vec{j}_\ell = 0$$

Comment 2: In a vacuum without charge or current, Eqs. (13) and (14) reduce to:

$$\Delta V_r - \frac{1}{c^2}\frac{\partial^2 V_r}{\partial t^2} = 0 \quad \text{and} \quad \Delta\vec{A}_r - \frac{1}{c^2}\frac{\partial^2 \vec{A}_r}{\partial t^2} = 0 .$$

By convention, the following operators are noted thus:

$$\Delta - \frac{1}{c^2}\frac{\partial^2}{\partial t^2} \equiv \square , \quad \text{called the d'Alembertian,} \; \Rightarrow \square\, V_r = 0$$

$$\vec{\Delta} - \frac{1}{c^2}\frac{\partial^2}{\partial t^2} \equiv \overrightarrow{\square} , \quad \text{called the d'Alembertian vector} \Rightarrow \overrightarrow{\square}\, \vec{A}_r = 0 .$$

Comment 3: A Lorentz condition is imposed on the gauge function, in which case, we still find that $\mathrm{div}\vec{A}_r + \varepsilon\mu \dfrac{\partial V_r}{\partial t} = 0$.

With $\vec{A}'_r = \vec{A}_r + \overrightarrow{\mathrm{grad}f}$ and $V'_r = V_r - \dfrac{\partial f}{\partial t}$ we also have:

$$\mathrm{div}\,\vec{A}'_r = \mathrm{div}\,\vec{A}_r + \mathrm{div}\overrightarrow{\mathrm{grad}f} \quad \text{(a); and}$$

$$\varepsilon\mu\frac{\partial V'_r}{\partial t} = \varepsilon\mu\frac{\partial V_r}{\partial t} - \varepsilon\mu\frac{\partial^2 f}{\partial t^2} \quad \text{(b).}$$

By adding each successive member in (a) and (b), we obtain

$$0 = \underbrace{\mathrm{div}\,\vec{A}'_r + \varepsilon\mu\frac{\partial V'_r}{\partial t}}_{= 0 \text{ as the Lorentzian gauge}} = \mathrm{div}\,\vec{A}_r + \varepsilon\mu\frac{\partial V_r}{\partial t} + \mathrm{div}\,\overrightarrow{\mathrm{grad}f} - \varepsilon\mu\frac{\partial^2 f}{\partial t^2}$$

$$\Rightarrow \quad \mathrm{div}\overrightarrow{\mathrm{grad}f} - \varepsilon\mu\frac{\partial^2 f}{\partial t^2} = 0$$

$$\rightleftharpoons \Delta f - \frac{1}{v^2}\frac{\partial^2 f}{\partial t^2} = 0$$

Finally, Lorentz's gauge condition requires that f satisfies a wave equation.

Comment 4: This additional remark concerns Coulomb's gauge. If there are no charges or currents, then Eq. (9) becomes:

$$\Delta V_r + \frac{\partial}{\partial t}(\mathrm{div}\vec{A}_r) = 0$$

If V_r is chosen so that $\Delta V_r = 0$ (Coulomb's gauge), we than have $\mathrm{div}\ \vec{A}_r = 0$ as in the stationary regime.

9.2.2. Equation for the propagation of potentials and retarded potentials

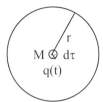

Figure 9.1. *Potential generated by a charge [q(t)] throughout* $d\tau$.

Equations (13) and (14) are the equations for propagation and they can be integrated by hypothesizing that at a given instant (t), a charge [q(t)] is in an elementary volume (dτ), which is assumed to be spherical given the symmetry caused by the isotopic propagation and is situated at a point (M) as described in Figure 9.1. The action of the charge depends only on the distance (r) from which its effect is studied. Within the spherical symmetry of the problem, the Laplacian of V_r is in the form:

$$\Delta V_r = \frac{1}{r} \frac{\partial^2 (rV_r)}{\partial r^2},$$

And Eq. (13) gives:

$$\frac{1}{r} \frac{\partial^2 (rV_r)}{\partial r^2} - \frac{1}{v^2} \frac{\partial^2 (V_r)}{\partial t^2} + \frac{\rho_\ell}{\varepsilon} = 0 \qquad (13')$$

Outside of the element dτ, the singularity of which can be considered to be a quasipoint, we have $\rho = 0$, and therefore can write for the outside that:

$$\frac{1}{r} \frac{\partial^2 (rV_r)}{\partial r^2} - \frac{1}{v^2} \frac{\partial^2 (V_r)}{\partial t^2} = 0 \quad \Leftrightarrow \quad \frac{\partial^2 (rV_r)}{\partial r^2} - \frac{1}{v^2} \frac{\partial^2 (rV_r)}{\partial t^2} = 0. \qquad (15)$$

Equation (15) is for propagation, and the function rV_r at t and r has a solution in the form:

$$rV_r(r,t) = G(t - \frac{r}{v}) + F(t + \frac{r}{v}).$$

As the source is at M and it is only waves emitted toward $r > 0$ that have physical solutions which are in the form $G(t - \frac{r}{v})$, we therefore find:

$$V_r(r,t) = \frac{G(t - \frac{r}{v})}{r}. \qquad (16)$$

In order to determine G, a limiting scenario might be used where $r \rightarrow 0$ and in Eq. (13) we find that $\dfrac{1}{r}$ becomes very large, just as $\dfrac{1}{r}\dfrac{\partial^2(rV_r)}{\partial r^2}$, which is such that:

$$\frac{1}{r}\frac{\partial^2(rV_r)}{\partial r^2} >> \frac{1}{v^2}\frac{\partial^2(V_r)}{\partial t^2} .$$

Equation (13') thus tends toward $\dfrac{1}{r}\dfrac{\partial^2(rV_r)}{\partial r^2} + \dfrac{\rho_\ell}{\varepsilon} = 0$, so that :

$$\Delta V_r + \frac{\rho_\ell}{\varepsilon} = 0 . \qquad (13'')$$

Equation (13') therefore tends toward the more usual Poisson equation, which has a solution that is for r $\rightarrow$ 0 (point charge at M):

$$V_r(r,t) = \frac{q(t)}{4\pi\varepsilon r} , \text{ so that also, } [r\,V_r(r,t)]_{r\rightarrow 0} = \frac{q(t)}{4\pi\varepsilon} .$$

From this can be determined that

$$\left[G\!\left(t - \frac{r}{v} \right) \right]_{r\rightarrow 0} = [r\,V_r(r,t)]_{r\rightarrow 0} = \frac{q(t)}{4\pi\varepsilon} \left.\right\} \Rightarrow G(t) = \frac{q(t)}{4\pi\varepsilon} .$$
$$= G(t)$$

And therefore, from Eq. (16), we find:

$$V_r(r,t) = \frac{q(t - \dfrac{r}{v})}{4\pi\varepsilon r} . \qquad (17)$$

If in a volume (V) there is a distribution of charges q(t), then there is an accumulation of potentials from the elementary charges such that $q(t) = \rho(t)\,d\tau$ and the general expression for the potential at t is:

$$V_r(r,t) = \frac{1}{4\pi\varepsilon} \iiint \frac{\rho\!\left(t - \dfrac{r}{v} \right)}{r} d\tau . \qquad (18)$$

In addition, Eq. (14) yields potential vector components similar to those of Eq. (13), with the condition that $1/\varepsilon$ is replaced by μ. The components of the potential vector thus have solutions of the type given by Eq. (18) in which $1/\varepsilon$ must be replaced by μ. These solutions can be brought together in a single equation:

$$\vec{A}_r(r,t) = \frac{\mu}{4\pi} \iiint \frac{\vec{j}\!\left(t - \dfrac{r}{v} \right)}{r} d\tau . \qquad (19)$$

For a distribution of rectilinear currents, we have:

$$\vec{A}_r(r,t) = \frac{\mu}{4\pi} \int \frac{I\left(t - \dfrac{r}{v}\right)}{r} d\vec{l} . \qquad (19')$$

To conclude, we thus obtain as a function of the active charges and currents at an instant $(t - \dfrac{r}{v})$ the potentials V_r and $\vec{A}_r$ for t and r with a delay of $\dfrac{r}{v}$, which takes into account the duration of the propagation. The potentials V_r and $\vec{A}_r$ are delayed potentials and substituted into Eqs. (7) and (8) yield $\vec{E}$ and $\vec{B}$.

9.3. Dipole Field at a Great Distance

9.3.1. Expression for the potential vector $\vec{A}$ generated by a domain D
9.3.1.1. General formula

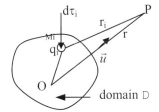

Figure 9.2. *Distribution of charges (q_i) placed around O by* $\overrightarrow{OM_i} = \vec{\delta_i}$.

Given an element, with a volume ($d\tau_i$) and an associated charge (q_i) such that $q_i = \rho\, d\tau_i$, which is within a domain (D) that has a distribution of q_i charges placed around O by $\overrightarrow{OM_i}$, we can associate with D a nonzero dipole moment given by $\vec{p} = \sum_i q_i \vec{\delta_i} \neq 0$ where $\vec{\delta_i} = \overrightarrow{OM_i}$.

Of interest is the form of the potential vector $\vec{A}$ at a point (P) located with respect to the origin (O) at the heart of D and the distribution of charges, by $\overrightarrow{OP} = \vec{r} = r\vec{u}$. Given Eq. (19) and that the system is in a vacuum, with $r_i = M_iP$, its possible to write that:

$$\vec{A}_r(r,t) = \vec{A}(P,t) = \frac{\mu_0}{4\pi} \iiint \frac{\vec{j_i}\left(t - \dfrac{r_i}{c}\right)}{r_i} d\tau_i .$$

It is possible to state that if $\vec{v}_i(t)$ is the velocity of q_i associated with $d\tau_i$ in D, with $\vec{j_i} = \rho\vec{v}_i$, we have $\vec{j_i}d\tau_i = \rho\vec{v}_i d\tau_i = \vec{v}_i\rho d\tau_i = \vec{v}_i q_i$, and can then write:

$$\vec{A}(P,t) = \frac{\mu_0}{4\pi} \sum_i \frac{q_i \vec{v}_i (t - \frac{M_i P}{c})}{M_i P} , \qquad (20)$$

where the summation is overall the charges within D.

In order to calculate the electromagnetic field that comes from this potential vector, the space can be divided into three:

- a zone defined by $r \ll \lambda$, inside which the field can be thought of as quasistationary;
- an intermediate zone where $r \approx \lambda$; and
- an external zone for which $r \gg \lambda$ and where the radiation is the concern of this study.

9.3.1.2. The form of $\vec{A}$ in the radiation zone

In the outer zone, it can be assumed that $r_i = M_i P \approx r$, as all the points in D are, in practical terms, a distance r from P. From this, Eq. (20) can be rewritten so that

$$\vec{A}(P,t) = \frac{\mu_0}{4\pi r} \sum_i q_i \vec{v}_i (t - \frac{r}{c}). \qquad (21)$$

With $\vec{v}_i = \frac{d\overrightarrow{OM_i}}{dt}$, we have $\sum_i q_i \vec{v}_i (t) = \dot{\vec{p}}(t)$, where $\dot{\vec{p}}(t)$ is the derivative with respect to time of the resultant dipole moment due to all charges in D. This finally gives:

$$\vec{A}(P,t) = \frac{\mu_0}{4\pi} \frac{\dot{\vec{p}}(t - \frac{r}{c})}{r}. \qquad (21')$$

It is interesting to note that the form of $\vec{A}$ therefore corresponds to a spherical wave, which at a great distance from D would be considered to be a source generating a plane wave (see Section 6.2.2.3). Indeed, at a great distance from the source, the surface of such a sphere can be associated with a tangential plane wave. With $\tau = t - \frac{r}{c}$, and as $\frac{d\tau}{dt} = 1$, it can also be noted that:

$$\frac{d\vec{p}(\tau)}{dt} = \frac{d}{d\tau}\vec{p}(\tau)\frac{d\tau}{dt} = \frac{d}{d\tau}\vec{p}(\tau).$$

The derivative therefore can be taken indifferently in respect of t or τ.

9.3.2. Expression for the electromagnetic field in the radiation zone (r >>λ)

With $\vec{B} = \overrightarrow{rot}\vec{A}$, and placing by notation, $\dot{\vec{p}}(t - \dfrac{r}{c}) = \dot{\vec{p}}(\tau) = \dot{\vec{p}}_\tau$ (which in effect means placing into lower magnitudes with respect to the index τ):

$$\vec{B}(P,t) = \overrightarrow{rot} \ \ \vec{A}(P,t) = \frac{\mu_0}{4\pi}\overrightarrow{rot}\left[\frac{\dot{\vec{p}}(t - \dfrac{r}{c})}{r}\right] = \frac{\mu_0}{4\pi}\overrightarrow{rot}\frac{\dot{\vec{p}}_\tau}{r}.$$

By using $\overrightarrow{rot}(a\vec{A}) = a\overrightarrow{rot}\vec{A} + \overrightarrow{grad}\,a \times \vec{A}$, it is possible to write:

$$\vec{B} = \frac{\mu_0}{4\pi r}\overrightarrow{rot}\dot{\vec{p}}_\tau + \frac{\mu_0}{4\pi}\overrightarrow{grad}\frac{1}{r}\times\dot{\vec{p}}_\tau,$$ the second term of which brings in

$\overrightarrow{grad}\ \dfrac{1}{r} = -\dfrac{\vec{r}}{r^3} = -\dfrac{\vec{u}}{r^2}$, a term that varies with respect to $\dfrac{1}{r^2}$ and at a great distance

becomes negligible with respect to the first term which only varies as $\dfrac{1}{r}$. Therefore:

$$\vec{B} \approx \frac{\mu_0}{4\pi r}\overrightarrow{rot}\dot{\vec{p}}_\tau. \qquad (22)$$

A quite involved calculation then gives $\overrightarrow{rot}\dot{\vec{p}}_\tau = -\dfrac{\vec{u}}{c}\times\ddot{\vec{p}}_\tau$, where $\ddot{\vec{p}} = \dfrac{\partial^2}{\partial\tau^2}\vec{p}(\tau)$. In order to get this result, one can verify that:

• on the one hand, we have

$$\left[\overrightarrow{rot}\dot{\vec{p}}_\tau\right]_x = \frac{d}{dy}\left[\frac{\partial p_z(\tau)}{\partial\tau}\right] - \frac{d}{dz}\left[\frac{\partial p_y(\tau)}{\partial\tau}\right]$$

$$= \frac{\partial^2 p_z}{\partial\tau^2}\frac{\partial\tau}{\partial y} - \frac{\partial^2 p_y}{\partial\tau^2}\frac{\partial\tau}{\partial z} = \left[\overrightarrow{grad}\tau \times \frac{\partial^2\vec{p}(\tau)}{\partial\tau^2}\right]_x$$

so that by reproducing the calculation in three dimensions, we find:

$$\overrightarrow{rot}\dot{\vec{p}}_\tau = \overrightarrow{grad}\tau \times \frac{\partial^2\vec{p}(\tau)}{\partial\tau^2}; \text{ and}$$

• on the other hand, $\overrightarrow{grad}\tau = -\dfrac{\vec{r}}{cr} = -\dfrac{\vec{u}}{c}$. This result can be obtained by calculating,

for example, that $[\overrightarrow{grad}\tau]_x = \dfrac{\partial\tau}{\partial x} = -\dfrac{1}{c}\dfrac{\partial}{\partial x}(x^2 + y^2 + z^2)^{1/2} = -\dfrac{x}{cr}$, from which we

have:

$$\overrightarrow{grad}\tau = -\frac{1}{cr}[x\vec{i} + y\vec{j} + z\vec{k}] = -\frac{\vec{r}}{cr}.$$

On taking the result $\overrightarrow{\text{rot}}\ddot{\vec{p}}_\tau = -\dfrac{u}{c} \times \ddot{\vec{p}}_\tau$ into Eq. (22), we finally reach:

$$\vec{B} = \frac{\mu_0}{4\pi c} \frac{\ddot{\vec{p}}(\tau) \times \vec{u}}{r} . \qquad (23)$$

In the plane wave approximation, for an electric field, $\vec{E} = c\vec{B} \times \vec{u}$, so that :

$$\vec{E} = \frac{\mu_0}{4\pi} \frac{1}{r}\left[\ddot{\vec{p}}(\tau) \times \vec{u}\right] \times \vec{u} . \qquad (24)$$

To conclude, if $\ddot{\vec{p}}(\tau) = 0$, the Eqs. (23) and (24) show that the radiated magnetic field is zero. By consequence, only accelerated charges radiate, as $\ddot{\vec{p}} = \sum\limits_i q_i \dfrac{d\vec{v}_i}{dt}$ and the wave corresponding to $\ddot{\vec{p}} \neq 0$ is termed the "acceleration wave".

9.3.3. Power radiated by a dipole

In vacuum, the Poynting vector is defined as : $\vec{S} = \dfrac{\vec{E} \times \vec{B}}{\mu_0}$ (Chapter 7, Section 7.3.).

Staying with the planar wave approximation, so that $\vec{E} = c\vec{B} \times \vec{u}$, $\vec{S}$ can be written as :

$$\vec{S} = \frac{cB^2}{\mu_0}\vec{u} .$$

By taking Eq. (23) and substituting it into this equation, for an angle (θ) between $\ddot{\vec{p}}(\tau)$ and $\vec{u}$ as shown in Figure 9.3(a), we find:

$$\vec{S} = \frac{\mu_0}{16\pi^2 c} \frac{\ddot{\vec{p}}^2(\tau)}{r^2} \sin^2\theta \quad \vec{u} . \qquad (25)$$

If $\theta = 0$, then in other words if the radiation intensity is observed in the direction of $\ddot{\vec{p}}$ it would be found to be zero.

However, the radiation maximum is found when $\theta = \dfrac{\pi}{2}$.

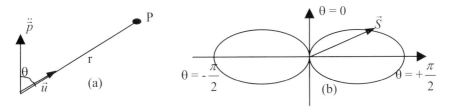

Figure 9.3. *Definition of θ (a) and the variation of S with θ (b).*

The variation in S as a function of θ is represented in Figure 9.3.b, and it can be seen that $|\vec{S}|$ is at a maximum when $\theta = \pm\dfrac{\pi}{2}$, and S = 0 when θ = 0.

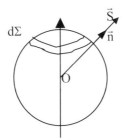

Figure 9.4. *Calculation of the flux ($\vec{S}$) through a sphere (Σ) with center O.*

The power radiated through a surface can be calculated from the flux of $\vec{S}$ through the total surface (see Section 7.3.1.2). For the spherical surface under consideration, the total radiation is given by a calculation of the flux $\vec{S}$ across the sphere (Σ) around a center O of radius r, as detailed in Figure 9.4.

As $\vec{S} \mathbin{/\mkern-5mu/} \vec{n}.d^2\Sigma$), we have $P = \oiint_{\Sigma} S.d^2\Sigma$. With $d^2\Sigma = r^2\sin\theta\ d\theta\ d\varphi$, so that also $d\Sigma = 2\pi\ r^2 \sin\theta\ d\theta$, we find that

$$P = \int_{\theta=0}^{\theta=\pi} S\ 2\pi r^2 \sin\theta\ d\theta = \frac{\mu_0}{8\pi c}\ddot{p}^2(\tau) \int_{\theta=0}^{\theta=\pi} \sin^3\theta\ d\theta.$$

With: $\displaystyle\int_{\theta=0}^{\theta=\pi} \sin^3\theta\ d\theta = -\int_{\cos\theta=1}^{\cos\theta=-1} \sin^2\theta\ d(\cos\theta) = \frac{4}{3}$, we obtain the Larmor equation:

$$P = \frac{\mu_0}{6\pi c}\ddot{p}^2(\tau) = \frac{1}{4\pi\varepsilon_0}\frac{2}{3c^3}\ddot{p}^2(\tau). \qquad (26)$$

9.4. Antennas

9.4.1. Principle: a short antenna where $\ell << \lambda$

9.4.1.1. Oscillating charge (or current) and oscillating dipole equivalency: Hertz's dipole

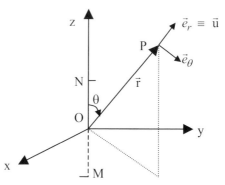

Figure 9.5. *Positions taken up by an oscillating charge.*

A charge (Q_m) is moved from a point (M) toward N, as shown in Figure 9.5. This movement also means that in going from its initial position, through the middle point (O), there is a dipole moment represented by:

$$\vec{p}_m = Q_m\ \overrightarrow{MN} = Q_m\ \vec{s}_m\ (\text{- } Q_m \text{ at M and } Q_m \text{ at N}).$$

If we now impose upon Q_m an oscillating elongation that is harmonic and linear represented by $\vec{s} = \vec{s}_m \cos\omega t = \vec{s}_m \exp(j\omega t)$ (by convention the notation Re is omitted in front of the complex term), then in a manner similar to the generation of a dipole with an instantaneous moment, it can be considered that:

$$\vec{p}(t) = Q_m\ \vec{s} = Q_m\ \vec{s}_m \cos\omega t = Q_m\ \vec{s}_m \exp(j\omega t),\qquad(27)$$

where $\vec{p}(t) = Q_m\ \vec{s}_m \cos\omega t = \vec{p}_m \cos\omega t = \vec{p}_m\ \exp(j\omega t)$ is the sinusoidal electric dipole moment created by charge oscillation. This dipole is called Hertz's dipole, and given Eq. (27) describing its moment, it can also be seen as a two oscillating charges, placed at M and N such that $\overrightarrow{MN} = \vec{s}_m$, and with an instantaneous value given by:

$$|Q| = |Q_m\ \cos \omega t| = |Q_m\ \exp(j\omega t)|.$$

It also is possible therefore to think that between the two charges at M and N there is a alternating current, as shown in Figure 9.6, and of a value given by:

$$I = \frac{dQ}{dt} = j\omega Q_m \exp(j\omega t) = I_m\ \exp(j\omega t),\qquad(28)$$

where $I_m = j\omega\ Q_m$, and with $j = \exp[j\pi/2]$ the dephasing of $\pi/2$ is brought in.

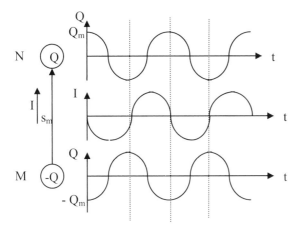

Figure 9.6. *Variations in charges and current with time (t).*

Dividing Eq. (28) by (27), it is possible to form a relation between p(t) and I(t) , as in

$$p(t) = \frac{s_m}{j\omega} I(t) = -\frac{js_m}{\omega} I(t) . \qquad (29)$$

9.4.1.2. Practical elements of a Hertzian dipole antenna

The left-hand side of Figure 9.6 means that in practical terms an antenna can be represented by two wires of length $\dfrac{\ell}{2} = \dfrac{s_m}{2}$ connected by a coaxial cable, as in Figure 9.7, so that the resultant current is always zero. The antenna is traversed by a sinusoidal current given by $I = I_m \exp(j\omega t)$.

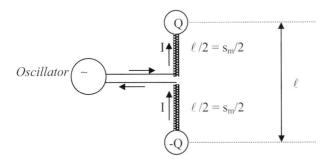

Figure 9.7. *Hertzian dipole.*

9.4.1.3. Radiation field due to a Hertzian dipole

Given that the two charges are separated by a small distance (ℓ), seen from P in the radiation zone, they can be assumed equivalent to a point charge. Evidently, Figure 9.5 is not to scale as in reality MN= ℓ is much greater than OP = r. The two charges are in effect the same distance (r) from P and the wave is pretty much a plane wave at P. These assumptions were used in Section 9.3 to perform the calculations, leading in particular to an expression for $\vec{E}$ in Eq. (24).

From Figures 9.5 and 9.7, $\vec{p}(t) = Q_m \, \vec{s}_m \exp(j\omega t) = Q_m \, \ell \exp(j\omega t) \, \vec{e}_z$, so we have

$$\ddot{\vec{p}}(\tau) = -\omega^2 \vec{p}(\tau) = -\omega^2 p(\tau) \, \vec{e}_z . \qquad (30)$$

In terms of spherical coordinates, and with $\vec{u} \equiv \vec{e}_r = \dfrac{\overrightarrow{OP}}{OP} = \dfrac{\vec{r}}{r}$ (from Figure 9.5), the double vectorial product given in Eq. (24) is such that:

$$[\vec{e}_z \times \vec{e}_r] \times \vec{e}_r = \left\{ \begin{pmatrix} \cos\theta \\ -\sin\theta \\ 0 \end{pmatrix} \times \begin{pmatrix} 1 \\ 0 \\ 0 \end{pmatrix} \right\} \times \begin{pmatrix} 1 \\ 0 \\ 0 \end{pmatrix} = \begin{pmatrix} 0 \\ 0 \\ \sin\theta \end{pmatrix} \times \begin{pmatrix} 1 \\ 0 \\ 0 \end{pmatrix} = \begin{pmatrix} 0 \\ \sin\theta \\ 0 \end{pmatrix} = \sin\theta \, \vec{e}_\theta$$

From this can be deduced that:

$$\vec{E} = \frac{\mu_0}{4\pi} \frac{1}{r} \left[\ddot{\vec{p}}(\tau) \times \vec{u} \right] \times \vec{u} = -\frac{\mu_0 \omega^2}{4\pi r} p(\tau) \sin\theta \, \vec{e}_\theta , \qquad (31)$$

so that in addition, with $\omega^2 = k^2 c^2 = \dfrac{k^2}{\varepsilon_0 \mu_0}$, and $k = \dfrac{2\pi}{\lambda}$, we find:

$$\vec{E} = -\frac{k^2}{4\pi\varepsilon_0 r} p(\tau) \sin\theta \, \vec{e}_\theta = -\frac{\pi}{\varepsilon_0 \lambda^2 r} p(\tau) \sin\theta \, \vec{e}_\theta . \qquad (32)$$

It is possible to see that a field radiated by a Hertzian dipole has only one component with respect to $\vec{e}_\theta$, is inversely proportional to r, but also depends on r and t through the term $p(\tau)$.

9.4.1.4. Radiation power of a Hertzian dipole when $\ell = s_m \ll \lambda$

In supposing that $\ell = s_m \ll \lambda$, for any given moment the intensity (I) can be assumed constant over the entire length (ℓ) of the dipole, which is itself considered in practical terms much as a single point from the external position (P) situated outside the radiation zone, so that r >> λ. Larmor's formula given in Eq. (26), obtained by removing $\ddot{p}$ from the integral, therefore can be applied. Thus we find that with $Q = Q_m \cos\omega t$, and that $I = \dfrac{dQ}{dt} = -\omega Q_m \sin\omega t = I_m \sin\omega t$, then

$p = Q_m \ell \cos\omega t$, and $\ddot{p} = -Q_m \ell\omega^2 \cos\omega t$, from which $\ddot{p}^2 = Q_m^2 \ell^2 \omega^4 \cos^2\omega t$ and we now have:

$$\langle P \rangle = \frac{\mu_0}{6\pi c} \ddot{p}^2(\tau) = \frac{\mu_0 Q_m^2 \ell^2 \omega^4}{6\pi c} \left\langle \cos^2\omega t \right\rangle.$$

By taking $I^2_m = \omega^2 Q^2_m$ and given that $<\cos^2\omega t> = \frac{1}{2}$, we finally have:

$$\langle P \rangle = \frac{\mu_0 \ell^2 \omega^2}{12\pi c} I^2_m = \frac{1}{2} R_{ray}\, I^2_m. \qquad (33)$$

This relationship gives a definition of R_{ray}, the so-called radiation resistance, which has resistance as a dimension and is directly obtained from:

$$R_{ray} = \frac{\mu_0 \omega^2}{6\pi c} \ell^2 = \frac{2\pi \mu_0 c}{3} \left(\frac{\ell}{\lambda} \right)^2. \qquad (34)$$

With $\mu_0 c = \mu_0 \dfrac{1}{\sqrt{\varepsilon_0 \mu_0}} = \sqrt{\dfrac{\mu_0}{\varepsilon_0}} = Z_0 = 378\ \Omega$, we can also write that:

$$R_{ray} = \frac{2\pi}{3} Z_0 \left(\frac{\ell}{\lambda} \right)^2 \approx 780 \left(\frac{\ell}{\lambda} \right)^2. \qquad (35)$$

9.4.2. General remarks on various antennae: half-wave and "whip" antennae

Antennae can be seen everywhere in modern day life. They can be found in portable telephones, televisions, cars, and automatic doors, just to give a few examples. According to Eq. (35), short antennae radiate poorly, and it is for this reason that antennae with sizes of the order of the wavelength are preferred. The most basic versions are still made up of a simple dipole, as shown in Figure 9.8a.

When the antenna is emitting, it is powered by an oscillator and parallel wire cables as detailed above (Section 9.4.1.2). If the antenna has a length λ/2 and establishes a stationary current so that there are nodes at the two ends and an anti-node at the origin in the middle ($z = \pm\dfrac{\lambda}{4}$), then the current has the form:

$$I(t) = I_m \cos kz \sin \omega t.$$

Each part along z with a length dz thus radiates as if it were a small dipole. This concept is used in a problem at the end of the chapter.

When the antenna is in the receptive mode, the electric field around it induces a current (see more precisely Section 12.3.3.3), which propagates through the cable up to the receiver.

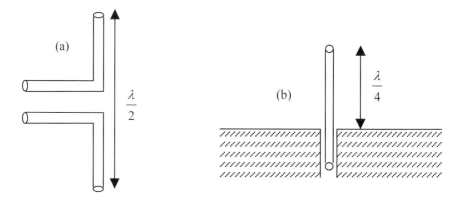

Figure 9.8. (a) Half-wave antenna; and (b) quarter wave or "whip" antenna.

One of the more common types of antennae are the "whip" antennae often seen on cars. They are electrical monopoles, as shown in Figure 7.8a, and form an electrical image in the plane of the conductor so that a positive charge that moves along the antenna toward the top produces a "reflected" image negative charge that moves symmetrically toward the bottom. Typically, the length of an antenna is λ/4, but its actual length needs to be determined quite precisely before determining its exact radiation diagram. When its length is equal to λ/4, the quarter-wave length antenna acts in effect as a half-wave antenna as it can generate its symmetrically opposite image in the plane of the conductor.

There are many other types of antennae. To cite just a few:
- "disk" antennae formed from microstrips that are applied for example onto airplane wings;
- frame antennae, also called a magnetic dipole antenna, based on a ferrite iron core, which concentrates the magnetic field;
- horn antennae (used with wave guides); and
- parabolic antennae that concentrate waves at a focal point where a horn antenna is placed.

9.5. Problem
Radiation from a half-wave antenna

For a half-wave antenna, that is with a length (ℓ) such that $\ell = \lambda/2$, and ignoring any energy losses such as the radiation, it can be assumed that it carries a current (I) which can be written as:

$$I = I_m \cos kz \exp j\omega t = I_m \cos\frac{2\pi}{\lambda} z \exp j\omega t .$$

Assuming that the radiation is isotropic, it is independent of the angular coordinate (φ). This problem concerns the radiation zone where the radius or distance from the antenna (r) is such that $r \gg \lambda$.

1. Determine the expression for the electric field at a point (P), which is such that $P \equiv P(r,\theta)$, i.e., has spherical coordinates (r,θ) and $\varphi = 0$ in the plane of the paper.

2. Assuming that when $r \gg \lambda$, the vectors $\bar{E}$ and $\bar{H}$ are orthogonal as in a plane wave and are related in a vacuum by $\dfrac{E}{B} = c$. Given that $\dfrac{E}{H} = \sqrt{\dfrac{\mu_0}{\varepsilon_0}} = Z_0 \approx 378\ \Omega$,

calculate $\vec{H}$.

3. Calculate the average Poynting vector.

Answers
1.

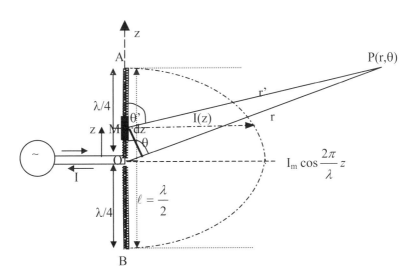

It can be assumed that $\theta' \approx \theta$ at P in the radiation zone where $r \gg \lambda \approx \ell$. The actual dipole can be considered to be a resultant of many small dipoles of a length dz centered on a point M such that $MP = r' = r - z \cos \theta$, as schematized in the diagram above.

An elementary Hertzian dipole with a moment denoted by dp radiates, as detailed in Section 9.4.1.3, an elementary electric field $d\vec{E}$ that follows the form:

$$d\vec{E} = -\frac{\mu_0 \omega^2}{4\pi\, r'} dp(\tau)\, \sin\theta\, \vec{e}_\theta \,, \text{ where } r' = MP, \text{ [see Eq. (31)].}$$

From Eq. (29), we can also write that:

$$d\vec{E} = \frac{\mu_0\, j\omega\, I(\tau)}{4\pi\, r'} dz\, \sin\theta\, \vec{e}_\theta \,, \text{ where } s_m = dz, \text{ so that in this problem,}$$

$$I(\tau) = I\left(t - \frac{r'}{c}\right) = I_m \cos kz \exp\left(j\omega\left[t - \frac{r'}{c}\right]\right).$$

As $r \gg \lambda$, and $r \gg \ell$, the denominator r' can be replaced by r. This cannot be said to be true for the numerator as the exponential phase varies rapidly with r'. So, integrating over the length of the antenna and using $r' = r - z\cos\theta$ to clean up, we end up with:

$$\vec{E} = \frac{\mu_0}{4\pi r} j\omega I_m \exp\left(j\omega\left[t - \frac{r}{c}\right]\right)\sin\theta \int_{-\lambda/4}^{+\lambda/4} \cos kz \exp\frac{j\omega z\cos\theta}{c} dz\, \vec{e}_\theta \,.$$

By using $\cos kz = \dfrac{e^{ikz} + e^{-ikz}}{2}$ and $\dfrac{\omega}{c} = \dfrac{2\pi}{\lambda} = k$, we obtain:

$$\vec{E} = \frac{\mu_0\, j\omega I_m}{8\pi r} \exp\left(j\omega\left[t - \frac{r}{c}\right]\right)\sin\theta$$

$$\times \int_{-\lambda/4}^{+\lambda/4} \left\{\exp\left(jkz[\cos\theta + 1]\right) + \exp\left(jkz[\cos\theta - 1]\right)\right\} dz\, \vec{e}_\theta$$

The integral gives:

$$\int_{-\lambda/4}^{+\lambda/4} \left\{\exp\left(jkz[\cos\theta + 1]\right) + \exp\left(jkz[\cos\theta - 1]\right)\right\} dz$$

$$= \frac{1}{jk}\left\{ \frac{e^{j\frac{2\pi}{\lambda}\frac{\lambda}{4}[\cos\theta+1]}}{\cos\theta + 1} - \frac{e^{-j\frac{2\pi}{\lambda}\frac{\lambda}{4}[\cos\theta+1]}}{\cos\theta + 1} + \frac{e^{j\frac{2\pi}{\lambda}\frac{\lambda}{4}[\cos\theta-1]}}{\cos\theta - 1} - \frac{e^{-j\frac{2\pi}{\lambda}\frac{\lambda}{4}[\cos\theta-1]}}{\cos\theta - 1} \right\}$$

$$= \frac{1}{jk}\left\{ \frac{2j\sin\left\{\frac{\pi}{2}[\cos\theta + 1]\right\}}{\cos\theta + 1} + \frac{2j\sin\left\{\frac{\pi}{2}[\cos\theta - 1]\right\}}{\cos\theta - 1} \right\}$$

With $\dfrac{\mu_0 j\omega I_m}{8\pi r}\dfrac{2}{k} = \dfrac{\mu_0 j\omega I_m}{4\pi r}\dfrac{c}{\omega} = \dfrac{j I_m}{4\pi\varepsilon_0 c r}$, we obtain:

$$\vec{E} = \frac{j I_m}{4\pi\varepsilon_0 c r}\sin\theta\exp(j\omega\tau)\left\{\frac{\sin\left\{\dfrac{\pi}{2}[\cos\theta+1]\right\}}{\cos\theta+1} + \frac{\sin\left\{\dfrac{\pi}{2}[\cos\theta-1]\right\}}{\cos\theta-1}\right\}\vec{e}_\theta$$

As:

$$\sin\left\{\frac{\pi}{2}[\cos\theta+1]\right\} = \cos\left(\frac{\pi}{2}\cos\theta\right)$$

$$\sin\left\{\frac{\pi}{2}[\cos\theta-1]\right\} = -\cos\left(\frac{\pi}{2}\cos\theta\right)$$

we have $\vec{E} = \dfrac{j I_m}{4\pi\varepsilon_0 c r}\sin\theta\exp(j\omega\tau)\left\{\dfrac{2}{\sin^2\theta}\cos\left(\dfrac{\pi}{2}\cos\theta\right)\right\}\vec{e}_\theta$ so that finally:

$$\vec{E} = \frac{j}{2\pi\varepsilon_0 c r}\frac{\cos\left(\dfrac{\pi}{2}\cos\theta\right)}{\sin\theta}I_m\exp(j\omega\tau)\vec{e}_\theta\,.$$

It is worth noting that this expression is indeterminate when $\sin\theta = 0$, i.e., when $\theta = 0$ or $\theta = \pi$.

2. Assuming that at great distance the wave still assumes a plane wave structure, the vector $\vec{H}$ will be orthogonal to $\vec{E}$. So with the trihedral $\vec{E}, \vec{H}, \vec{e}_r$ where $\vec{e}_r$ indicates the direction and sense of propagation, $\vec{H}$ is collinear with $\vec{e}_\varphi$.

Therefore with $H = E\sqrt{\varepsilon_0/\mu_0}$ we obtain

$$\vec{H} = \frac{j}{2\pi r}\frac{\cos\left(\dfrac{\pi}{2}\cos\theta\right)}{\sin\theta}I_m\exp(j\omega\tau)\vec{e}_\varphi\,.$$

3. By applying Eq. (33) of Section 7.3.3, we can write that

$$\langle \vec{S} \rangle = \frac{1}{2} \operatorname{Re}\left(\vec{E} \times \vec{H}^{*}\right) \text{ where } B = \mu_0 H,$$

from which can be obtained

$$\langle \vec{S} \rangle = \frac{1}{\left(2\pi r\right)^{2}\, \varepsilon_0 c} \frac{\cos^{2}\left(\dfrac{\pi}{2}\cos\theta\right)}{\sin^{2}\theta}\, \frac{I_{m}^{2}}{2}\, \left(\vec{e}_\theta \times \vec{e}_\varphi\right)$$

$$= \frac{1}{4\pi\varepsilon_0} \frac{1}{\pi c} \frac{\cos^{2}\left(\dfrac{\pi}{2}\cos\theta\right)}{\sin^{2}\theta}\, \frac{I_{eff}^{2}}{r^{2}}\, \vec{e}_{r}$$

$$\approx 9.5\, \frac{\cos^{2}\left(\dfrac{\pi}{2}\cos\theta\right)}{\sin^{2}\theta}\, \frac{I_{eff}^{2}}{r^{2}}\, \vec{e}_{r} \quad \text{(S.I.), that is in W/m}^2.$$

Integration over the sphere of radius r gives the total radiated power, much as the same method used to obtain the Larmor relation given in Section 9.3.3. The result that can be obtained from the ensuing numerical calculation is $P \approx 73\, I_{eff}^{2}$ and gives a final radiation resistance for the half-wave antenna of the order of 73 Ω.

Chapter 10

Interactions between Materials and Electromagnetic Waves, and Diffusion and Absorption Processes

10.1. Introduction

Having looked at the way in which electromagnetic (EM) waves are emitted by a particle accelerating in a vacuum in Chapter 9, this chapter is concerned with how EM waves interact with materials, and in particular:

- the form of wave emitted by a material subject to an incident EM wave (Rayleigh diffusion);
- how EM waves are formed by charged particles interacting with a material as the so-called Rutherford or Bremsstrahlung (German for "braking radiation") radiation; and
- the absorption and emission phenomena of EM waves interacting with excited (more exactly "perturbed" as detailed below by the intervention of Hamiltonian perturbation) materials. The description used will be semiclassical, or possibly semiquantic, depending on whether the glass is half full or half empty with regards to a knowledge of quantum mechanics! That is to say that EM radiation can be written in a classic form while the material can be understood in quantic terms generally known by readers. The quantic terms used however, generally are for students following their first university degree course, and although Section 10.4 details in the simplest terms this semiquantic theory, it can be skipped over during an initial read. The essential results have been set out in Section 10.5 and these can be compared in a second read with the results given in chapter 3 of the volume, "Applied Electromagnetism and Materials" (notably the last Figure of that chapter).

10 .2. Diffusion Mechanisms

10.2.1. Rayleigh diffusion: radiation diffused by charged particles

An electron with a charge denoted by q is subject to a monochromatic planar polarized electromagnetic (MPPEM) wave which is polarized along Ox (complex amplitude of the incident wave given by $\underline{\vec{E}}_{inc}//\overrightarrow{Ox}$) and propagates along Oz, as shown in Figure 10.1. The electric wave can be described by $\vec{E} = \underline{\vec{E}}_{inc}\exp(j\omega t)$.

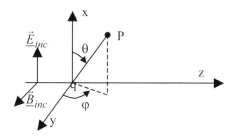

Figure 10.1. *MPPEM wave incident on an electron with charge denoted by q.*

If the electron is within a dense material, following its displacement along Ox by the electric field of the incident wave, it can be assumed that it will be subject to a returning force in the direction of its equilibrium position and in the form $f_r = - k\,x$ while simultaneously being subject to frictional forces of the form $f_t = - f\,\dfrac{dx}{dt}$.

Rigorously speaking, the interactive force between the EM wave and the velocity ($\vec{v}$) of the electron is given by $\vec{F}_{em} = q(\vec{E} + \vec{v} \times \vec{B})$

With $B = \dfrac{E}{c}$, we have $vB = v\dfrac{E}{c} << E$ assuming that $v << c$, so that the electromagnetic force can be reduced simply to Coulomb's force, as in :

$$\vec{F}_{em} \approx \vec{F}_C = q\vec{E}.$$

The fundamental dynamic equation, $\sum\vec{F} = m\vec{\gamma}$, given in terms of Ox thus gives:

$$q\underline{E}_{inc}e^{j\omega t} - k\,x - f\frac{dx}{dt} = m\,\frac{d^2x}{dt^2}$$

The solution for a steady state can be looked for in the form $x = \underline{x}_0\,e^{j\omega t}$, and the placing of this into the preceding fundamental equation gives:

$$x = \frac{q\underline{E}_{inc}}{k - m\omega^2 + j\omega f}e^{j\omega t} = \underline{x}_0\,e^{j\omega t}.$$

The result is that q displaced by x generates a dipole moment:

$$p(t) = q\, x(t) = q\, \underline{x}_0\, e^{j\omega t}$$

and radiates according to the result of Section 9.3.2 concerning an EM field such that $\vec{E}$ and $\vec{B}$ are expressed as a function of $\ddot{\vec{p}}(\tau) = q\, \ddot{\vec{x}}(\tau) = q\, \vec{\gamma}$, so that

$$\ddot{p}(\tau) = -\omega^2 q\, x = -\omega^2 q\, \underline{x}_0\, e^{j\omega\tau}.$$

The power radiated and given by Eq. (26) of Section 1.3.3 is proportional to $\ddot{p}^2(\tau)$, and thus $\ddot{p}(\tau) = -\omega^2 q\, \underline{x}_0\, e^{j\omega\tau}$.

Two simple examples can now be considered.

10.2.1.1. Diffusion by bound electrons (valence electrons of the atmospheric molecules O_2 and N_2)

The electrons are well bound to the molecules, and at the level of the forces involved, the constants introduced are such that $k \gg m\omega^2$ and $k \gg \omega f$. The result is that the dynamic fundamental equation, as in

$$x = \frac{q\underline{E}_{inc}}{k - m\omega^2 + j\omega f}\, e^{j\omega t},$$

is reduced to $x \approx \dfrac{q\underline{E}_{inc}}{k}\, e^{j\omega t}$, so that $\underline{x}_0 = \dfrac{q\underline{E}_{inc}}{k}$, where $\underline{x}_0$ thus appears independent of ω. With $\ddot{p}(\tau) = -\omega^2 q\, \underline{x}_0\, e^{j\omega\tau}$, we have:

$$\ddot{p}^2(\tau) \propto \omega^4.$$

The radiated power is proportional to $\ddot{p}^2(\tau)$ and is in the form $P \propto \omega^4$, that is the power radiates to the fourth power of the angular frequency (ω).

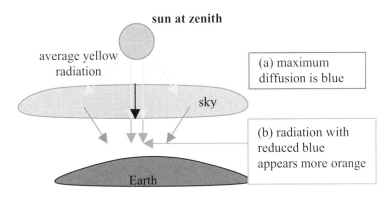

Figure 10.2. *Observations made when the sun is at its zenith.*

Given that it is radiation with a high angular frequency that dominates:

- When we look at the sky, but not at the sun, the sky appears blue as shown in Figure 10.2 a. The power radiated, or rather diffused, by electrons in the molecules that make up the atmosphere is at a maximum for the highest angular frequencies, i.e., those that correspond to the color blue in the incident optical radiation from the sun.
- When we look at the sun when at its zenith (an observation worth avoiding due to possible eye damage!) as shown in Figure 10.2 b, the sun appears on earth more orange than the yellow observed by a satellite outside of earth's atmosphere. This is because the transmitted light is equal to the incident light minus light lost to diffusion during its passage through the atmosphere. In effect, the light we see has an "impoverished" blue region, especially if the sky is cloudy, resulting in an apparent color shift toward red.
- When we see the setting sun, as shown in Figure 10.3, the observed waves have had to travel through a long distance in the atmosphere, much longer than that when the sun is at its zenith, with the result that there is an greater impoverishment of blue light diffused all along the light's pathway. This makes the light look red as all other light has been diffused.

setting sun

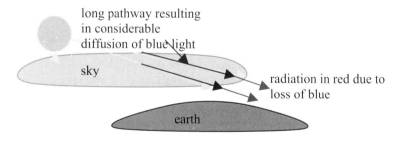

Figure 10.3. *Observation of the setting sun.*

10.2.1.2. Diffusion by free electrons

This section looks at the study of free electrons using, for example, Laue type diffusions in crystalline solids.

Free electrons, which are little or even not at all bound to atoms, are not subject to returning forces toward a given atomic nucleus. Thus, $k = 0$ and the frictional forces are also negligible, so that $f = 0$.

The solution to the differential equation, $x = \dfrac{qE_{inc}}{k - m\omega^2 + j\omega f} e^{j\omega t}$ thus is reduced to $x = -\dfrac{q\underline{E}_{inc}}{m\omega^2} e^{j\omega t}$; $\ddot{\underline{p}}(\tau) = -\omega^2 q\,\underline{x_0}\,e^{j\omega\tau}$ then gives

$$\ddot{\underline{p}}(\tau) = \dfrac{q^2}{m}\underline{E}_{inc}e^{j\omega\tau}\;.$$

In addition, according to Eq. (24) of Section 9.3.2, we have:

$$\vec{E} = \dfrac{\mu_0}{4\pi}\dfrac{1}{r}\Big[\ddot{\vec{p}}(\tau)\times\vec{u}\Big]\times\vec{u}$$

and as in Figure 10.4, we have $\ddot{\vec{p}}$ // $\vec{e}_x$.

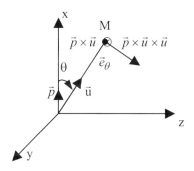

Figure 10.4. *Orientation of the vectors used in the text.*

For a point (M) located by the direction of a vector ($\vec{u}$), it can be written that $\vec{E} = E\vec{e}_\theta$, so that in terms of moduli, $E = E_\theta$.

As $\Big|\big[\ddot{\vec{p}}(\tau)\times\vec{u}\big]\times\vec{u}\Big| = \Big|\big[\ddot{\vec{p}}(\tau)\times\vec{u}\big]\Big|\sin\dfrac{\pi}{2} = \Big|\big[\ddot{\vec{p}}(\tau)\times\vec{u}\big]\Big| = \ddot{p}\sin\theta$

we have $E_\theta = \dfrac{\mu_0}{4\pi r}\ddot{p}\sin\theta = \dfrac{\mu_0}{4\pi r}\dfrac{q^2}{m}\underline{E}_{inc}\sin\theta\;\;e^{j\omega\tau}$.

With $\tau = t - \dfrac{r}{c}$, and $\underline{F} = m\,\underline{\gamma} = q\,\underline{E}_{inc}$, so that $\underline{\gamma} = \dfrac{q\underline{E}_{inc}}{m}$, finally we have

$$\underline{E}_\theta = \dfrac{q\underline{\gamma}}{4\pi\varepsilon_0 c^2}\dfrac{\sin\theta}{r}\,e^{-j\omega\frac{r}{c}}e^{j\omega t} = \underline{E}_{00}\,e^{j\omega t}\;.$$

The wave appears to be well proportional to the acceleration denoted by γ; hence the term "accelerating wave".

10.2.2. Radiation due to Rutherford diffusion

10.2.2.1. A general description of Rutherford diffusion

Here the incident particle is an electron denoted $-e$ with a mass given by m that is a great distance from a proton that is assumed to be fixed and at a reference origin (O). The trajectory of the electron is considered rectilinear with a velocity v_∞. If there are no forces acting on the incident electron, then it will have a trajectory given by the straight line (D), as shown in Figure 10.5, which passes at a minimum distance (b) from the target (b is called the impact parameter).

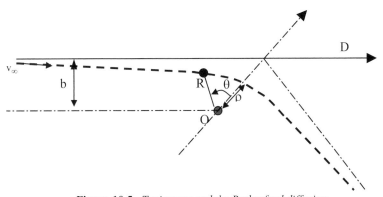

Figure 10.5. *Trajectory and the Rutherford diffusion.*

In reality, when an electron nears a proton, the Coulombic interactive force is no longer negligible. The electron is subject to a Coulombic force ($\vec{f}$) that is directed along the line between the proton and electron distance (R). If $\vec{\sigma}$ denotes the electron's kinetic moment, the theory of kinetic moment is written (with $\vec{f}$ and $\vec{R}$ being collinear) as

$$\frac{d\vec{\sigma}}{dt} = \vec{R} \times \vec{f} = 0.$$

The result is that $\vec{\sigma} = \vec{R} \times m\vec{v}$ is a first integral of the movement where:
- the direction of $\vec{\sigma}$ is fixed and the movement is thus plane; and
- the modulus (σ) of the vector $\vec{\sigma}$ is a constant.

With the electron being located in the plane of its movement by the polar angle θ, it is possible to state that:

$$\sigma = mRv = mR^2 \frac{d\theta}{dt} \, ,$$

from which the surface law can be deduced, as in:

$$R^2 \frac{d\theta}{dt} = 2\frac{dS}{dt} = C = \frac{\sigma}{\mu} \, .$$

By denoting the potential energy of an electron by $E_p(R)$ and its total energy by E, the application of the law of energy conservation results in the equation:

$$\frac{d\theta}{dR} = \pm \frac{\sigma}{mr^2 \sqrt{\frac{2}{m}\left[E - E_p(R)\right] - \frac{\sigma^2}{m^2 R^2}}} \Leftrightarrow \frac{dR}{d\theta} = \pm \frac{mR^2 \sqrt{\frac{2}{m}\left[E - E_p(R)\right] - \frac{\sigma^2}{m^2 R^2}}}{\sigma}$$

The derivative $\dfrac{dR}{d\theta}$ changes sign and R abruptly changes from a decreasing to an increasing region, or the inverse, when the numerator of the last equation cancels out so that for a value ρ for R is such that:

$$E_p(\rho) = E - \frac{\sigma^2}{2m\rho^2} \, .$$

In addition, by taking the origin of the potential energies at infinity, it also is possible to write for the two constants for the movement that the kinetic moment (σ) and the energy (E) that

$$\sigma = mbv_\infty \quad \text{and} \quad E = \frac{1}{2}mv_\infty^2 \, ,$$

from which can be deduced for $E_p(\rho)$ that

$$E_p(\rho) = \frac{mv_\infty^2}{2}\left(1 - \frac{b^2}{\rho^2}\right) \, .$$

Here the interaction potential $[E_p(\rho)]$ is negative and the equation shows that $\rho < b$.

It is equally feasible to introduce a parameter (δ) that corresponds to a distance between the two particles at which the Coulombic energy of the electron, given by $\dfrac{e^2}{4\pi\varepsilon_0 R}$, is equal to its mass energy as in mc^2, so that $\delta = \dfrac{e^2}{4\pi\varepsilon_0 mc}$.

10.2.2.2. Rutherford diffusion and radiation

During the deviation caused by electrostatic interactions, the electron – proton pair constitutes a dipole that has a dipole moment given by $\vec{p} = e(-\vec{R})$, which varies continuously, as $\vec{R}$ varies just as well in terms of direction as in terms of modulus. The sign changes [in $(-\vec{R})$] due to the fact that $\vec{R}$ is orientated with respect to the origin (the proton) toward the electron, while the dipole moment goes from the negative (the electron) to the positive (the proton) charge.

The consequence of this is that the dipole radiates, which is assumed, to a first approximation, to not modify the movement. It thus is possible to calculate the electromagnetic field radiated at a point (P), which is in the radiation zone. Also, OP $\gg \lambda$, where the origin of the dipole also is at O (by supposing that R/2 $\ll$ OP = r).

The field ($\vec{B}$) thus is given simply by Eq. (23) in Chapter 9, as in

$$\vec{B} = \frac{\mu_0}{4\pi c} \frac{\ddot{\vec{p}}(\tau) \times \vec{u}}{r},$$

into which, here, $\vec{p} = e(-\vec{R})$, so that $\ddot{\vec{p}} = e(-\ddot{\vec{R}})$.

The field ($\vec{E}$) is given, for its part, by Eq. (24) in Chapter 9, by supposing that at a great distance from the radiation zone the wave is plane and is such that:

$$\vec{E} = c\vec{B} \times \vec{u}, \text{ where } \vec{u} = \frac{\overrightarrow{OP}}{r}.$$

Across a sphere with a given radius (r) the radiation power is given by Eq. (26) in Chapter 9, i.e.,

$$P(r,t) = \frac{1}{4\pi\varepsilon_0} \frac{2}{3c^3} \ddot{p}^2 (t - \frac{r}{c}).$$

With $\ddot{\vec{R}}$ given by the fundamental dynamic relationship:

$$-m\ddot{\vec{R}} = -\frac{e^2}{4\pi\varepsilon_0 R^2} \vec{e}_R, \text{ we have } \ddot{\vec{p}} = e\left(-\ddot{\vec{R}}\right) = -\frac{e^3}{4\pi\varepsilon_0 mR^2} \vec{e}_R,$$

from which, it can deduced that:

$$P(r,t) = \frac{1}{4\pi\varepsilon_0} \frac{2}{3c^3} \ddot{p}^2 (t - \frac{r}{c}) = \frac{e^6}{\left(4\pi\varepsilon_0 c\right)^3 m^3} \frac{2m}{3R^4}$$

so that with $R = R(t - \frac{r}{c})$, we have

$$P(r,t) = \frac{2}{3} \frac{mc^3 \delta^3}{R^4}.$$

To conclude, the electron and the associated dipole radiate essentially when the electron passes with the neighborhood of the nucleus, at which point $\vec{R}$ and thus $\vec{p}$ vary the most both in terms of modulus and of direction. The expression for the radiated power shows that as the electron distances itself from the proton (R increases), the power of the radiation decreases very rapidly, as it is proportional to R^{-4}.

10.3. Radiation Produced by Accelerating Charges: Synchrotron Radiation and Bremsstrahlung

10.3.1. Synchrotron radiation

Spectroscopy is an essential method in "divining" materials when the most precise structural representations are required. Sensitivity in these methods therefore is of primary importance and can be addressed properly by increasing the signal to noise ratio. This can be done by increasing the power of the radiation source, which also permits a reduction in sampling time. Synchrotron radiation displays exactly this advantage. It was first observed at the end of the 1940s and exhibited the additional benefit of being easily controlled. A considerable number of synchrotrons since have been constructed, and they remain in heavy demand.

An electron synchrotron is made up of a large vacuum chamber, in the form of an annular, into which are pulsed fast electrons that have undergone an acceleration in an electric field. Magnetic fields, resulting from magnets placed around the ring, oblige electrons to follow the ring's curve.

At speeds close to light, the accelerating electrons emit, at a tangent to their trajectory, a fine beam of radiation. The modification of the acceleration field permits a wide variation in this so-called synchrotron radiation. This controllability and sensitivity of the source has allowed a large number of experimental difficulties to be overcome. Experimental techniques such as extended X-ray absorption fine structure (EXAFS) spectroscopy were developed using synchrotron sources.

10.3.2. Bremsstrahlung: electromagnetic stopping radiation

When electrons or ions penetrate a material, their movement is stopped by their interaction with the constituent electrons or ions of the target material. The depth of penetration is governed by electronic and nuclear-stopping strengths. The deceleration of the particles results in an emission of radiation called Bremsstrahlung radiation, given in German to mean "braking radiation" and signifying stopping radiation.

The loss in kinetic energy occurs progressively and is fractionated by the order of successive interactions. The resulting electromagnetic radiation is polychromatic. With total energy being conserved, the highest frequency radiation corresponds to a particle that has lost its energy in a single and unique interaction. An application of this type of radiation is the production of polychromatic X-rays via X-ray tubes.

10. 4. Process of Absorption or Emission of Electromagnetic Radiation by Atoms or Molecules (to Approach as Part of a Second Reading)

10.4.1. The problem

The study here considers the interaction of an atom or a molecule, in terms of a quantic description that takes on board the discreet energy levels associated with the positions of electrons in their orbitals, with an incident wave (here a luminous wave) assumed to be periodic and described in classical terms by a periodic vector potential.

By using a Coulomb gauge (see comment 4 from Section 9.2.1.3), the vector potential accords with the gauge condition, that is $\mathrm{div}\vec{A}_r = 0$, and the associated scalar potential ($U(r,t)$) can be taken as zero, as in $U(r,t) = 0$.

In order to deal with the problem, we will consider the effect of an incident wave as a perturbation to the state of an atom or molecule. This perturbation can be characterized by a Hamiltonian that must first be evaluated to then show that it typically can be reduced to a electric dipole Hamiltonian. Finally, by applying the theory of time-dependent perturbations, we will be brought to studying the electric dipole transitions generated inside an atom that interact with the incident radiation.

10.4.2. Form of the interaction Hamiltonian

10.4.2.1. Hamiltonian in analytical mechanics: form for a particle interacting with a wave

In analytical mechanics (see also, for example, "Elements of analytical mechanics" by J. W. Leech, Dunod 1961), it is simple enough to show that the Lagrangian associated with the movement of a particle, with a charge denoted by q and a mass denoted by m placed in an electromagnetic field derived from a scalar potential (V) where V is the only potential (V_{Coul}) generated by the atom, and the vector potential ($\vec{A}$) associated with the incident wave has the form:

$$L = \frac{1}{2}mv^2 + q\left[\vec{v}.\vec{A} - V\right]. \quad (1)$$

The limiting conditions of the integral $I = \int L\, dt$ give the movement equation obtained from Eq. (2):

$$\vec{F} = q\left[\vec{E} + \vec{v}\times\vec{B}\right] = m\vec{\gamma}. \quad (2)$$

This result can be obtained by using the fact that there is an equivalence between $\partial I = 0$ with Euler's equation:

$$\frac{d}{dt}\left(\frac{\partial L}{\partial \dot{q}_i}\right) - \frac{\partial L}{\partial q_i} = 0, \quad (3)$$

which gives rise to, for example, for the variable $q_i = x$ [Eq. (4) identical to Eq. (2) depending on values of x]:

$$m\ddot{x} = -q\left[\frac{\partial A_x}{\partial t} + \frac{\partial V}{\partial x}\right] + q\left[\frac{dy}{dt}\left(\frac{\partial A_y}{\partial x} - \frac{\partial A_x}{\partial y}\right) - \frac{dz}{dt}\left(\frac{\partial A_x}{\partial z} - \frac{\partial A_z}{\partial x}\right)\right]. \qquad (4)$$

With $p_i = \dfrac{\partial L}{\partial \dot{q}_i}$ (conjugate moment), we obtain with L given by Eq. (1) and in a Cartesian frame $p_x = m\dot{x} + qA_x$, so that in terms of vectors

$$\vec{p} = m\vec{v} + q\vec{A} . \qquad (5)$$

Calculating then the Hamiltonian given by:

$$H = \sum_i p_i \dot{q}_i - L \qquad (6)$$

where L is given by Eq. (1), an expression in which according to Eq. (5):

$$\frac{1}{2}m\vec{v}^2 = \frac{1}{2m}\left(m\vec{v}\right)^2 = \frac{1}{2m}\left(\vec{p} - q\vec{A}\right)^2 .$$

From this can be deduced that $\vec{v} = \dfrac{1}{m}\left(\vec{p} - q\vec{A}\right)$, so that Eq. (1) can be written:

$$L = \frac{1}{2m}\left(\vec{p} - q\vec{A}\right)^2 + \frac{q}{m}\left(\vec{p} - q\vec{A}\right)\vec{A} - q\,V_{Coul}$$

$$= \frac{\vec{p} - q\vec{A}}{2m}\left(\vec{p} - q\vec{A} + 2q\vec{A}\right) - q\,V_{Coul}$$

so that finally:

$$L = \frac{\left(\vec{p} - q\vec{A}\right)\left(\vec{p} + q\vec{A}\right)}{2m} - q\,V_{Coul} = \frac{\vec{p}^2 - q^2\vec{A}^2}{2m} - q\,V_{Coul} . \qquad (7)$$

In additional terms, from Eq. (5) we have:

$$p_i = m\frac{dq_i}{dt} + qA_i \quad \Rightarrow \quad \frac{dq_i}{dt} = \dot{q}_i = \frac{1}{m}\left(p_i - qA_i\right),$$

which moved with Eq. (7) into Eq. (6) gives:

$$H = \frac{1}{m}\vec{p}\left(\vec{p} - q\vec{A}\right) - L$$

$$= \frac{1}{2m}\left(\vec{p} - q\vec{A}\right)\left(2\vec{p} - \vec{p} - q\vec{A}\right) + qV_{Coul}$$

so that:

$$H = \frac{1}{2m}\left(\vec{p} - q\vec{A}\right)^2 + qV_{Coul}. \qquad (8)$$

10.4.2.2. Hamiltonian operator and the perturbation Hamiltonian

In terms of operators, $\vec{p}$ corresponds to the operator $\hat{\vec{p}} \equiv -j\hbar\vec{\nabla}$, so that on development, we have as an operator associated with Eq. (8):

$$\hat{H} = \frac{\vec{p}^2}{2m} - \frac{q}{2m}\left(\hat{\vec{p}}\vec{A} + \vec{A}\hat{\vec{p}}\right) + \frac{q^2\vec{A}^2}{2m} + qV_{Coul} \qquad (9)$$

To calculate the commutator $\left[\hat{\vec{p}}, f(r)\right]$ by applying the function $\psi(r)$ we have:

$$\left[\hat{\vec{p}}, f(r)\right]\psi(r) = -j\hbar\left[\vec{\nabla}, f(r)\right]\psi(r) = -j\hbar\left(\vec{\nabla}f(r)\psi(r) - f(r)\vec{\nabla}\psi(r)\right)$$

from which

$$\left[\hat{\vec{p}}, f(r)\right]\psi(r) = -j\hbar\left(\vec{\nabla}f(r) \cdot \psi(r) + f(r)\vec{\nabla}\psi(r) - f(r)\vec{\nabla}\psi(r)\right)$$

$$= -j\hbar\vec{\nabla}f(r) \cdot \psi(r).$$

So definitely:

$$\left[\hat{\vec{p}}, f(r)\right] = -j\hbar\vec{\nabla}f(r), \qquad (10)$$

and by making $\vec{A}$ take on the role of $f(r)$, we have:

$$\left[\hat{\vec{p}}, \vec{A}\right] = -j\hbar\vec{\nabla}.\vec{A} = -j\hbar \; \text{div } \vec{A} \;,$$

so that with Coulomb's gauge, for which $\operatorname{div} \vec{A} = 0$:

$$\left[\hat{\vec{p}}, \vec{A} \right] = 0. \qquad (11)$$

By using the result of Eq. (11) in Eq. (9), we have:

$$\hat{H} = \frac{\hat{\vec{p}}^2}{2m} - \frac{q}{m} \hat{\vec{p}} \, \vec{A} + \frac{q^2 \vec{A}^2}{2m} + qV_{Coul},$$

which we can write in the form:

$$\hat{H} = \hat{H}^0 + \hat{H}^{(1)}$$

with

$$\left\{ \begin{array}{l} \hat{H}^0 = \dfrac{\hat{\vec{p}}^2}{2m} + qV_{Coul} \\[4mm] \hat{H}^{(1)} = -\dfrac{q}{m} \hat{\vec{p}} \, \vec{A} + \dfrac{q^2 \vec{A}^2}{2m}. \end{array} \right. \qquad (12)$$

The first term ($\hat{H}^0$) is the atomic Hamiltonian that describes the particle (the electron undergoing an interaction with the wave) in the presence of a Coulombic potential in the atom.

For its part, $\hat{H}^{(1)}$ represents the interaction of the electron with the wave characterized by the vector potential denoted as $\vec{A}$.

In an approximation of the scale of the wavelengths (optical waves with wavelengths of the order of 600 nm), we can assume that the wave is practically uniform over all of the atom (which has a dimension of the order of Bohr's radius, 0.053 nm) or indeed of the molecule. Under such conditions, $\vec{A}$ depends only on the position ($\vec{r}_a$) of the atom or molecule and the second term of the Hamiltonian perturbation ($\hat{H}^{(1)}$) is a scalar, for which the atomic states between the two different states of the atom are zero. The result of this is that the second term cannot undergo a transition and consequently is ignored in the following calculations. The perturbation Hamiltonian thus is reduced to (with q = -e as the charge of the interacting electron)

$$\widehat{H}^{(1)} = -\frac{q}{m}\widehat{p}\,\vec{A} = \frac{e}{m}\widehat{p}\,\vec{A}. \quad (13)$$

This Hamiltonian sometimes is called the "$\vec{A}.\vec{p}$ Hamiltonian" in which the vector potential of the Coulomb gauge intervenes, evaluated for the atom's position.

10.4.3. Transition rules

10.4.3.1. Preliminary introduction to quantum mechanics (see also, for example, C. Cohen-Tannoudji, *Quantum Mechanics*, Chapter XIII, Hermann, Paris, 1973)

By looking at the course notes of most courses in quantum mechanics, we can see the expression for the probability of obtaining a state (m) from a state (n) after an instant (t), where $\Delta t = t - t_0$ and $t_0 = 0$, for a particle subject to a time-dependent perturbation $(\widehat{H}^{(1)})$. Thus found, the probability is in the form:

$$P_{nm} = a^{*}_{\,m}(t)a_m(t) \quad (14)$$

where

$$a_m(t) = \frac{1}{j\hbar}\int_0^t H^{(1)}_{mn}(t')dt' \quad (15)$$

and equation in which:

$$H^{(1)}_{mn}(t') = \left\langle \Psi^0_m(t') \,|\, \widehat{H}^{(1)} \,|\, \Psi^0_n(t') \right\rangle = \left\langle \psi^0_m \, e^{-j\frac{E^0_m}{\hbar}t'} \,|\, \widehat{H}^{(1)} \,|\, \psi^0_n \, e^{-j\frac{E^0_n}{\hbar}t'} \right\rangle.$$

10.4.3.2. Application to a wave perturbation

In the approximation of an electric dipole, the size of the atom is smaller than the wavelength of the luminous perturbation ($\vec{k}.\vec{r} \approx \frac{2\pi}{\lambda}r$ is low, whereas the value of λ is high) and the vector potential, which has the general form $\vec{A}(r,t) = \vec{A}_0 \cos(\omega t - \vec{k}.\vec{r})$, can be reduced to $\vec{A}(t) = \vec{A}_0 \cos\omega t$.

$\widehat{H}^{(1)}$ given by Eq. (13) thus takes on the form

$$\widehat{H}^{(1)} = \frac{e}{m}\widehat{p}\,\vec{A}_0 \cos\omega t = -j\hbar\frac{e}{m}\vec{A}_0\vec{\nabla}\cos\omega t. \quad (16)$$

By making

$$\Omega = -j\hbar \frac{e}{m} \vec{A}_0 \vec{\nabla} \qquad (17)$$

definitively, we find that:

$$\widehat{H}^{(1)} = \Omega \cos \omega t = \frac{\Omega e^{i\omega t} + \Omega e^{-i\omega t}}{2}. \qquad (18)$$

By making $\Omega_{mn} = \left\langle \psi_m^0 | \Omega | \psi_n^0 \right\rangle \overset{notation}{=} \left\langle m | \Omega | n \right\rangle$ and by remarking that Ω is Hermitian and time independent, we have according to Eq. (15):

$$a_m(t) = \frac{\Omega_{mn}}{2j\hbar} \int_0^t e^{\frac{j}{\hbar}\left(E_m^0 - E_n^0\right)t'} \left[e^{j\omega t'} + e^{-j\omega t} \right] dt'. \qquad (19)$$

With:

$$\omega_0 = \frac{E_n^0 - E_m^0}{\hbar} \qquad (20)$$

and following integration, Eq. (19) gives:

$$a_m(t) = \frac{\Omega_{mn}}{2j\hbar} \left[\frac{e^{-j(\omega_0 - \omega)t} - 1}{\omega_0 - \omega} + \frac{e^{-j(\omega_0 + \omega)t} - 1}{\omega_0 + \omega} \right]. \qquad (21)$$

This result indicates that $a_m(t)$ is large only when the denominator tends toward zero, that is when $\omega = \pm \omega_0$.

If $\omega = \omega_0$, the first term is large with respect to the second, and Eq. (21) gives:

$$\left| a_m(t) \right|^2 = a_m^*(t) a_m(t) = \frac{\left| \Omega_{mn} \right|^2}{\hbar^2 (\omega_0 - \omega)^2} \frac{\left(e^{-j[\omega_0 - \omega]t} - 1 \right)^* \left(e^{-j[\omega_0 - \omega]t} - 1 \right)}{4}$$

so that in addition,

$$|a_m(t)|^2 = \frac{|\Omega_{mn}|^2}{2\hbar^2}\frac{[1 - \cos(\omega_0 - \omega)t]}{(\omega_0 - \omega)^2}. \tag{22}$$

The term shown in Eq. (22) represents the probability of finding a particle in the state m at an instant t, and the probability of the transition $P_{n\to m}$ is defined by the variation as a function of time of the probability $|a_m(t)|^2$ of finding the particle in the final state m, as in:

$$P_{n\to m} = \frac{d}{dt}|a_n(t)|^2 = \frac{\sin(\omega_0 - \omega)t}{(\omega_0 - \omega)}\frac{1}{2\hbar^2}|\langle m \mid \Omega \mid n \rangle|^2 .$$

If we suppose that the time t is large with respect to the period of the electromagnetic field, which is f the order of 10^{-15} sec in the optical domain, and if we make $x = \omega_0 - \omega$, we obtain:

$$P_{n\to m} = \frac{1}{2\hbar^2}|\langle m \mid \Omega \mid n \rangle|^2 \lim_{t\to\infty}\frac{\sin xt}{x} .$$

Dirac's function is defined by

$$\delta x = \frac{1}{\pi}\lim_{t\to\infty}\frac{\sin xt}{x}$$

and the probability of the transition thus is written as

$$P_{n\to m} = \frac{\pi}{2\hbar^2}|\langle m \mid \Omega \mid n \rangle|^2 \delta(\omega_0 - \omega). \tag{23}$$

This indicates that for the transition probability from one state n to another state m is zero, then

$$\delta(\omega_0 - \omega) \neq 0 \quad \Rightarrow \quad \omega = \omega_0 .$$

Therefore, the angular frequency (ω) of the incident electromagnetic wave must be such that:

$$\hbar\omega = \hbar\omega_0 = E_n^0 - E_m^0 . \tag{24}$$

This expression confirms in some way the principle of the conservation of energy and can be considered as a "resonance condition".

With $\omega \simeq \omega_0$, we have $E_n^0 = E_m^0 + \hbar\omega$. There is an absorption of energy ($\hbar\omega$) by the atom.

If we had taken as the more important term in Eq. (21) that which corresponded to $\omega \simeq -\omega_0$, we would have obtained $E_n^0 = E_m^0 - \hbar\omega$. Therefore, there is an emission of an energy $\hbar\omega$ induced by the perturbation.

10.4.3.3. Determination of the transition rule, and dipole transition moment

The probability of transition between the state n and the state m is given by Eq. (23), in which the resonance condition intervenes through the intermediary of Dirac's function ($\delta(\omega_0 - \omega)$). With the resonance condition being fulfilled ($\omega \simeq \omega_0$), the transition probability thus brings in the term $P_0 = \dfrac{\pi}{2\hbar^2}|\langle m \mid \Omega \mid n\rangle|^2$, which evaluated with the help of Eq. (17) is of the form:

$$P_0 = \frac{\pi}{2\hbar^2}|\langle m \mid \Omega \mid n\rangle|^2 = \frac{\pi}{2\hbar^2}\left(\frac{\hbar e}{m}\right)^2 \vec{A}_0^2|\langle m \mid \vec{\nabla} \mid n\rangle|^2. \qquad (25)$$

In order to evaluate the term $\langle m \mid \vec{\nabla} \mid n\rangle$ the following mathematical term (described in the problems at the end of this Chapter) can be used:

$$\frac{\hbar^2}{m}\langle \psi_m \mid \vec{\nabla} \mid \psi_n\rangle = \langle \psi_m \mid \vec{r} \mid \psi_n\rangle(E_n - E_m). \qquad (26)$$

Under these condition, Eq. (25) which can be written as:

$$P_0 = \frac{\pi}{2\hbar^4}\vec{A}_0^2\left[\left(\frac{\hbar^2 e}{m}\right)\langle m \mid \vec{\nabla} \mid n\rangle\right]\left[\left(\frac{\hbar^2 e}{m}\right)\langle m \mid \vec{\nabla} \mid n\rangle\right]^*$$

brings in the terms in the form (where $\vec{\mu} = e\vec{r}$, the dipole moment)

$$\left(\frac{\hbar^2 e}{m}\right)\langle m \mid \vec{\nabla} \mid n\rangle = \langle \psi_m \mid e\vec{r} \mid \psi_n\rangle(E_n - E_m) = \vec{\mu}_{mn}(E_n - E_m).$$

To conclude, the probability of a transition is proportional to the transition dipole moment defined by:

$$\mu_{mn} = \langle \psi_m \mid \mu \mid \psi_n\rangle = \int \psi_m^* \mu\, \psi_n\, d\tau \qquad (27)$$

Once the resonance condition is established, it is the result of the calculation for this integral that finally indicates the permitted transitions between the various possible m and n atomic states.

10.5. Conclusion: Introduction to Atomic and Molecular Spectroscopy

In this last part, which gives the most important results from Section 10.4 on the semiclassic calculations for dipolar approximations, we will underline the importance of dipolar radiation and apply it to the description of the possible different internal transitions (for absorption or emission) in atoms and molecules.

10.5.1. Result concerning the dipole approximation

In very general terms, the dipolar approximation, which states that the electromagnetic field is practically constant at the level of an atom or a molecule, means assuming that there is only a weak coupling between an electromagnetic field and an atom or a molecule. In effect, it is inefficient because its length is way below that of the length of the electromagnetic waves.

We thus can treat the effect of an electromagnetic field as that of a perturbation. By carrying out a development of the interaction energy of the electromagnetic field in terms of various contributions in an order of decreasing importance (analogous to that of a distribution of charges at multipoles, with successive terms due to total charge, dipole moment, quadripolar moment, etc.) we find that the term for the electric dipole interaction is greater than the term for the magnetic dipole interaction, which in turn is greater than the term for the quadripolar electric interaction and so on.

In physical terms, the luminous perturbation, characterized by its electric field ($\vec{E}$) moves electrons through a coupling energy in the form $W = -\vec{\mu}.\vec{E}$, where

$$\vec{E} = -\frac{\partial \vec{A}}{\partial t} = \omega \vec{A}_0 \sin \omega t \quad \text{in the Coulombic gauge, where the scalar potential}$$

associated with the luminous wave field is zero (we can see that this coupling energy intervenes in Eq. (25) taking Eq. (26) into account).

If the following electronic redistribution is symmetric, there is no overall variation in the electric dipole and the transition is forbidden. Inversely, if the redistribution is asymmetric, then the transition is possible even if the atom (or molecule) showed no initial permanent dipole moment.

10.5.2. Different transitions possible in an electromagnetic spectrum

Various types of transitions (for emissions or absorptions) can be written, depending on the wavelength of the electromagnetic field.

10.5.2.1. Rotational transitions

A molecule with a permanent magnetic dipole can appear as a carrier of a variable dipole moment when touring on its own axis (figure 10.6). Inversely, if the molecule is symmetric (without a permanent moment), there is no apparent variation in the dipole moment with rotation. Only molecules that possess a permanent electric dipole can emit or absorb radiation (in the far infrared or microwave region) by making a transition between rotational states.

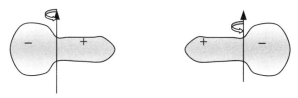

Figure 10.6. *Variation of the dipole moment of a rotating polar molecule.*

Figure 10.7a shows a rotating dipolar molecule, such as HCl, being influenced by an alternating electric field (E) along Ox at an initial time t = 0. Figure 10.7b represents in the plane Oxy the rotation of the dipole at the various instant when t = 0, t_1, t_2, t_3 etc., and Figure 10.7c shows the variation of a component of the dipole in the direction Ox with respect to time.

As shown in Figure 10.7b, the dipole periodically changes position with the electric field due to the variation of the orientation of E and the variation in the coupling energy, which is of the form $W = -\vec{\mu}\vec{E}$ and when t = 0, $W_{min} = -\mu_x E$ as E is directed along Ox at the initial time. As a consequence, the component of the dipole in a given direction (shown for Ox in Figure 10.7c) fluctuates regularly, and thus the component μ_x exhibits a fluctuation with a form resembling that of the electric field from the electromagnetic wave. (In emission, the radiation would be of the same frequency as long as the rotation is not slowed by the interaction of the dipole with neighboring molecules in the material.)

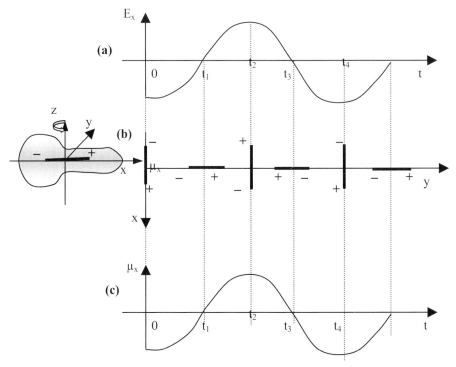

Figure 10.7. *Variation of (a) a component (E_x) of a varying electric field; (b) the position of a permanent dipole moment (μ) in the plane Oxy; and (c) the value of μ_x with time.*

10.2.2. Vibrational transitions

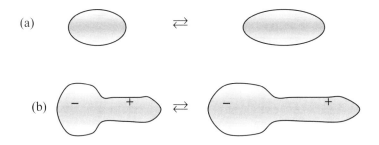

Figure 10.8. *A vibration results in a stretching of the bond, and for (a) the nonpolar molecule shows no variation in dipole moment, but (b) a polar molecule exhibits a change in dipole moment due to the oscillation.*

These transitions appear when there is a vibration associated with, for example, a stretching or a torsion of the bonds between the atoms of a molecule, resulting in a variation in the dipole moment of the molecule. Only those vibrations that are associated with a modification of the dipole moment of a molecule are accompanied with an emission of absorption of electromagnetic radiation. Given the energies brought into play, the mechanism is apparent in the infrared.

10.5.2.3. Electronic transitions
Electronic transitions are apparent if, following an electromagnetic excitation or emission, the resulting redistribution of electrons changes the dipole moment of the molecule. For this to happen, the redistribution must be asymmetric; if the redistribution is symmetric, then the corresponding transition is forbidden. The energies brought into play are in the optical region.

10.5.3. Conclusion
This chapter has described different phenomena of absorption and emission of EM waves by materials. Absorption phenomena will be detailed further in Chapter 3 of the second volume called "Applied Electromagnetism and Materials" using a more phenomenological approach based simply on the interaction of electromagnetic waves with molecules. The deformation that the molecules undergo will be described as a function of the region in which the frequency of the EM wave falls and as part of a more classic mechanical treatment, as detailed in the last figure of Chapter 3 of Volume 2.

It is worth adding that if the energies are of the order of radio frequencies, then it is the magnetic spin moment that interacts with the electromagnetic wave. The reversing magnetic spin dipole thus is at the origin of electronic paramagnetic resonance (EPR) that follows the reversing electron spin. If the effect involves the proton spin, then phenomena associated with nuclear magnetic resonance (NMR) are then observed.

10.5. Problems

10.5.1. Diffusion due to bound electrons
This problem concerns a bound electron with a charge denoted by $-q$, which is initially situated at the origin of a trihedral defined by Oxyz of an atom or molecule subject to:

- a monochromatic plane polarized (along Ox) electromagnetic wave incident along
  Oz and with an electric field in the form $\vec{E} = \underline{E}_0 \exp(i\omega t)\, \vec{u}_x$; and

- and a returning force toward its equilibrium position of the form $\vec{f} = -m\omega_0{}^2 \, x \, \vec{u}_x$.

It is assumed that the angular frequency (ω) of the incident wave is well below that of ω_0. The atom or molecule thus is an oscillating dipole.

1. It is also assumed that the velocity acquired by the electron is negligible with respect to that of the speed of light.

(a) Give the fundamental dynamic equation.

(b) From this deduce the solution for the permanent steady state along the x abscissa of the electron.

(c) What is the value of the dipole moment (p(t)) of the system?

2. The equation for an electromagnetic field radiated at a great distance by a dipole at a point (M) such that $\overline{OM} = \vec{r}$ is:

$$\vec{B} = \frac{\mu_0}{4\pi c} \frac{\ddot{\vec{p}}(\tau) \wedge \vec{u}_r}{r} \quad \text{and} \quad \vec{E} = \frac{\mu_0}{4\pi} \frac{1}{r} (\ddot{\vec{p}}(\tau) \wedge \vec{u}_r) \wedge \vec{u}_r .$$

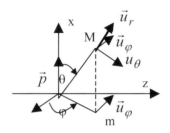

What does τ correspond to? Write the equation for $\vec{p}(\tau)$. Thus calculate $\vec{B}$ and $\vec{E}$.

3. If $\underline{\vec{E}}_W$ and $\underline{\vec{B}}_W$ are the complex amplitudes of the electric field and the magnetic field, respectively, given by $\underline{\vec{E}} = \underline{\vec{E}}_W \exp(i\omega t)$ and $\underline{\vec{B}} = \underline{\vec{B}}_W \exp(i\omega t)$, in general terms, the average value of the Poynting vector ($\vec{S}$) is of the form

$$\langle \vec{S} \rangle = \frac{1}{2} \text{Re} \left(\frac{\underline{\vec{E}}_W \wedge \underline{\vec{B}}^*_W}{\mu_0} \right), \quad \text{where } \underline{\vec{B}}^*_W \text{ is the conjugate complex of } \underline{\vec{B}}_W .$$

Determine the average Poynting vector for the problem.

4. Determine from this the total power radiated by the electron and from the result makes a conclusion.

Answers

1.

(a) The displacement with respect to Ox can be found by considering that for this axis:

$\vec{F}_{em} + \vec{F}_{rappel} = m\vec{\gamma}$, so that

$$-q\left(\vec{E} + \vec{v} \times \vec{B}\right) + F_{return} = m\vec{\gamma}$$

With $B = \dfrac{E}{c}$, we have $v\,B = \dfrac{v\,E}{c} << E$ if $\dfrac{v}{c} << 1$.

With respect to Ox, we therefore have $-q\underline{E}_0 e^{i\omega t} - m\omega_0^2 x = m\dfrac{d^2 x}{dt^2}$.

(b) Under a permanent steady state, we are looking for a solution of the form:

$x = \underline{x}_0 e^{i\omega t}$, that with $\dfrac{d^2 x}{dt^2} = -\omega^2 x$ and by simplifying with $e^{i\omega t}$, then

$-q\underline{E}_0 - m\omega_0^2 \underline{x}_0 = -m\omega^2 \underline{x}_0 \Rightarrow \underline{x}_0 = -\dfrac{q\underline{E}_0}{m\left[\omega_0^2 - \omega^2\right]}$, so that with $\omega_0 >> \omega$, we find

$\underline{x}_0 \approx -\dfrac{q\underline{E}_0}{m\omega_0^2}$, and finally

$$x = -\dfrac{q\underline{E}_0}{m\omega_0^2} e^{i\omega t}$$

In terms of vectors,

$$\vec{x} = x\vec{u}_x = \underline{x}_0 e^{i\omega t} \vec{u}_x$$

where $\vec{u}_x$ is the unit vector along Ox.

(c) Moving the charge –q is equivalent to applying a dipole moment of p = - q x, for which $p(t) = -qx = \dfrac{q^2 \underline{E}_0}{m\omega_0^2} e^{i\omega t}$, so that in terms of vectors,

$$\vec{p}(t) = \frac{q^2 E_0}{m\omega_0^2} e^{i\omega t} \vec{u}_x .$$

2. We have $\vec{B}(M) = \frac{\mu_0}{4\pi c} \frac{\overset{\cdot\cdot}{\vec{p}}(\tau) \wedge \vec{u}_r}{r}$, where $\tau = t - \dfrac{r}{c}$ (change in variable associated

with the propagation) . With $\overset{\cdot\cdot}{\vec{p}}(\tau) = -\omega^2 p(\tau)$ and $\vec{p}(\tau) = \frac{q^2 E_0}{m\omega_0^2} e^{i\omega\left(t-\frac{r}{c}\right)} \vec{u}_x$, we find

that for $\vec{B}(M)$

$$\vec{B}(M) = -\frac{\mu_0 q^2 \omega^2 E_0}{4\pi crm\omega_0^2} \sin\theta \, \vec{u}_\varphi \, e^{i\omega\left(t-\frac{r}{c}\right)} = \underline{\vec{B}}_w \, e^{i\omega t} .$$

Similarly, with $\vec{u}_x \times \vec{u}_r = \sin\theta \, \vec{u}_\varphi$ and $\sin\theta \, \vec{u}_\varphi \times \vec{u}_r = \sin\theta \, \vec{u}_\theta$, the equation for $\vec{E}$
results in:

$$\vec{E}(M) = -\frac{\mu_0 q^2 \omega^2 E_0}{4\pi rm\omega_0^2} \sin\theta \, \vec{u}_\theta \, e^{i\omega\left(t-\frac{r}{c}\right)} = \vec{E}_w \, e^{i\omega t} .$$

3. The average Poynting vector thus is (where $\vec{u}_\theta \times \vec{u}_\varphi = \vec{u}_r$) given by

$$\langle \vec{S} \rangle = \frac{1}{2} \mathrm{Re}\left(\frac{\vec{E}_w \wedge \vec{B}^*_w}{\mu_0} \right) = \frac{\mu_0 q^4 \omega^2 E_0^2}{32\pi^2 r^2 m^2 \omega_0^4 c} \sin^2\theta \, \vec{u}_r .$$

4. The total power radiated is given by:

$$P = \oiint_\Sigma \langle \vec{S} \rangle \overline{d^2\Sigma} .$$

With $\overline{d^2\Sigma} = r^2 \sin\theta \, d\theta \, d\varphi \, \vec{u}_r$, we have:

$$P = \frac{\mu_0 q^4 \omega^2 E_0^2}{32\pi^2 r^2 m^2 \omega_0^4 c} r^2 \iint_{\theta,\varphi} \sin^3\theta d\theta d\varphi = \frac{\mu_0 q^4 \omega^2 E_0^2}{32\pi^2 m^2 \omega_0^4 c} 2\pi \int_{\theta=0}^{\pi} \sin^3\theta d\theta .$$

With $\int\limits_{\theta=0}^{\pi} \sin^3\theta d\theta = \dfrac{4}{3}$, we finally obtain:

$$P = \frac{\mu_0 q^4 E_0^2}{12\pi m^2 c}\left(\frac{\omega^4}{\omega_0^4}\right)$$

Comment: With $\omega = \dfrac{2\pi c}{\lambda}$ and $\omega_0 = \dfrac{2\pi c}{\lambda_0}$, we also can write that:

$$P = \frac{\mu_0 q^4 E_0^2}{12\pi m^2 c}\left(\frac{\lambda_0^4}{\lambda^4}\right).$$

As detailed in the course, the smaller λ (0.4 μm in the visible spectrum, that is blue light), the greater the power radiated (hence the blue color diffused by electrons in molecules present in the sky).

10.5.2. Demonstration of the relationship between matrix elements
[*see Eq. (26) of Section 10.4.3.3 as in*

$$\frac{\hbar^2}{m}\langle \psi_m \mid \vec{\nabla} \mid \psi_n \rangle = \langle \psi_m \mid \vec{r} \mid \psi_n \rangle (E_n - E_m)]$$

For this, we will establish the following equation:

$$\frac{d\langle A \rangle}{dt} = \frac{1}{i\hbar}\langle [A, H] \rangle$$

which is a relative equation for the physical magnitude A, which is not explicitly time dependent, so that the associated operator ($\hat{A}$) is Hermitian (proper real values) and is such that $\langle A \rangle = \langle \psi \mid \hat{A} \mid \psi \rangle$.

Answers

Calculate $\dfrac{d\langle A \rangle}{dt} = \dfrac{d}{dt}\langle \psi \mid \hat{A} \mid \psi \rangle = \left\langle \dfrac{d\psi}{dt} \mid \hat{A} \mid \psi \right\rangle + \left\langle \psi \mid \hat{A} \mid \dfrac{d\psi}{dt} \right\rangle$.

However, according to Schrödinger, $\hat{H}\psi = j\hbar\dfrac{d\psi}{dt}$, for which $\dfrac{d\psi}{dt} = \dfrac{1}{j\hbar}\hat{H}\psi$.

We thus obtain:

$$\frac{d\langle A\rangle}{dt} = \left\langle \frac{1}{j\hbar}\widehat{H}\psi \mid \widehat{A} \mid \psi \right\rangle + \left\langle \psi \mid \widehat{A} \mid \frac{1}{j\hbar}\widehat{H}\psi \right\rangle$$

$$= -\frac{1}{j\hbar}\left\langle \widehat{H}\psi \mid \widehat{A} \mid \psi \right\rangle + \frac{1}{j\hbar}\left\langle \psi \mid \widehat{A} \mid \widehat{H}\psi \right\rangle.$$

H and A are Hermitian (being equal to their associated terms and are such that, for example, $\langle \psi \mid H \mid \varphi \rangle = \langle H\psi \mid \varphi \rangle$), so we also find that:

$$\frac{d\langle A\rangle}{dt} = \frac{1}{j\hbar}\left[-\left\langle \psi \mid \widehat{H}\widehat{A} \mid \psi \right\rangle + \left\langle \psi \mid \widehat{A}\widehat{H} \mid \psi \right\rangle \right]$$

$$= \frac{1}{j\hbar}\left\langle \psi \mid \left[\widehat{A},\widehat{H} \right] \mid \psi \right\rangle = \frac{1}{j\hbar}\left\langle [A,H] \right\rangle = \frac{j}{\hbar}\left\langle [H,A] \right\rangle.$$

With $\widehat{A} \equiv m\widehat{\vec{r}}$, in terms of operators we also can write that (see also, for example, E. Durand, *Quantum Mechanics*, Masson, p.69, 1970):

$$m\frac{d\widehat{\vec{r}}}{dt} = \frac{jm}{\hbar}\left[\widehat{H},\widehat{\vec{r}} \right].$$

As $m\dfrac{d\widehat{\vec{r}}}{dt}$ corresponds to an operator of a quantity of movement, $-j\hbar\vec{\nabla}$, we also have:

$$-j\hbar\vec{\nabla} = \frac{jm}{\hbar}\left[\widehat{H},\widehat{\vec{r}} \right], \text{ from which } \frac{\hbar^2}{m}\vec{\nabla} = \left[\widehat{\vec{r}},\widehat{H} \right].$$

Multiplication of the left side of the last expression by ψ_m^* and the right side by ψ_n , and then carrying out with a scalar product gives:

$$\frac{\hbar^2}{m}\left\langle \psi_m \mid \vec{\nabla} \mid \psi_n \right\rangle = \left\langle \psi_m \mid \widehat{\vec{r}}\widehat{H} \mid \psi_n \right\rangle - \left\langle \psi_m \mid \widehat{H}\widehat{\vec{r}} \mid \psi_n \right\rangle.$$

$\widehat{H}$ is Hermitian (just as is $\widehat{\vec{r}}$), so that:

$$\left\langle \psi_m \mid \widehat{H}\widehat{\vec{r}} \mid \psi_n \right\rangle = \left\langle \psi_m \mid \widehat{H} \mid \widehat{\vec{r}}\psi_n \right\rangle = \left\langle \widehat{\vec{r}}\psi_n \mid \widehat{H} \mid \psi_m \right\rangle^* = \int \widehat{\vec{r}}\psi_n \widehat{H}\psi_m^* dr .$$

Similarly, we obtain:

$$\left\langle \psi_m \mid \hat{\vec{r}}\hat{H} \mid \psi_n \right\rangle = \left\langle \psi_m \mid \hat{\vec{r}} \mid \hat{H}\psi_n \right\rangle = \left\langle \hat{H}\psi_n \mid \hat{\vec{r}}\psi_m \right\rangle^* = \int \hat{\vec{r}}\psi_m^* \hat{H}\psi_n \, dr \; .$$

The result is that:

$$\frac{\hbar^2}{m} \left\langle \psi_m \mid \vec{\nabla} \mid \psi_n \right\rangle = \int \hat{\vec{r}}[\psi_m^* \hat{H}\psi_n - \psi_n \hat{H}\psi_m^*] \, dr \; .$$

As $\hat{H}\psi_m^* = E_m\psi_m^*$ and $\hat{H}\psi_n = E_n\psi_n$, we have:

$$\frac{\hbar^2}{m} \left\langle \psi_m \mid \vec{\nabla} \mid \psi_n \right\rangle = E_m \int \psi_m^* \hat{\vec{r}}\psi_n \, dr - E_n \int \psi_m^* \hat{\vec{r}}\psi_n \, dr \; ,$$

and hence, the looked-for equation:

$$\frac{\hbar^2}{m} \left\langle \psi_m \mid \vec{\nabla} \mid \psi_n \right\rangle = \left\langle \psi_m \mid \hat{\vec{r}} \mid \psi_n \right\rangle \left(E_n - E_m \right) .$$

Chapter 11

Reflection and Refraction of Electromagnetic Waves in Absorbent Materials of Finite Dimensions

11.1. Introduction

In practical terms, electromagnetic (EM) waves in materials of limited dimensions correspond to those in systems such as coaxial cables, optical fibers, and wave guides. It is by this process that signals, and therefore information, are transmitted with as low a degree of attenuation and parasitic phenomena as possible.

This chapter will look at the influence of discontinuity, at the interface between two materials, on the propagation of EM waves. Typically, the two materials are:

• linear, homogeneous, and isotropic (lhi) dielectrics and not magnetic so that their magnetization intensity can be written as $\vec{I} = 0$, so that $\vec{B} = \mu_0\vec{H} + \vec{I} = \mu\vec{H}$, $\mu = \mu_0$ with $\vec{I} = 0$;

• uncharged, as in $\rho_\ell = 0$, and not traversed by real currents, as in $\vec{j}_\ell = 0$; and

• described by their complex dielectric permittivity as $\underline{\varepsilon} = \varepsilon' - j\varepsilon''$ (electrokinetic notation), or as $\underline{\varepsilon} = \varepsilon' + j\varepsilon''$ (optical notation).

In this chapter, and then in Chapter 12, which mostly looks at applications using the optical region, the notation used is that classically used elsewhere in optics as in sinusoidal planar progressive EM waves are written in the form (see also Section 6.4.2):

$$\underline{\vec{E}} = \underline{\vec{E}}_m \exp\left(j[\vec{k}\vec{r} - \omega t]\right). \quad (1)$$

11.2. Law of Reflection and Refraction at an Interface between Two Materials
11.2.1. Representation of the system

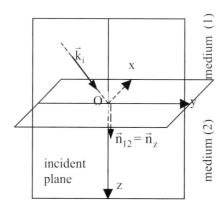

Figure 11.1. *Interface between media (1) and (2).*

As a general system, the volume under consideration is made up of two media, denoted (1) and (2) which have permittivities ε_1 and ε_2, respectively, and are separated by an interface on the plane Oxy. The incident wave has an angular frequency denoted by ω_i, is written in the form $\underline{\vec{E}}_i = \underline{\vec{E}}_{im} \exp\left(j[\vec{k}_i \vec{r} - \omega_i t]\right)$, and thus by notation for the incident wave $\underline{\vec{E}}_m = \underline{\vec{E}}_{im}$. In addition, by choosing the origin of phases on this wave such that $\underline{\vec{E}}_{im} = \vec{E}_i^0 \exp(j\varphi_i) = \vec{E}_i^0$ when $\varphi_i = 0$, then it is possible to state:

$$\underline{\vec{E}}_i = \vec{E}_i^0 \exp\left(j[\vec{k}_i \vec{r} - \omega_i t]\right). \qquad (2)$$

The incident plan is defined by the plane that contains the incident wavevector ($\vec{k}_i$) as well as the normal to the interface ($\vec{n}_{12} = \vec{n}_z$). In order to simplify the representations, the incident plane can be taken as Oxy in which consequently, is the wavevector $\vec{k}_i$. Also, we will assume that the first material, medium (1), is nonabsorbent so that $\varepsilon_{r1} = n_1^2$ so that the modulus of $\vec{k}_i$ is $\left|\vec{k}_i\right| = k_1 = k_0 n_1$.

At the reflection on the plane of the interface Oxy, the incident wave will observe a modification in one part of its amplitude (which now will be denoted $\vec{E}_{r0}$), while the other part will be subject to a dephasing by a quantity denoted as φ_r. The reflected wave then can be written in a general form where ω_r designates the angular frequency following reflection:

$$\vec{E}_r = \vec{E}_r^0 \exp(j\varphi_r) \exp\left(j[\vec{k}_r \vec{r} - \omega_r t]\right) = \vec{E}_r^0 \exp\left(j[\vec{k}_r \vec{r} - \omega_r t]\right). \quad (3)$$

Similarly, for the transmitted wave, self-evident notations are used:

$$\vec{E}_t = \vec{E}_t^0 \exp(j\varphi_t) \exp\left(j[\vec{k}_t \vec{r} - \omega_t t]\right) = \vec{E}_t^0 \exp\left(j[\vec{k}_t \vec{r} - \omega_t t]\right). \quad (4)$$

In these equations, $\vec{k}_r$ and $\vec{k}_t$ are the wavevector relating to the reflected and transmitted wave, respectively. As detailed below, $\vec{k}_t$ can be a complex magnitude (thus denoted $\underline{\vec{k}}_t$).

11.2.2. Conservation of angular frequency on reflection or transmission in linear media

Two different types of reasoning can be used in this case: the first is by a physical analysis of the problem, and the second via an equation using the conditions of continuity at an interface.

11.2.2.1. Physical reasoning

As shown in Chapter 10 dipoles associated with a displacement of charges inside a dielectric that are subject to an incident wave ($\vec{E}_i$) with a given angular frequency ($\omega_i = \omega$), and within an approximation of a linear oscillator (returning force proportional to the induced displacement), emit in turn a wave with the same angular frequency as the incident wave. The result is that little by little, in media (1) and (2), the incident wave with angular frequency is reemitted with the same angular frequency, so that

$$\omega_i = \omega_r = \omega_t = \omega. \quad (5)$$

11.2.2.2. Reasoning with the help of equations of continuity

By denoting a_i, a_r and a_t as the proper projections of $\vec{E}_i^0$, $\vec{E}_r^0$, and $\vec{E}_t^0$ at the plane of the interface Oxy, the continuity condition of the tangential component of the electric field gives rise to:

$a_i \exp\left(j[\vec{k}_i \vec{r} - \omega_i t]\right) + a_r \exp\left(j[\vec{k}_r \vec{r} - \omega_r t]\right) = a_t \exp\left(j[\vec{k}_t \vec{r} - \omega_t t]\right)$. By multiplying the two members by $\exp\left(-j[\vec{k}_i \vec{r} - \omega_i t]\right)$, we obtain:

$$a_i + a_r \, e^{j[(\vec{k}_r - \vec{k}_i)\vec{r} - (\omega_r - \omega_i)t]} = a_t \, e^{j[(\vec{k}_t - \vec{k}_i)\vec{r} - (\omega_t - \omega_i)t]}. \quad (6)$$

This equation must be true for all instants (t), and each term of the preceding equation should have the same temporal dependence, so that $0 = \omega_r - \omega_i = \omega_t - \omega_i$. The same result is found as that given by Eq. (5), that is $\omega_i = \omega_r = \omega_t = \omega$.

11.2.3. Form of the wavevectors with respect to the symmetry of the media
Once again, the same two types of reasoning may be applied.

11.2.3.1. Reasoning based on physics of the symmetry of a system
The EM field of the incident wavevector is given by:

$$\vec{k}_i \begin{cases} k_{ix} = 0 \\ k_{iy} \neq 0 \\ k_{iz} \neq 0 \end{cases}$$

Taking into account the disposition of the interface, which is such that the discontinuity only occurs in the direction Oz, the propagation in the directions x and y is not perturbed at the interface. This results in perturbation being only in the direction Oz, where the wavevector thus is modified on reflection and transmission. Physically then we should have:

$$\begin{aligned} k_{ix} &= k_{rx} = k_{tx} = 0 \\ k_{iy} &= k_{ry} = k_{ty} \qquad (7) \\ k_{iz} &\neq k_{rz} \neq k_{tz} \end{aligned}$$

11.2.3.2. Mathematically based reasoning and the use of the continuity equation
Again using the result given by Eq. (6) and also from the imposition of the condition of continuity of the tangential component of the electric field, the equation is valid only at the level of the interface where the vector ($\vec{r}$) has the components given by

$$\vec{r} \begin{cases} x \neq 0 \\ y \neq 0 \\ z = 0 \end{cases}$$

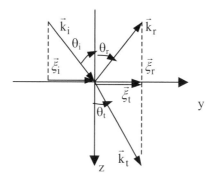

Figure 11.2. *Representation of the projected vectors ($\vec{\xi}$) at the interface.*

By denoting $\vec{\xi}_i, \vec{\xi}_r,$ and $\vec{\xi}_t$, the projected vectors at the interface Oxy of the vectors $\vec{k}_i, \vec{k}_r, \vec{k}_t$, respectively, we can state that they have components given by:

$$\vec{\xi}_i \begin{cases} k_{ix} \\ k_{iy} \\ 0 \end{cases} \qquad \vec{\xi}_r \begin{cases} k_{rx} \\ k_{ry} \\ 0 \end{cases} \qquad \vec{\xi}_t \begin{cases} k_{tx} \\ k_{ty} \\ 0 \end{cases}$$

Eq. (6) for the vectors $\vec{r}$ located at the interface thus is in the form:

$$a_i + a_r \, e^{j[(\vec{\xi}_r - \vec{\xi}_i)\vec{r} - (\omega_r - \omega_i)t]} = a_t \, e^{j[(\vec{\xi}_t - \vec{\xi}_i)\vec{r} - (\omega_t - \omega_i)t]} . \qquad (6')$$

This Eq. (6') should be true for all $\vec{r}$ vectors at the interface, and we should find for each term in Eq. (6') the same spatial dependence, which can be written: $0 = \left(\vec{\xi}_r - \vec{\xi}_i\right)\vec{r} = \left(\vec{\xi}_t - \vec{\xi}_i\right)\vec{r}$, so that $\vec{\xi}_i = \vec{\xi}_r = \vec{\xi}_t$.

With $\vec{k}_i$ in the plane of the wave $(k_{ix} = 0)$, it can be determined that:

$$k_{ix} = k_{rx} = k_{tx} = 0 \text{ and } k_{iy} = k_{ry} = k_{ty} . \qquad (7')$$

Following reflection and transmission of the EM wave at the level of the interface, there remains only k_{rz} and k_{tz} to be determined. This can be done only through a use of the equations for the propagation in each of the media.

11.2.4. Symmetry and linear properties of the media and the form of the related field
11.2.4.1. In medium (1)

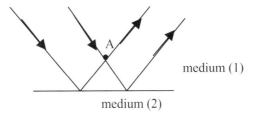

Figure 11.3. *Representation of a wave at point A in medium 1.*

A wave at any point (A) in medium (1) is in fact a superposition of the incident and reflected wave. Taking the results of Sections 11.2.2 and 11.2.3 into account, which made it possible to state that $\omega_r = \omega$, that $k_{rx} = k_{ix} = 0$ and that $k_{ry} = k_{iy}$, along with that of the form of Eq. (2) for the incident wave, the resultant wave in medium (1) therefore must be of the form:

$$\underline{\vec{E}}_1(\vec{r},t) = \underline{\vec{E}}_1^0(z) \exp\left[j\left(k_{iy}y - \omega t\right)\right] . \qquad (8)$$

The exponential term contains the well-determined (and unchanged) part of the wave, where the temporal variation is $\omega_r = \omega$ and the spatial variation with respect to x is given by $k_{rx} = k_{ix} = 0$ and with respect to y by $k_{ry} = k_{iy}$.

The term $\vec{E}_1^0(z)$ contains the behavior with respect to Oz, which remains for the moment unknown; the resolution of the propagation equation in medium (1) will allow a determination of $\vec{E}_1^0(z)$ and k_{rz} (as a function of k_{iz}).

11.2.4.2. In the medium (2)

Once again, $\omega_t = \omega$ and $k_{ty} = k_{iy}$, while only k_{tz} remains unknown. In medium (2) the wave thus is in the form:

$$\vec{E}_2(\vec{r},t) = \vec{E}_2^0(z) \exp\left[j\left(k_{iy}y - \omega t\right)\right]. \qquad (9)$$

The exponential contains the known terms, while $\vec{E}_2^0(z)$ encloses that which is unknown for the moment (behaviour with respect to Oz) and it is the resolution of the propagation equation for medium (2) which will permit a determination of $\vec{E}_2^0(z)$ and $k_{tz} \equiv k_{2z}$.

11.2.5. Snell-Descartes law for the simple scenario where media (1) and (2) are nonabsorbent and k_1 and k_2 are real

11.2.5.1.   The wavevectors associated with incident, reflected, and transmitted waves are all in the same plane (incidence plane)

The result given in the subtitle can be determined immediately from the fact that we have $k_{ix} = k_{rx} = k_{tx} = 0$ from the first expressions of Eq. (7). The result is that the wavevectors $\vec{k}_i, \vec{k}_r$, and $\vec{k}_t$ are all in the plane Oyx, which is their hypothetical plane of incidence.

11.2.5.2. Law of reflection: $\theta_i = -\theta_r$

The first equivalence of the second expression of the equations given in (7) is that $k_{ry} = k_{iy}$.   With $k_i = \dfrac{\omega}{c}n_1$ and also $k_r = \dfrac{\omega}{c}n_1$, the relation $k_{ry} = k_{iy}$, which represents the equality of the projection of vectors $\vec{k}_i$ and $\vec{k}_r$ with respect to Oy, gives in moduli (see Figure 11.2) $\dfrac{\omega}{c}n_1 \sin\theta_i = \dfrac{\omega}{c}n_1 \sin\theta_r$, so that $\theta_i = \theta_r$, and if we orient the angles with respect to the normal, then $\vec{n}_{12} = \vec{n}_z$, and

$$\theta_i = -\theta_r. \qquad (10)$$

11.2.5.3. Law of refraction: $n_1 \sin \theta_i = n_2 \sin \theta_t$

Still with respect to the second of the expressions in Eq. (7), we also have $k_{ty} = k_{iy}$. When the materials are nonabsorbing, the wavevectors are real just as

the optical indices. We thus have $k_i = \dfrac{\omega}{c} n_1$ and $k_t = \dfrac{\omega}{c} n_2$ and the relation

$k_{ty} = k_{iy}$, which brings in a projection of the vectors $\vec{k}_i$ and $\vec{k}_t$ with respect to Oy,

gives rise to (Figure 11.2) $\dfrac{\omega}{c} n_1 \sin \theta_i = \dfrac{\omega}{c} n_2 \sin \theta_t$, so that :

$$n_1 \sin \theta_i = n_2 \sin \theta_t . \qquad (11)$$

11.2.6. Equation for the electric field in medium (1): the law of reflection

As above indicated, medium (1) is not magnetic or absorbent, n_1 and

$k_i = \dfrac{\omega}{c} n_1 = k_0 n_1$ are real, and the components of $\vec{k}_i$ are (0, $k_{iy} \neq 0$ and $k_{iz} \neq 0$).

In medium (1), the monochromatic wave is in accordance with the wave equation [Eq. (7") in Chapter 7]:

$$\Delta \underline{\vec{E}}_1 + \frac{\omega^2}{c^2} n_1^2 \, \underline{\vec{E}}_1 = 0 \qquad (12)$$

where $\underline{\vec{E}}_1$ is the form given by Eq. (8), as in

$$\underline{\vec{E}}_1(\vec{r},t) = \underline{\vec{E}}_1^0(z) \exp\left[j\left(k_{iy} y - \omega t \right) \right] .$$

By making $f(y) = \exp\left[j\left(k_{iy} y - \omega t \right) \right]$, we have $\underline{\vec{E}}_1(\vec{r},t) = \underline{\vec{E}}_1^0(z) \, f(y)$, with the result that

$$\Delta \underline{\vec{E}}_1 = \vec{i} \, \Delta \underline{E}_{1x} + \vec{j} \, \Delta \underline{E}_{1y} + \vec{k} \, \Delta \underline{E}_{1z}$$

$$= \vec{i} \, \Delta \left[\underline{\vec{E}}_{1x}^0 (z) \, f(y) \right] + \vec{j} \, \Delta \left[\underline{\vec{E}}_{1y}^0 (z) \, f(y) \right] + \vec{k} \, \Delta \left[\underline{\vec{E}}_{1z}^0 (z) \, f(y) \right]$$

$$= f(y) \left[\frac{\partial^2}{\partial z^2} \underline{\vec{E}}_1^0 (z) \right] + \left[\vec{i}^2 \, k_{iy}^2 \left(\vec{i} \, E_{1x}^0 + \vec{j} \, E_{1y}^0 + \vec{k} \, E_{1z}^0 \right) f(y) \right]$$

$$= \left[\frac{\partial^2}{\partial z^2} \underline{\vec{E}}_1^0 (z) - k_{iy}^2 \underline{\vec{E}}_1^0 (z) \right] \exp\left[i\left(k_{iy} y - \omega t \right) \right].$$

Following simplification of the two latter terms with the term $f(y) = \exp\left[j\left(k_{iy} y - \omega t \right) \right]$, Eq. (12) becomes:

$$\frac{d^2\vec{E}_1^0(z)}{dz^2} + \left(\frac{\omega^2}{c^2}n_1^2 - k_{iy}^2\right)\vec{E}_1^0(z) = 0 .$$ (13)

By making $k_{1z}^2 = \frac{\omega^2}{c^2}n_1^2 - k_{iy}^2$, then Eq. (13) can be written as

$$\frac{d^2\vec{E}_1^0(z)}{dz^2} + k_{1z}^2\ \vec{E}_1^0(z) = 0 .$$ (13')

It is notable that:

$$\left.\begin{array}{l} k_i^2 = k_{ix}^2 + k_{iy}^2 + k_{iz}^2 = k_{iy}^2 + k_{iz}^2 \\[2mm] k_i^2 = (k_0\ n_1)^2 = \dfrac{\omega^2}{c^2}n_1^2 \end{array}\right\} \Rightarrow k_{iz}^2 = \frac{\omega^2}{c^2}n_1^2 - k_{iy}^2 = k_{1z}^2 .$$

Where $k_{iz} = k_{1z} > 0$ so that the incident propagation can occur for a positive value of Oz, the general solution for $\vec{E}_1^0(z)$ is given by

$$\vec{E}_1^0(z) = \vec{E}_i^0\exp(i\ k_{1z}\ z) + \vec{E}_r^0\exp(- i\ k_{1z}\ z) .$$ (14)

$$\underbrace{\qquad\qquad}_{\substack{\text{incident} \\ \text{propagation}}} \qquad \underbrace{\qquad\qquad}_{\substack{\text{reflected} \\ \text{propagation}}}$$

By substituting $\vec{E}_1^0(z)$ into Eq. (8), we obtain:

$$\begin{aligned} \vec{E}_1(\vec{r},t) &= \vec{E}_1^0(z)\exp\left[j\left(k_{iy}y - \omega t\right)\right] \\ &= \vec{E}_i^0\exp\left[i\left(k_{iy}y + k_{1z}z - \omega t\right)\right] + \vec{E}_r^0\exp\left[i\left(k_{iy}y - k_{1z}z - \omega t\right)\right] \\ &= \vec{E}_i + \vec{E}_r . \end{aligned}$$ (15)

The first term details the incident wave for which the wavevector is given by:

$$\vec{k}_i \begin{cases} k_{ix} = 0 \\ kiy \\ k_{iz} = k_{1z} > 0. \end{cases}$$

The second term represents the reflected wave at the level of the interface and is the monochromatic plane progressive wave for which the vector $\vec{k}_r$ has the component:

$$\vec{k}_r \begin{cases} k_{rx} = k_{ix} = 0 \\ k_{ry} = k_{iy} \neq 0 \\ k_{rz} = -k_{1z} = -k_{iz} < 0 \end{cases}$$

From the relations $k_{rx} = k_{ix} = 0$ and $k_{ry} = k_{iy}$, we find once again the laws of reflection obtained in Section 11.2.5. The incident and reflected rays are in the incident plane Oyz and $\theta_i = -\theta_r$. It is worth noting that these laws do not depend on the nature of medium (2), as no hypothesis has been formulated concerning it, and it could be nonabsorbing or absorbent and even a perfect insulator or conductor.

11.2.7. The Snell-Descartes law for reflection a system where medium (2) can be absorbent: n_2 and k_2 are complex

Here medium (1) is nonabsorbent and n_1 and k_1 are real.

11.2.7.1. Form of the field in medium (2)

Here medium (2) can be absorbent and is characterized by a complex dielectric permittivity, so that $\underline{\varepsilon}_{r2} = \underline{n}_2^2$.

The wave equation in medium (2) is $\overline{\Delta \vec{E}}_2 + \dfrac{\omega^2}{c^2} \underline{n}_2^2 \, \vec{E}_2 = 0$.

Into this the substitution of the form of $\vec{E}_2$ given by Eq. (9), i.e.,
$\vec{E}_2(\vec{r},t) = \vec{E}_2^0(z) \exp\left[j\left(k_{iy}y - \omega t\right) \right]$, gives

$$\frac{d^2 \vec{E}_2^0}{dz^2} + \left(\frac{\omega^2}{c^2} \underline{n}_2^2 - k_{iy}^2 \right) \vec{E}_2^0 = 0 . \qquad (16)$$

By making $\underline{k}_{2z}^2 = \left(\dfrac{\omega^2}{c^2} \underline{n}_2^2 - k_{iy}^2 \right)$, so that with $k_{iy} = k_0 \, n_1 \sin\theta_i$, then

$$\underline{k}_{2z}^2 = k_0^2 \, (\underline{n}_2^2 - n_1^2 \sin^2\theta_i) . \qquad (17)$$

Then the equation for the propagation of the wave takes on its more normal form [see for example Eq. (15) in Chapter 7]:

$$\frac{d^2 \vec{E}_2^0}{dz^2} + \underline{k}_{2z}^2 \, \vec{E}_2^0 = 0 . \qquad (16')$$

The general solution is thus of the form:

$$\vec{E}_2^0(z) = \vec{E}_t^0 \exp(i\underline{k}_{2z}z) + \vec{E}_t^{\prime 0} \exp(-i\underline{k}_{2z}z) . \qquad (18)$$

Physically, in medium (2) there is no source of light and the only wave that actually can be found is a wave transmitted and distancing itself from the interface. In this case, the second component of Eq. (18) must be equal to zero, as it represents the effect due to a wave that nears the interface (the $\vec{E}\,'^0_t$ also must be taken as equal to zero in order that the condition is held). Only the first component of Eq. (18), which clearly represents the effect of a wave leaving from the interface, can be retained on a physical basis, so that finally we have

$$\vec{\underline{E}}^0_2(z) = \vec{\underline{E}}^0_t \ \exp(i\underline{k}_{2z}z) . \qquad (18')$$

For its part, $\underline{k}_{2z}$ defined by Eq. (17), generally has a complex form and can be written using a notation in the form:

$$\underline{k}_{2z} = \underline{k}_{tz} = k'_{tz} + ik''_{tz} . \qquad (19)$$

On substituting Eq. (18') into Eq. (9), we obtain:

$$\vec{E}_2(\vec{r},t) = \vec{\underline{E}}^0_2(z) \exp\left[i\left(k_{iy}y - \omega t\right)\right] = \vec{\underline{E}}^0_t \ \exp\left[i\left(k_{iy}y + \underline{k}_{2z} - \omega t\right)\right], \qquad (20)$$

so that with Eq. (19):

$$\vec{E}_2(\vec{r},t) = \vec{\underline{E}}^0_t \ \exp(-k''_{tz}z) \ \exp\left[i\left(k_{iy}y + k'_{tz}z - \omega t\right)\right]. \qquad (21)$$

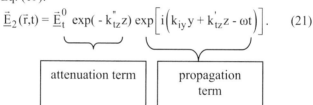

| attenuation term | propagation term |

11.2.7.2. Form of the wavevector in the medium (2)

Equation (21) for the wave in medium (2) can be rewritten in the form:

$$\vec{E}_2(\vec{r},t) = \vec{\underline{E}}^0_t \ \exp(-\vec{k}''_t \vec{r}) \ \exp\left[i\left(\vec{k}'_t \vec{r} - \omega t\right)\right]. \qquad (21')$$

This relation defines the vectors $\vec{k}'_t$ and $\vec{k}''_t$ for which the components are given from Eq. (21):

$$\vec{k}'_t \begin{cases} k'_{tx} = 0 \ (= k_{ix}) & (22) \\[2mm] k'_{ty} = k_{iy} = k_0 \, n_1 \, \sin\theta_i \\[2mm] k'_{tz} \overset{(19)}{=} R(\underline{k}_{2z}) \overset{(17)}{=} k_0 R\left\{\left(n_2^2 - n_1^2\sin^2\theta_i\right)^{1/2}\right\} \end{cases}$$

$$\vec{k}''_t \begin{cases} k''_{tx} = 0 \\ k''_{ty} = 0 \\ k''_{tz} \overset{(19)}{=} \mathrm{Im}(\underline{k}_{2z}) \overset{(17)}{=} k_0 \mathrm{Im}\left\{\left(\underline{n}_2^2 - n_1^2\sin^2\theta_i\right)^{1/2}\right\} \end{cases} \qquad (23)$$

Finally, the wave can be written in the form:

$$\underline{E}_2(\vec{r},t) = \underline{\vec{E}}_t^0(z)\,\exp\left[i\left(\underline{\vec{k}}_t\vec{r} - \omega t\right)\right],$$

where $\underline{\vec{k}}_t$ is defined by $\underline{\vec{k}}_t = \vec{k}'_t + i\vec{k}''_t$, such that:

$$\underline{\vec{k}}_t \begin{cases} \vec{k}'_t = k_{iy}\,\vec{e}_y + k_0 R\left\{\left(\underline{n}_2^2 - n_1^2\sin^2\theta_i\right)^{1/2}\right\}\vec{e}_z = k'_{ty}\,\vec{e}_y + k'_{tz}\,\vec{e}_z \\ \vec{k}''_t = k_0 \mathrm{Im}\left\{\left(\underline{n}_2^2 - n_1^2\sin^2\theta_i\right)^{1/2}\right\}\vec{e}_z. \end{cases}$$

$$(24)$$

Consequences: inhomogeneous and homogeneous waves

Generally speaking, the wavevector for the transmitted wave can be written in accord with Eq. (24) in the form:

$$\underline{\vec{k}}_t = k'_{ty}\,\vec{e}_y + k'_{tz}\,\vec{e}_z + i\,(k''_{tz}\,\vec{e}_z). \qquad (25)$$

According to Eq. (21), the propagation in medium (2), which is determined by the real component of $\underline{\vec{k}}_t$, is along Oy and Oz, while the attenuation associated with the imaginary component of $\vec{k}''_t$ is only in Oz. The directions of propagation and attenuation thus are different and the wave is termed inhomogeneous.

In the specific case where $k'_{ty} = 0$, so that when $k'_{ty} = k_1\sin\theta_1$ we have $\theta_1 = 0$ (normal incidence), we have:

$$\underline{\vec{k}}_t = k'_{tz}\,\vec{e}_z + i\,(k''_{tz}\,\vec{e}_z)$$

and the propagation and attenuation are in the same direction (Oz), and thus the wave is termed homogeneous.

11.2.7.3. First law of refraction

Here we have $k'_{tx} = 0$ and $k''_{tx} = 0$, so that $\left[\vec{\underline{k}}_t\right]_x = 0$. The $\underline{k}_t$ has no component along Ox and the transmitted wave is propagated in the plane of incidence, i.e., Oyz.

11.2.7.4. Second law of refraction

When $k'_{tz} \neq 0$, the transmitted wave effectively propagates in medium (2); supposing $\theta_2 = \theta_t = (\vec{n}_{12}, \vec{k}'_t)$, so that $k'_{ty} = k'_t \sin \theta_2$ (see Figure 11.2). In addition, it also is possible to state, from Eq. (22), that $k'_{ty} = k_{iy} = k_0 n_1 \sin \theta_i$ which can written with $k_1 = k_0 n_1$ and by making $\theta_1 = \theta_i$ by notation, that $k'_{ty} = k_1 \sin \theta_1$. As a consequence, we have the equation:

$$k_1 \sin \theta_1 = k'_t \sin \theta_2, \qquad (26)$$

which can be rewritten using the notations introduced, as

$$k_i \sin \theta_i = k'_t \sin \theta_t. \qquad (26')$$

11.2.7.4.1. Case where $\vec{\underline{k}}_t$ is real $\rightleftarrows \vec{k}''_t = 0$

Here, according to Eq. (23), the term $\left(\underline{n}_2^2 - n_i^2 \sin^2 \theta_i\right)$ must have a positive and real magnitude: if it were negative, its square root would be purely imaginary and k''_t would be nonzero. Looking at the following conditions:

• first condition, $\left(\underline{n}_2^2 - n_i^2 \sin^2 \theta_i\right)$ has a real magnitude, imposes that $\underline{n}_2^2$ is real, so that $\underline{n}_2^2 = n_2^2 = \varepsilon_{r2}$ where ε_{r2} has a real magnitude. Medium (2) thus is intrinsically [that is to say by itself without medium (1)] nonabsorbent as $\underline{k}_2^2 = \dfrac{\omega^2}{c^2} \underline{n}_2^2 = \dfrac{\omega^2}{c^2} \varepsilon_{r2}$ is real if $\underline{\varepsilon}_{r2} = \varepsilon_{r2}$ is real (and positive)

$$\underline{n}_2^2 - n_1^2 \sin^2 \theta_1 = n_2^2 - n_1^2 \sin^2 \theta_1 \, ;$$

• second condition, the term $n_2^2 - n_1^2 \sin^2 \theta_1$ has a positive magnitude if

$$n_2^2 > n_1^2 \sin^2 \theta_1. \qquad (27)$$

This is the supplementary condition to $\underline{n}_2^2$ being real [medium (2) nonabsorbent and ε_{r2} real] so that $\vec{\underline{k}}_t$ can be real $(\rightleftarrows \vec{\underline{k}}_t'' = 0)$ when medium (2) is in the presence of medium (1).

When $k_t'' = 0$, we have $k_t = k_t'$, which is such that according to Eq. (22):

$$k_t^2 = k_t'^2 = k_{ty}'^2 + k_{tz}'^2 = (k_0 \ n_1 \ \sin \ \theta_1)^2 + (k_0^2 \ [n_2^2 - n_1^2 \ \sin^2 \ \theta_1]) = k_0^2 n_2^2 = \frac{\omega^2}{c^2} n_2^2 = k_2^2 .$$

Also, when $k_t'' = 0$, Eq. (26) is written as $k_1 \ \sin \ \theta_1 = k_2 \ \sin \ \theta_2$, so that in turn:

$$\boxed{n_1 \ \sin \ \theta_1 = n_2 \ \sin \ \theta_2} . \qquad (28)$$

The conditions required so that $n_2^2 > n_1^2 \sin^2\theta_1$ $(\rightleftarrows \vec{\underline{k}}_t'' = 0)$ are:
- by simple evidence, that $n_2 > n_1$ (see also Figure 11.4);
- or $n_2 < n_1$ and $\theta_1 < \theta_\ell$ where θ_ℓ is defined by $n_1 \ \sin \ \theta_\ell = n_2$. $\qquad (29)$

In effect, so that $n_2 < n_1$ we have $\theta_2 > \theta_1$ (Figure 11.5). The θ_2 cannot exceed $\theta_{2\ell} = \frac{\pi}{2}$ and θ_ℓ also cannot exceed a limiting value given by $\theta_1 = \theta_\ell$ and such that $n_1 \ \sin\theta_\ell = n_2 \ \sin \ \frac{\pi}{2}$. The limiting angle (θ_ℓ) above which there can be no refracted radiation is associated with the total reflection and thus is defined by the equation $\sin\theta_\ell = \frac{n_2}{n_1}$.

The condition $\theta_1 < \theta_\ell$ yields $\sin\theta_1 < \sin\theta_\ell = \frac{n_2}{n_1}$, which otherwise can be stated as $n_1 \ \sin\theta_1 < n_2$ and corresponds well to Eq. (27), for which $\vec{\underline{k}}_t$ is real $(\rightleftarrows \vec{\underline{k}}_t'' = 0)$.

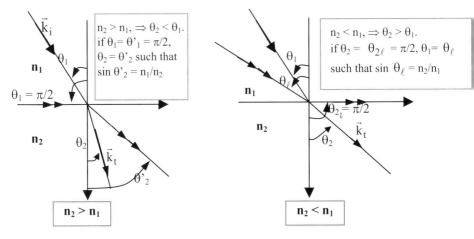

Figure 11.4. *Refraction when $n_2 > n_1$.*   **Figure 11.5.** *Refraction when $n_2 < n_1$.*

11.2.7.4.2. When $\underline{\vec{k}}_t$ is complex: $\rightleftharpoons \vec{k}_t'' \neq 0$

When $\underline{\vec{k}}_t$ is complex, then $k_{tz}'' \neq 0$ must be true. The attenuation occurs in the Oz direction. According to the third relation given in Eq. (23), the following must be true:

- either $\underline{n}_2^2 = \underline{\varepsilon}_{r2}$ is imaginary and medium (2) is absorbent; or

- $\underline{n}_2^2 = \underline{\varepsilon}_{r2} = \varepsilon_{r2}$ is real [medium (2) is not absorbent] and $n_2 < n_1$ and $\theta_1 > \theta_\ell$ (which correspond to $n_1 \sin\theta_1 > n_1 \sin\theta_\ell = n_2$).

We thus have $n_2^2 - n_1^2 \sin^2\theta_1 < 0$, so that $n_2^2 - n_1^2 \sin^2\theta_1 = i^2(n_1^2\sin^2\theta_1 - n_2^2)$ and, according to Eqs. (22) and (23), we can deduce that, respectively:

$$
\begin{cases}
k_{tz}' = k_0 \, R\left\{\left[\; i^2(n_1^2\sin^2\theta_1 - n_2^2)\right]^{1/2}\right\} = 0 \\[2mm]
k_{tz}'' = k_0 \, Im\left\{\left[\; i^2(n_1^2\sin^2\theta_1 - n_2^2)\right]^{1/2}\right\} = \pm k_0\left(n_1^2\sin^2\theta_1 - n_2^2\right)^{1/2} \neq 0.
\end{cases}
$$

$$(30)$$

So that the wave undergoes an exponential "braking", Eq. (21) must have $k_{tz}''z > 0$. As $z > 0$ in medium (2), the positive solution from Eq. (30) for k_{tz}'' is retained, and in terms of vectors, we have [from Eq. (24)]:

$$\begin{cases} \vec{k}_t' = k_{iy} \; \vec{e}_y = k_0 \; n_1 \; \sin \theta_1 \; \vec{e}_y \\[2mm] \vec{k}_t'' = k_0 \left(n_1^2 \sin^2\theta_1 - n_2^2 \right)^{1/2} \vec{e}_z \end{cases} \qquad (30')$$

$\Rightarrow$ the wave is inhomogeneous. It is as shown in Figure 11.6.

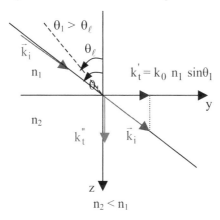

Figure 11.6. *Inhomogeneous wave where* $\vec{k}_t'' \neq 0$, $n_2 < n_1$ *and* $\theta_1 > \theta_\ell$.

11.3. Coefficients for Reflection and Transmission of a Monochromatic Plane Progressive EM Wave at the Interface between Two Nonabsorbent lhi Dielectrics (n_1 and n_2 are real), and the Fresnel Equations

11.3.1. Hypothesis and aim of the study

We thus can suppose that $\underline{n}_1^2 = \varepsilon_{r1} = \varepsilon_{r1}$ are real, just as $\underline{n}_2^2 = \varepsilon_{r2} = \varepsilon_{r2}$. Once again it is assumed that the wave source imposes an angular frequency (ω) and that the state of the polarization of the incident wave, which is always a combination of two orthogonal polarization states, is either such that:

• $\vec{E}_i^0$ is perpendicular to the plane of incidence ($\vec{E}_{i\perp}^0$) in which case it is termed a transverse electric (TE) polarization, as the electric field is orthogonal to the plane of incidence (it can also be stated that the wave is in an orthogonal polarization as in Figure 11.7 a); or

• $\vec{E}_i^0$ is parallel to the plane of incidence ($\vec{E}_{i\parallel}^0$) and in which case the term is one of transverse magnetic (TM) polarization as the magnetic field thus is orthogonal to the plane of incidence. Another term used is that of parallel polarization, as shown in Figure 11.7b.

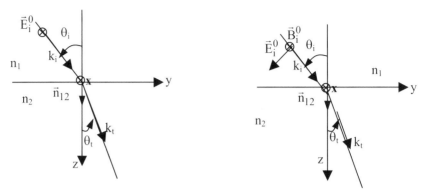

Fig. 11.7(a). *TE polarization.*   **Fig. 11.7(b)** *TM polarization.*

Given the incident wave, as in $\vec{\underline{E}}_i = \vec{E}_i^0 \exp(j[k_i \sin\theta_i \, y + k_i \cos\theta_i \, z - \omega t])$, it should be possible to determine either the vectors $\vec{\underline{E}}_r^0$ and $\vec{\underline{E}}_t^0$ for the reflected and transmitted waves, respectively, or the coefficients for the amplitudes of reflection ($\underline{r}$) and transmission ($\underline{t}$) defined by:

$$\underline{r} = \frac{E_r^0}{E_i^0} \quad \text{and} \quad \underline{t} = \frac{E_t^0}{E_i^0}. \qquad (31)$$

The $\underline{r}$ and $\underline{t}$ are *a priori* complex magnitudes that can take into account any possible dephasing between the reflection and the transmission.

We thus have two unknowns to determine, which can be done with the help of two equations established from the limiting conditions. With the media assumed to be nonmagnetic, the following two relations for continuity at the interface are used: $E_{1t} = E_{2t}$ and $B_{1t} = B_{2t}$ (as $B_{1t} = \mu_0 \, H_{1t} = \mu_0 H_{2t} = B_{2t}$ in nonmagnetic media where $\mu = \mu_0$). The indices "1" and "2" denote the media with refractive indices n_1 and n_2, respectively. Given that the study concerns monochromatic planar progressive electromagnetic (MPPEM) waves, between $\vec{E}$ and $\vec{B}$ there now is the well-used equation $\vec{B} = \dfrac{\vec{k} \times \vec{E}}{\omega}$.

11.3.2. Fresnel equations for perpendicular polarizations (TE)

11.3.2.1. When k_t is real ($n_2 > n_1$ or $n_2 > n_1$ and $\theta_1 < \theta_\ell$)

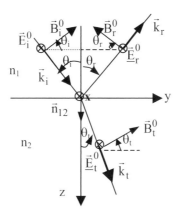

Figure 11.8 *In TE mode where $n_2 > n_1$ or $n_2 < n_1$ and $\theta_1 < \theta_\ell$.*

By symmetry, the reflection and transmission fields conserve the same polarization directions as the incident wave. It is supposed that the fields thus are placed as indicated in Figure 11.8. The eventual dephasing with respect to the reflection or the transmission will be determined through an argument around the coefficients for reflection or transmission.

For the configuration then it is possible to write equations for the continuity at the interface:

• with respect to Ox: $\underline{E}_i^0 + \underline{E}_r^0 = \underline{E}_t^0$, so that on dividing the two members by $\underline{E}_i^0$, we have:

$$1 + \underline{r}_\perp = \underline{t}_\perp \qquad (32);$$

• with respect to Oy: $\underline{B}_i^0 \cos\theta_i - \underline{B}_r^0 \cos\theta_i = \underline{B}_t^0 \cos\theta_t$, so that also

$$\frac{k_i}{\omega} \underline{E}_i^0 \cos\theta_i - \frac{k_r}{\omega} \underline{E}_r^0 \cos\theta_i = \frac{k_t}{\omega} \underline{E}_t^0 \cos\theta_t \text{ from which with}$$

$k_i = k_r = k_0 n_1$ and $k_t = k_0 n_2$ and division of two members by $\underline{E}_i^0$:

$$(1 - \underline{r}_\perp) n_1 \cos\theta_1 = \underline{t}_\perp n_2 \cos\theta_2 . \qquad (33)$$

By substituting $r_\perp = t_\perp - 1$ from Eq. (32) into Eq. (33), and with $|\theta_i| = |\theta_r| = \theta_1$ and $\theta_t = \theta_2$, it can be immediately deduced that:

$$t_\perp = \frac{2n_1\cos\theta_1}{n_1\cos\theta_1 + n_2\cos\theta_2} \quad \text{and} \quad r_\perp = \frac{n_1\cos\theta_1 - n_2\cos\theta_2}{n_1\cos\theta_1 + n_2\cos\theta_2}. \quad (34)$$

With the help of Eq. (28) ($n_1 \sin\theta_1 = n_2 \sin\theta_2$), we can eliminate n_1 and n_2 from Eq. (34), and then by multiplying the top and bottom of the preceding equations by $\dfrac{\sin\theta_2}{n_1}$ yields:

$$t_\perp = \frac{2\sin\theta_2 \cos\theta_1}{\sin(\theta_1 + \theta_2)} \quad \text{and} \quad r_\perp = \frac{\sin(\theta_2 - \theta_1)}{\sin(\theta_1 + \theta_2)}. \quad (34')$$

The angles θ_1 and θ_2 vary at most between 0 and $\dfrac{\pi}{2}$, and $\sin\theta_2$ and $\cos\theta_1$ are always positive just as is $\sin(\theta_1 + \theta_2)$ as $(\theta_1 + \theta_2)$ also has a variation limited to between 0 and π. The $t_\perp$ is in fact always real and positive here, and the transmitted wave does not exhibit a dephasing with respect to the incident wave.

10.3.2.1.1. When $n_2 > n_1$

With k_t being real, in agreement with the relation $n_1 \sin\theta_1 = n_2 \sin\theta_2$, we have $\theta_2 < \theta_1$, and $r_\perp$ is always real and negative as the maximum variation in $(\theta_2 - \theta_1)$ is given by $-\dfrac{\pi}{2} < (\theta_2 - \theta_1) < 0$. We can conclude therefore that a TE wave reflected by the more refractive material undergoes at the reflection a dephasing ($\varphi_\perp$) such that $\exp(j\varphi_\perp) = -1$, so that:

$\varphi_\perp = \pi$ (as $\exp(j\pi) = \cos\pi + j \sin\pi = -1$).

In the limiting case, Eq. (34) shows that:

• for a normal incidence, where $\theta_1 = 0$, $r_\perp = \dfrac{n_1 - n_2}{n_1 + n_2} < 0$, so that in terms of moduli:

$$|r_\perp| = \frac{n_2 - n_1}{n_1 + n_2} \quad (35)$$

• for a glancing incidence, where $\theta_1 = \dfrac{\pi}{2}$ then $r_\perp = -1$.

As θ_1 increases, the modulus of $r_\perp$ continuously increases from a value given by Eq. (35) up to unity. Finally, we obtain the representation given in Figure 4.9 for when $n_2 > n_1$.

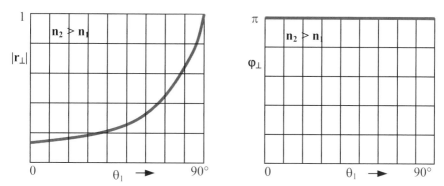

Figure 11.9. *Modulus and phase angle of the reflection coefficient as a function of the angle of incidence (θ_1) for a TE wave with $n_2 > n_1$.*

11.3.2.1.2. When $n_2 < n_1$ and $\theta_1 < \theta_\ell$ (k_t real)

According to Eq. (28), $\theta_2 > \theta_1$ and $0 < (\theta_2 - \theta_1) < \dfrac{\pi}{2}$ so that $\sin(\theta_2 - \theta_1) > 0$, and according to Eq. (34'), $r_\perp$ is real and positive, so that when medium (2) is less refractive than medium (1) and when $\theta_1 < \theta_\ell$, the reflected wave rests in phase with the incident wave. According to Eq. (34), we can see that as θ_1 increases, the modulus of $r_\perp$ steadily increases from a value given by $r_\perp = \dfrac{n_1 - n_2}{n_1 + n_2} > 0$, obtained for a normal incident wave ($\theta_1 = 0$), up to the unit value given for the limit of incidence ($\theta_1 = \theta_\ell$), as at this point θ_2 takes on the value $\dfrac{\pi}{2}$. Figure 11.10 gives a representation of this zone.

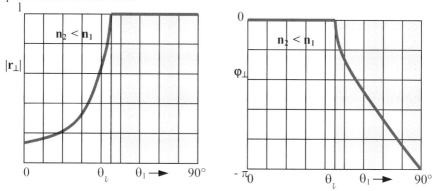

Figure 11.10. *Modulus and phase angle of the reflection coefficient as a function of the angle of incidence (θ_1) for a TE wave and with $n_2 < n_1$.*

11.3.2.2. When k_t is complex $(n_2 < n_1$ and $\theta_1 > \theta_\ell)$, the wave is inhomogeneous and is as shown in Figure 11.6

The relations for the continuity at the interface (Figure 11.8) give:

- with respect to Ox, we again find Eq. (32); and
- with respect to Oy, we have:

$$\frac{k_i}{\omega} E_i^0 \cos\theta_i - \frac{k_r}{\omega} E_r^0 \cos\theta_i = \frac{1}{\omega}\left(\vec{k}_t \times \vec{E}_t^0\right)\vec{e}_y = \frac{1}{\omega}\left(\left[\vec{k}_t\right]_y + \left[\vec{k}_t\right]_z\right) \times \vec{E}_t^0 \vec{e}_y .$$

As $\vec{E}_t^0$ is directed along Ox (TE wave), the term on the right-hand side is reduced to

$\frac{1}{\omega}\left(\left[\vec{k}_t\right]_z \times \vec{E}_t^0\right)\vec{e}_y$. With $\left[\vec{k}_t\right]_z = ik_{tz}''\vec{e}_z$ and by simplifying with $\dfrac{E_i^0}{\omega}$ the two members of the continuity equation for Oy we have:

$$(1 - \underline{r}_\perp) k_1 \cos\theta_1 = i k_{tz}'' \underline{t}_\perp,$$

with according Eq. (30') : $k_{tz}'' = k_0\left(n_1^2\sin^2\theta_1 - n_2^2\right)^{1/2}$. We thus obtain

$$\underline{t}_\perp = \frac{2k_1\cos\theta_1}{k_1\cos\theta_1 + ik_{tz}''} \quad \text{and} \quad \underline{r}_\perp = \frac{k_1 \cos\theta_1 - ik_{tz}''}{k_1 \cos\theta_1 + ik_{tz}''} . \qquad (36)$$

For $\underline{r}_\perp$, the numerator is the conjugate complex of the denominator, for which $|\underline{r}_\perp| = 1$. The angle for dephasing at the reflection thus is given by:

$\varphi_\perp = \text{Arg}(\underline{r}_\perp)$ tel que $\underline{r}_\perp = |\underline{r}_\perp| \exp(i\varphi_\perp) = \exp(i\varphi_\perp)$.

We   therefore   have $\underline{r}_\perp = \exp(i\varphi_\perp) = \dfrac{\overline{z}}{z}$, and   then   by   making

$z = k_1 \cos\theta_1 + ik_{tz}'' = \rho\exp(i\,\alpha)$, we also find that

$$\underline{r}_\perp = \exp(i\varphi_\perp) = \frac{\overline{z}}{z} = \frac{\rho \exp(-i\alpha)}{\rho \exp(i\alpha)} = \exp(-i\,2\alpha) , \text{ so that}$$

$$\tan\left(-\frac{\varphi_\perp}{2}\right) = \tan\alpha = \frac{k_{tz}''}{k_1 \cos\theta_1} .$$

By using $n_2 = n_1 \sin\theta_1$ (which makes it possible to replace n_2^2 in Eq. (30') by $n_1^2 \sin^2\theta_\ell$), we find that (in the shaded zone of Figure 11.10):

$$\tan\left(\frac{\varphi_\perp}{2}\right) = = -\frac{\left(\sin^2\theta_1 - \sin^2\theta_\ell\right)^{1/2}}{\cos\theta_1} . \qquad (37)$$

If $\theta_1 = \theta_\ell$, $\varphi_\perp = 0$, and if $\theta_1 = \dfrac{\pi}{2}$, $\dfrac{\varphi_\perp}{2} = -\dfrac{\pi}{2}$, thus $\varphi_\perp = -\pi$ (see the plot on the shaded part of Figure 11.10).

10.3.3. *Fresnel's equations for parallel magnetic field polarizations*

10.3.3.1. When k_t is real $(n_2 > n_1$ or $n_2 < n_1$ and $\theta_1 < \theta_\ell)$.

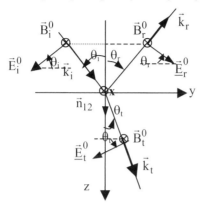

Figure 11.11. *TM wave with $n_2 > n_1$ or $n_2 < n_1$, and $\theta_1 < \theta_\ell$.*

As detailed in Section 10.3.2, the fields retain the same polarization directions at reflection or transmission through symmetry. We thus can assume that the fields, once reflected and transmitted, are laid out as described in Figure 11.11. For this configuration, the equations for continuity at the interface (still with $|\theta_i| = |\theta_r| = \theta_1$ and $\theta_t = \theta_2$) give:

• with respect to Ox: $B_i^0 + \underline{B}_r^0 = \underline{B}_t^0$, so in addition,

$$\frac{k_0 \; n_1 \; E_i^0}{\omega} + \frac{k_0 \; n_1 \; \underline{E}_r^0}{\omega} = \frac{k_0 \; n_2 \; \underline{E}_t^0}{\omega} \quad \text{from which can be deduced that:}$$

$$n_1(1 + \underline{r}_{\parallel}) = n_2 \; \underline{t}_{\parallel} \quad (38); \text{ and}$$

- with respect to Oy: $- E_i^0 \cos\theta_1 + E_r^0 \cos\theta_1 = - E_t^0 \cos\theta_2$, from which:

$$\cos\theta_1(1 - \underline{r}_{\parallel}) = \underline{t}_{\parallel} \; \cos\theta_2 . \quad (39)$$

From this can be deduced that:

$$\underline{r}_{\parallel} = \frac{n_2 \; \cos\theta_1 - n_1 \; \cos\theta_2}{n_2 \; \cos\theta_1 + n_1 \; \cos\theta_2} \quad \text{and} \quad \underline{t}_{\parallel} = \frac{2n_1 \; \cos\theta_1}{n_2 \; \cos\theta_1 + n_1 \; \cos\theta_2} . \quad (40)$$

By again using Eq. (28), it is possible to eliminate n_1 and n_2 from the above equations, so that they become:

$$r_\parallel = \frac{\sin\theta_1 \cos\theta_1 - \sin\theta_2 \cos\theta_2}{\sin\theta_1 \cos\theta_1 + \sin\theta_2 \cos\theta_2} = \frac{\sin2\theta_1 - \sin2\theta_2}{\sin2\theta_1 + \sin2\theta_2} = \frac{2\cos(\theta_1+\theta_2)\sin(\theta_1-\theta_2)}{2\sin(\theta_1+\theta_2)\cos(\theta_1-\theta_2)},$$

from which $r_\parallel = \dfrac{\tan(\theta_1 - \theta_2)}{\tan(\theta_1 + \theta_2)}$.   (40')

Similarly, we obtain $t_\parallel = \dfrac{2\cos\theta_1 \sin\theta_2}{\sin(\theta_1 + \theta_2)\cos(\theta_2 - \theta_1)}$.   (40'')

Using Eq. (40), it is immediately evident that $t_\parallel$ is always real and positive as $\cos\theta_1$ and $\cos\theta_2$ are both always positive (θ_1 and θ_2 being between 0 and $\dfrac{\pi}{2}$). Hence the dephasing between the incident wave and the transmitted wave is always zero.

With regard to the reflected wave, according to Eq. (40), it is possible to state that $r_\parallel = r_\parallel$ is a real magnitude and the dephasing of the reflected wave is dependent on the sign of $r_\parallel$, which in turn is such that:

$$\operatorname{sgn} r_\parallel = \{\operatorname{sgn}[\tan(\theta_1 - \theta_2)]\} \times \{\operatorname{sgn}[\tan(\theta_1 + \theta_2)]\}.$$

11.3.3.1.1. When $n_2 > n_1$: k_t is real and $\theta_2 < \theta_1$

For a normal incidence ($\theta_1 = 0°$), according to Eq. (40) $r_\parallel = \dfrac{n_2 - n_1}{n_2 + n_1} > 0$ so that $r_\parallel = r_\parallel$ is a magnitude that is both real and positive, and the dephasing $\varphi_\parallel$ thus is zero.

In general terms though, $\theta_2 < \theta_1$ and $\theta_1 \in [0, \pi/2]$, $(\theta_1 - \theta_2) \in [0, \pi/2]$, and $\operatorname{sgn}[\tan(\theta_1 - \theta_2)]$ always is positive. The result is such that $\{\operatorname{sgn} r_\parallel\}$ changes sign with $\{\operatorname{sgn}[\tan(\theta_1 + \theta_2)]\}$ and $r_\parallel$ goes from being positive to negative when $\theta_1 = \theta_B$ so that:

$$(\theta_B + \theta_2) = \frac{\pi}{2}.   (41)$$

The angle of incidence, $\theta_1 = \theta_B$, is called Brewster's angle and is such that $\tan(\theta_B + \theta_2) = \tan\dfrac{\pi}{2} = \infty$, with $r_\parallel = 0$. The wave therefore is entirely transmitted.

When $\theta_1 = \theta_B$, so that $\theta_2 = \dfrac{\pi}{2} - \theta_B$, we have

$n_1 \sin \theta_B = n_2 \sin \theta_2 = n_2 \sin \left(\dfrac{\pi}{2} - \theta_B\right) = n_2 \cos\theta_B$, from which:

$$\boxed{\tan \theta_B = \dfrac{n_2}{n_1}.}\qquad (42)$$

At an air/water interface, we have $\theta_B = $ Arc tan $1.5 \approx 57°$.

With $\theta_1 = 90°$, we have $r_\parallel = -1$ and $|r_\parallel| = 1$ and $\varphi_\parallel = \pi$.

So, to display the results, the plots of $|r_\parallel| = f(\theta_1)$ and $\varphi_\parallel = g(\theta_1)$ are shown in Figure 11.12.

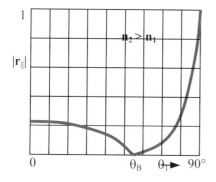
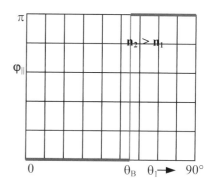

Figure 11.12. *Modulus and phase angle for the reflection coefficient as a function of θ_1 for a TM wave with $n_2 > n_1$.*

11.3.3.1.2. When $n_2 < n_1$ and $\theta_1 < \theta_\ell$ (k_t is real), thus here with $\theta_2 > \theta_1$

For   a   normal   incident   wave $(\theta_1 = 0°)$ according   to   Eq.   (40)

$\underline{r}_\parallel = \dfrac{n_2 - n_1}{n_2 + n_1} < 0$ and $\underline{r}_\parallel = r_\parallel$ is a real and negative magnitude. The dephasing ($\varphi_\parallel$) thus is equal to π.

When $\theta_2 > \theta_1, (\theta_1 - \theta_2) \in [0, -\pi/2]$ and $\text{sgn}\left[\tan(\theta_1 - \theta_2)\right]$ is always negative; the result is that $r_\parallel$ goes from being negative to positive when $\left[\tan(\theta_1 + \theta_2)\right]$

changes sign. This also happens when $\theta_1 = \theta_B$ such that $(\theta_B + \theta_2) = \dfrac{\pi}{2}$.

Simultaneously, $\varphi\|$ goes from π to 0.

We still have $\tan\theta_B = \dfrac{n_2}{n_1}$ though, and also can state that $\theta_B < \theta_\ell$. In effect,

$$\sin\theta_\ell = \dfrac{n_2}{n_1} = \tan\theta_B = \dfrac{\sin\theta_B}{\cos\theta_B} > \sin\theta_B \text{ , so that } \theta_\ell > \theta_B \,.$$

A representation of this is given in Figure 11.13.

It is worth noting that for the example of air and water, $n_1 = 1.5$ and $n_2 = 1$.

We therefore have $\theta_\ell = \text{Arc}\sin\left(\dfrac{1}{1.5}\right) \approx 42\,°$ and $\theta_B = \text{Arc}\tan\left(\dfrac{1}{1.5}\right) \approx 34\,°$.

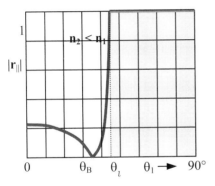 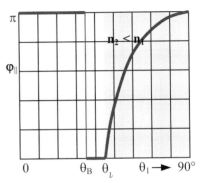

Figure 11.13. *Modulus and phase angle for the reflection coefficient as a function of θ_1 for a TM wave, with $n_2 < n_1$.*

11.3.3.2. When k_t is complex ($n_2 < n_1$ and $\theta_1 > \theta_\ell$)

In this situation it is not an easy task to reuse the reasoning given in Section 11.3.2.2; the projections made demand a symmetry around the angle $\theta_t \equiv \theta_2$, which has no real physical significance as the wave does not propagate only along the Oy axis. Mathematically, we can check that θ_2 is not a real angle, as in effect,

$$\sin\theta_2 = \dfrac{n_1}{n_2}\sin\theta_1 = \dfrac{\sin\theta_1}{\sin\theta_\ell}, \text{ so that with } \theta_1 > \theta_\ell \text{ we have } \sin\theta_2 > 1.$$

In Fresnel's equations, in the place of the usual equation, $\cos\theta_2 = \sqrt{1 - \sin^2\theta_2}$, now with $\sin^2\theta_2 > 1$ we must bring in for the term $\cos\theta_2$ a purely imaginary value given by $\cos\theta_2 = -i\sqrt{\sin^2\theta_2 - 1}$. The negative sign is required by the need to obtain at a later point an attenuation of the wave in medium (2). We also can state that $\cos\theta_2 = -i\sqrt{\dfrac{\sin^2\theta_1}{\sin^2\theta_\ell} - 1}$, and with $\dfrac{n_2}{n_1} = \sin\theta_\ell$, Eq. (40) gives:

$$\underline{r}_\| = \frac{\sin^2\theta_\ell\,\cos\theta_1 + j\sqrt{\sin^2\theta_1 - \sin^2\theta_\ell}}{\sin^2\theta_\ell\,\cos\theta_1 - j\sqrt{\sin^2\theta_1 - \sin^2\theta_\ell}} . \qquad (43)$$

Once again the numerator is the conjugate complex of the denominator, for which $|\,r_\||^2 = 1$. By writing $\underline{r}_\| = \exp(i\varphi_\|) = \dfrac{z}{\overline{z}}$, and by making

$$z = \sin^2\theta_\ell\,\cos\theta_1 + j\sqrt{\sin^2\theta_1 - \sin^2\theta_\ell} = \rho\,\exp(i\,\beta) , \text{ we obtain:}$$

$$\underline{r}_\| = \exp(i\varphi_\|) = \frac{z}{\overline{z}} = \frac{\rho\,\exp(\,i\beta)}{\rho\,\exp(-\,i\beta)} = \exp(\,i\,2\beta), \text{ so that with } \tan\left(\frac{\varphi_\|}{2}\right) = \tan\beta:$$

$$\tan\left(\frac{\varphi_\|}{2}\right) = \frac{\sqrt{\sin^2\theta_1 - \sin^2\theta_\ell}}{\cos\theta_1\,\sin^2\theta_\ell} . \qquad (44)$$

If $\theta_1 = \theta_\ell$, then $\tan\left(\dfrac{\varphi_\|}{2}\right) = 0$ and $\varphi_\| = 0$. If $\theta_1 = \dfrac{\pi}{2}$, then

$\cos\theta_1 = 0$, $\tan\left(\dfrac{\varphi_\|}{2}\right) = \infty$, and $\varphi_\| = \pi$ (see also the plot on the shaded area of Figure 11.13).

11.3.3.3. Comment

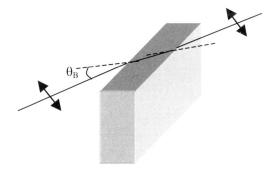

Figure 11.14. *A Brewster angle incidence for a TM wave at an air/glass window.*

To ensure that a ray passes through a glass/air interface, as shown in Figure 11.14, without loss by reflection, the wave should be TM polarized and undergo a Brewster type incidence as $\left[r_\parallel\right]_{\theta_1=\theta_B} = 0$ (such as is used in Brewster lasers with a Brewster window).

11.3.4. Reflection coefficients and energy transmission

11.3.4.1. Aide mémoire: energy flux through a surface

The energy flux, associated with the propagation of an electromagnetic field in a material, is equal to the flux traversing the surface given by the Poynting vector ($\vec{S}$).

Equation (33') of Chapter 7, Section 7.3.3, shows that for a MPPEM wave in nonmagnetic materials the average value for the Poynting vector is given by

$\langle\vec{S}\rangle = \dfrac{1}{2\mu_0}R(\underline{\vec{E}}_m \times \underline{\vec{B}}_m^*)$, so that with $\omega\underline{\vec{B}}_m = \vec{k}\times\underline{\vec{E}}_m$ [Eq. (13) of Chapter 7] we

have $\langle\vec{S}\rangle = \dfrac{1}{2\mu_0\omega}\vec{k}\left|\underline{\vec{E}}_m\right|^2$,   and   with $\dfrac{1}{\mu_0\omega} = \dfrac{\varepsilon_0 c}{k_0}$,   we   can   write   that

$\langle\vec{S}\rangle = \dfrac{\varepsilon_0 c}{2}\dfrac{\vec{k}}{k_0}\left|\underline{\vec{E}}_m\right|^2$. With our notations, and for incident, reflected, and transmitted

wave, the associated average value of the Poynting vector is, respectively:

$$\langle\vec{S}_i\rangle = \frac{\varepsilon_0 c}{2}\frac{\vec{k}_i}{k_0}\left|\vec{E}_i^0\right|^2, \quad \langle\vec{S}_r\rangle = \frac{\varepsilon_0 c}{2}\frac{\vec{k}_r}{k_0}\left|\vec{E}_r^0\right|^2, \quad \langle\vec{S}\rangle = \frac{\varepsilon_0 c}{2}\frac{\vec{k}_t}{k_0}\left|\vec{E}_t^0\right|^2.$$

The energy flux across the surface Σ is the energy transmitted per unit time through the surface. In effect, it also represents the power transmitted by the wave under consideration.

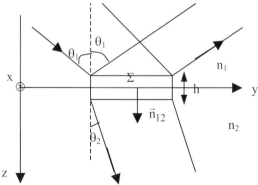

Figure 11.15. *Reflection and transmission of an incident flux traversing Σ.*

The radiated power of the wave traversing an interface with surface (Σ) from a medium with an index denoted by n_1 toward a medium with an index n_2 (Figure 11.15) thus is given by:

$$
\left.\begin{aligned}
<P_i> &= \left\langle \vec{S}_i \right\rangle \vec{n}_{12} \Sigma = \frac{\varepsilon_0 c}{2} \frac{\vec{k}_i}{k_0} \left|\vec{E}_i^0\right|^2 \vec{n}_{12} \Sigma = \frac{\varepsilon_0 c}{2} n_1 \left|\vec{E}_i^0\right|^2 \Sigma \cos\theta_1 \\
<P_r> &= \left\langle \vec{S}_r \right\rangle \vec{n}_{12} \Sigma = \frac{\varepsilon_0 c}{2} \frac{\vec{k}_r}{k_0} \left|\vec{E}_r^0\right|^2 \vec{n}_{12} \Sigma = -\frac{\varepsilon_0 c}{2} n_1 \left|\vec{E}_r^0\right|^2 \Sigma \cos\theta_1 \\
<P_t> &= \left\langle \vec{S}_t \right\rangle \vec{n}_{12} \Sigma = \frac{\varepsilon_0 c}{2} \frac{\vec{k}_t}{k_0} \left|\vec{E}_t^0\right|^2 \vec{n}_{12} \Sigma = \frac{\varepsilon_0 c}{2} \left|\vec{E}_t^0\right|^2 \Sigma \frac{k'_{tz}}{k_0}.
\end{aligned}\right\} \quad (45)
$$

11.3.4.2. Equation for the reflection(R) and transmission (T) coefficients: total energy

The titled coefficients are defined so as to give positive values, so that we have:

$$
R = - \frac{\langle P_r \rangle}{\langle P_i \rangle} \quad \text{and} \quad T = \frac{\langle P_t \rangle}{\langle P_i \rangle}. \quad (46)
$$

From this can be determined with the equations in Eq. (45) that:

$$
R = \frac{\left|\vec{E}_r^0\right|^2}{\left|\vec{E}_i^0\right|^2} = |\underline{r}|^2 \quad \text{and} \quad T = \frac{\left|\vec{E}_t^0\right|^2}{\left|\vec{E}_i^0\right|^2} \frac{k'_{tz}}{k_0\, n_1\, \cos\theta_1} = |\underline{t}|^2 \frac{k'_{tz}}{k_0\, n_1\, \cos\theta_1}. \quad (47)
$$

It is notable that in the case of the transmitted power, $T \neq |\underline{t}|^2$. This is due to the fact that the way in which the power is transported depends on the different cross sections, so that $\Sigma_i = \Sigma \cos\theta_1$ for an incident wave and $\Sigma_t = \Sigma \cos\theta_2$ for a transmitted wave (while for incident and transmitted powers, $\Sigma_i = \Sigma_r = \Sigma \cos\theta_1$).

With respect to the sum of the energy, we can calculate this by using a closed surface, such as a parallelepiped of a given height (h) such that $h \to 0$, and then drawn around the level of the surface (Figure 11.15). In physical terms, for an isolated system, the incoming radiation power must be equal to that leaving, so:
- the incoming radiation power is that of the incident wave, such that

$$
\vec{n}_{12}: \quad <P_i> = \left\langle \vec{S}_i \right\rangle \vec{n}_{12} \Sigma .
$$

- the outgoing power radiated has two components:

- power radiated by the reflected wave with respect to $-\vec{n}_{12}$

$$-\langle \vec{S}_r \rangle \vec{n}_{12} \Sigma = -<\text{Pr}>$$

- the power radiated by the transmitted wave with respect to $\vec{n}_{12}$

$$<P_t> = \langle \vec{S}_t \rangle \vec{n}_{12} \Sigma$$

We thus should have $<P_i> = -<P_r> + <P_t>$. Dividing through term by term by $<P_i>$ we obtain:

$$\boxed{R + T = 1} . (48)$$

Comment: The transmission coefficient for a homogeneous transmitted wave $(n_1 < n_2$ or $n_1 > n_2$ and $\theta_1 < \theta_\ell)$.

In this case, $\underline{\vec{k}}_t$ is real and equal to $\underline{\vec{k}}_t'$. With $k_{tz}' = k_0 n_2 \cos \theta_2$, we have

$$T = |\underline{t}|^2 \frac{n_2 \cos\theta_2}{n_1 \cos\theta_1}. (49)$$

11.3.4.3. Representation of the reflection coefficient as a function of the angle of incidence

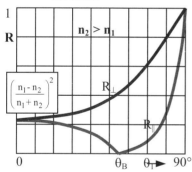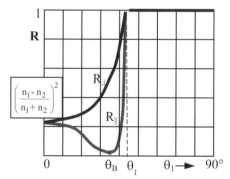

Figure 11. 16 a.

Factors in the reflection of energy:
$R_\perp = |r_\perp|^2$ and $R_{||} = |r_{||}|^2$ *as a function of*
θ_1, *for* $n_2 > n_1$ (*with* $r_\perp$ *and* $r_{||}$
from Figures 11.9 and 11.12)
$\theta_B = 57°$ *for the air/glass interface.*

Figure 11.16 b.

Factors in the reflection of energy:
$R_\perp = |r_\perp|^2$ and $R_{||} = |r_{||}|^2$
as a function of θ_1,
for $n_2 < n_1$ (*with* $r_\perp$ *and* $r_{||}$
from Figures 11.10 and 11.13).
$\theta_B = 34°$ *for an air/glass interface,*
and $\theta_\ell = 42°$.

Taking the equation $R = |\underline{r}|^2$ into account, we can use the representations given in Figure 11.15 for when $n_2 > n_1$, or $n_1 > n_2$.

For normal incidence, $\theta_1 = \theta_2 = 0$, and $R = \left(\dfrac{n_1 - n_2}{n_1 + n_2} \right)^2$.

At the same time, $T = |t|^2 \dfrac{n_2}{n_1} = \left(\dfrac{2n_1}{n_1 + n_2} \right)^2 \dfrac{n_2}{n_1} = \dfrac{4n_1 n_2}{(n_1 + n_2)^2} = 1 - R$.

It can be seen in the plots of Figure 11.16 that when $n_2 > n_1$ the reflection is always partial and relatively weak, except in the neighborhood of the glancing angle $(\theta_1 \approx \dfrac{\pi}{2})$. This is the reason why the air/water interface is always transparent $(n_2 \approx 1.33)$ except when there is a glancing angle, or under conditions of reflection.

Under a Brewster angle, only a wave polarized perpendicularly to the plane of incidence (TE waves) is partially reflected. At this incidence, the TM wave is entirely transmitted $(R_\| = 0$ when $\theta_1 = \theta_B)$.

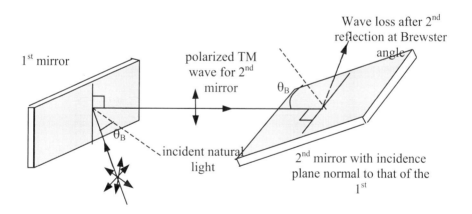

Figure 11.17. *Malus's experiment.*

Under natural light, the incident wave exhibits all polarization states, and for an incidence with the Brewster angle it is only waves polarized perpendicularly to the plane of incidence which is reflected (TE wave). It is this result that explains physically Malus's law, which demonstrates the polarization of light with the help of two mirrors set at a Brewster angle, so that the planes of incidence are orthogonal. The two successive reflections are at Brewster angles so that after the first reflection of natural light there remains only a component that is polarized perpendicularly to the plane of incidence of the first mirror (thus a TE wave for the first mirror). The incidence plane of the second mirror is perpendicular to that of the first, and the incoming wave is seen as a TM wave. Following a reflection at the Brewster angle on the second mirror, there is a complete extinction of the wave.

11.3.5. Total and frustrated total reflection

When $n_1 > n_2$ and $\theta_1 > \theta_\ell$, we have $k'_{tz} = 0$ and $\vec{k}'_t = k'_{ty}\,\vec{e}_y$; there no longer is a propagation along Oz and only propagation along Oy persists.

As $R = |r|^2 = 1$, we have $T = 1 - R = 0$, while at the same time with $t \neq 0$ (see Eq. (36), for example).

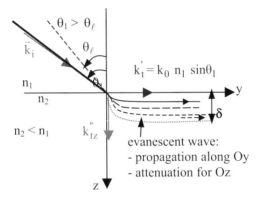

Figure 11.18. *The evanescent wave and frustrated total reflection.*

Therefore, even though there is a transmitted wave, as $t \neq 0$, the transmission coefficient for energy is zero. This is because the propagation through medium (2) can occur only at the interface (along $\vec{e}_y$). The wave itself is evanescent (Figure 11.18 being the "completed" Figure 11.6) with wavevector components ($\vec{k}_t$) given by Eq. (30'). By substituting the value for the components into Eq. (21), we find that the transmitted wave has the form:

$$\vec{E}_t(\vec{r},t) = \vec{E}^0_t\,\exp(-k''_{tz}z)\,\exp\left[i\left(k_{iy}y - \omega t\right)\right]\ ,$$

with $k''_{tz} = k_0(n_1^2\sin^2\theta_1 - n_2^2)^{1/2} = k_0 n_2\left(\dfrac{n_1^2}{n_2^2}\sin^2\theta_1 - 1\right)^{1/2}$.

The intensity of the transmitted wave, proportional to $\left|\vec{E}_t(\vec{r},t)\right|^2$, is thus of the form:

$$I_t(z) = \left|\vec{E}_t^0\right|^2 \exp(-2k_{tz}^{"}) = I_t(0)\exp(-\frac{2z}{\delta}), \text{ with } \delta = \frac{1}{k_{tz}^{"}} = \frac{1}{k_0 n_2}(\frac{n_1^2}{n_2^2}\sin^2\theta_1 - 1)^{-1/2}$$

where δ represents the degree of attenuation (when $z = \dfrac{\delta}{2}$, the wave intensity is divided by e).

The wavelength in medium (2) is given by $\lambda_2 = \dfrac{\lambda_0}{n_2}$, so that also we have:

$$\delta = \frac{\lambda_2}{2\pi}(\frac{n_1^2}{n_2^2}\sin^2\theta_1 - 1)^{-1/2}. \qquad (50)$$

If θ_1 increases, $\sin\theta_1$ also increases, as $(\dfrac{n_1^2}{n_2^2}\sin^2\theta_1 - 1)$, while $(\dfrac{n_1^2}{n_2^2}\sin^2\theta_1 - 1)^{-1/2}$ decreases with δ and the evanescent wave penetrates less medium (2).

Numerically, for the glass/air interface, and at what is practically a glancing angle, $(\theta_1 \approx \dfrac{\pi}{2})$, we find that $\delta \approx 0.14\,\lambda_2$.

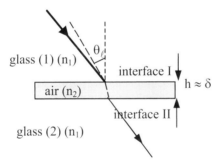

Figure 11.19. *Experimental demonstration of frustrated total reflection.*

The presence of an electromagnetic wave in a thickness of the order of δ of the second medium (such that T = 0) can be demonstrated by using the setup shown in Figure 11.19. At the level of the first interface (interface I) the wave penetrates to a depth of the order of δ and is then transmitted to the level of the second interface II that is such that its index [glass (2) with index n_1] is higher than that of the first medium (air with index n_2). In other terms, following its passage from interface I, the wave is attenuated over a depth of $h \approx \delta$ and what "remains'" at this depth is then transmitted to the level of the interface II, which in turn does not follow the conditions of total reflection. The result is that of frustrated total reflection.

In order to carry out such an experiment, a total reflection prism can be used with angles equal to 45°, which is greater than the 42° required for the glass/air interface ($\theta_1 = 45° > \theta_\ell$). By placing a second prism at a distance $h \approx \delta$ from the first, as described in Figure 11.20, a frustrated total reflection occurs exhibited by the presence of part of the incident wave in the right-hand side.

However, as mentioned above, the two prisms must be placed a distance apart given by $h \approx \delta \approx 0.14 \lambda_2$, which means that $h \approx 0.1$ μm for waves in the optical domain. This requires extremely delicate handling and can be made easier by the use of waves with wavelengths of the order of centimeters thus requiring $h \approx$ several millimeters (which is easy with paraffin prisms).

The results obtained for electromagnetic waves can be extended to waves associated with material particles. Frustrated total reflection can be considered that of a tunneling effect, in this case applied to photons.

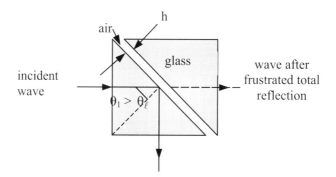

Figure 11.20. *A total reflection prism used to demonstrate frustrated total reflection.*

11.4. Reflection and Absorption by an Absorbing Medium

11.4.1. Reflection coefficient for a wave at a normal incidence to an interface between a nonabsorbent medium (1) (index of n_1) and an absorbent medium (2) (index of $\underline{n}_2$)

The index of medium (2) is complex and given by $\underline{n}_2 = n_2' + i\,n_2''$. For its part, the incident wave is assumed to be normal to the interface so that $\theta_1 = 0°$ and $\sin \theta_1 = 0$, and the wavevector of the transmitted wave thus can be written, with $k_{ty}' = k_0\,n_1\,\sin\theta_1 = 0$, as:

$$\vec{k}_t = \left(k_t' + i k_t'' \right) \vec{e}_z \text{ , where } \begin{cases} k_{tz}' = k_0 \ R\left(\underline{n}_2^2 - n_1^2 \sin^2\theta_1 \right)^{1/2} = k_0 n_2' \\ k_{tz}'' = k_0 \ Im\left(\underline{n}_2^2 - n_1^2 \sin^2\theta_1 \right)^{1/2} = k_0 n_2'' \end{cases}$$

so that $\vec{k}_t = k_0 \left(n_2' + n_2'' \right) \vec{e}_z = k_0 \ \underline{n}_2 \ \vec{e}_z$.

The EM field associated with the transmitted wave $\left(\vec{E}_t , \ \vec{B}_t \right)$ thus is given by:

$$\begin{cases} \vec{E}_t = \vec{E}_t^0 \exp\left(-k_0 n_2'' z \right) \exp\left(i[k_0 n_2' z - \omega t] \right) \\ \vec{B}_t = \dfrac{\vec{k}_t \times \vec{E}_t}{\omega} = \dfrac{\underline{n}_2}{c} \vec{e}_z \times \vec{E}_t . \end{cases}$$

With the index $\underline{n}_2$ being complex, $\vec{B}_t$ is no longer in phase with $\vec{E}_t$. Taking into account the form of the fields, we can equally state that this MPPEM wave propagates along Oz with a phase velocity given by $v_\varphi = \dfrac{\omega}{k_0 n_2'} = \dfrac{c}{n_2'}$ and is attenuated exponentially along Oz in the absorbent medium (2).

The condition of continuity at the interface of the tangential components for the fields $\vec{E}$ and $\vec{B}$ results in the same values as those cited previously for reflection and amplitude transmission coefficients, while nevertheless noting that $\underline{n}_2$ is in a complex form and that:

$$\underline{r} = \frac{n_1 - \underline{n}_2}{n_1 + \underline{n}_2} = \frac{n_1 - n_2' - i n_2''}{n_1 + n_2' + i n_2''} \text{ and } \underline{t} = \frac{2 n_1}{n_1 + \underline{n}_2} = \frac{2 n_1}{n_1 + n_2' + i n_2''} .$$

The arguments with respect to $\underline{r}$ and $\underline{t}$, respectively, give the dephasing of the reflected and the transmitted wave with respect to the incident wave.

For the reflection and transmission coefficients in terms of energy, we also obtain:

$$R = |\underline{r}|^2 = \frac{\left(n_1 - n_2' \right)^2 + n_2''^2}{\left(n_1 + n_2' \right)^2 + n_2''^2} \text{ and } T = 1 - R = \frac{4 n_1 n_2'}{\left(n_1 + n_2' \right)^2 + n_2''^2} .$$

11.4.2. Optical properties of a metal: reflection and absorption at low and high frequencies by a conductor

See Problem 1 of the present chapter.

11.5. The Antiecho Condition: Reflection from a Magnetic Layer; a Study of an Antiradar Structure; and a Dallenbach Layer

11.5.1. The antiecho condition: reflection from a nonconducting magnetic layer

11.5.1.1. Reflection from a magnetic layer

This follows on from the study in Section 11.3.2 on TE polarization by reflection at an interface between media (1) and (2) with the difference that here the media have a permeability different to that of a vacuum such that media (1) and (2), respectively, exhibit permeabilities μ_1 and μ_2.

It is assumed that the media are sufficiently insulating so that the density of free charges is negligible so that $\sigma_\ell = 0$ and $j_\ell = 0$.

Under such conditions, the continuity conditions for the interface mean that $\vec{E}_{1t} = \vec{E}_{2t}$ and $\vec{H}_{1t} = \vec{H}_{2t}$ and in turn $\dfrac{\vec{B}_{1t}}{\mu_1} = \dfrac{\vec{B}_{2t}}{\mu_2}$ with in addition $B_{1t} = B_{2t}$ as specifically to this case $\mu_1 \neq \mu_2$ (see also the end of Section 11.3.1). Equation (32) then becomes:

- along Ox: $\underline{E}_i^0 + \underline{E}_r^0 = \underline{E}_t^0$ in an unchanged relation due to the conservation of the continuity equation. By dividing the two terms by $\underline{E}_i^0$, we also find

$$1 + \underline{r}_\perp = \underline{t}_\perp, \qquad (51) \text{ and}$$

- along Oy: $\dfrac{k_i}{\omega\mu_1} \underline{E}_i^0 \cos\theta_i - \dfrac{k_r}{\omega\mu_1} \underline{E}_r^0 \cos\theta_i = \dfrac{k_t}{\omega\mu_2} \underline{E}_t^0 \cos\theta_t .$ (52)

With the media being nonconducting, i.e., $\sigma \approx 0$, we can state that in each medium, according to Eq. (5) of Section 8.5.1,

$k_i = k_r = \omega\sqrt{\varepsilon_1\,\mu_1}$ and $k_t = \omega\sqrt{\varepsilon_2\,\mu_2}$. The preceding equation with respect to

Oy becomes: $\sqrt{\dfrac{\varepsilon_1}{\mu_1}}\underline{E}_i^0 \cos\theta_i - \sqrt{\dfrac{\varepsilon_1}{\mu_1}}\underline{E}_r^0 \cos\theta_i = \sqrt{\dfrac{\varepsilon_2}{\mu_2}}\underline{E}_t^0 \cos\theta_t .$

With Eq. (6) from Section 8.5.2, i.e., $\dfrac{1}{Z_1} = \sqrt{\dfrac{\varepsilon_1}{\mu_1}}$, and $\dfrac{1}{Z_2} = \sqrt{\dfrac{\varepsilon_2}{\mu_2}}$, we have:

$Z_1^{-1}\underline{E}_i^0 \cos\theta_i - Z_1^{-1}\underline{E}_r^0 \cos\theta_i = Z_2^{-1}\underline{E}_t^0 \cos\theta_t .$

By dividing the two terms by $\underline{E}_i^0$, we obtain: $Z_1^{-1}\cos\theta_i - Z_1^{-1}\underline{r}_\perp \cos\theta_i = Z_2^{-1}\underline{t}_\perp\cos\theta_t ,$

so that with the same notations as those used in Section 11.3.1,

$$(1 - \underline{r}_\perp)\,Z_1^{-1}\cos\theta_1 = \underline{t}_\perp\,Z_2^{-1}\cos\theta_2 , \qquad (52')$$

From Eqs (51) and (52'), we can determine using a method analogous to that in Section 11.3.2 (where Z_1^{-1} was substituted for n_1 and Z_2^{-1} for n_2):

$$\underline{r}_\perp = \frac{\dfrac{\cos\theta_1}{Z_1} - \dfrac{\cos\theta_2}{Z_2}}{\dfrac{\cos\theta_1}{Z_1} + \dfrac{\cos\theta_2}{Z_2}},$$ so that following multiplication of top and bottom with $Z_1 Z_2$ we have:

$$\underline{r}_\perp = \frac{Z_2 \cos\theta_1 - Z_1 \cos\theta_2}{Z_2 \cos\theta_1 + Z_1 \cos\theta_2},$$ from which according to Eq. (47) we have:

$$R_\perp = \left| \frac{Z_2 \cos\theta_1 - Z_1 \cos\theta_2}{Z_2 \cos\theta_1 + Z_1 \cos\theta_2} \right|^2. \qquad (53)$$

As the incidence is normal (with directions $\vec{k}_i$ and $\vec{e}_z$ being merged) the notion of incidence plane loses its significance (otherwise defined by the vector directions $\vec{k}_i$ and $\vec{n}_{12} \equiv \vec{e}_z$), and we can state for the energy reflection factor for whatever polarization direction of the incident wave, that:

$$R = \left| \frac{Z_1 - Z_2}{Z_1 + Z_2} \right|^2. \qquad (54)$$

1.5.1.2. The antiecho condition

Here the reflected wave is annulled so that $R = 0$. For this to occur, taking Eq. (54) into account being applicable for a great distance where the wave exhibits a normal incidence on the target, it suffices that $Z_1 = Z_2$, that is, $\dfrac{\varepsilon_1}{\mu_1} = \dfrac{\varepsilon_2}{\mu_2}$, and hence:

$$\frac{\varepsilon_{1r}}{\mu_{1r}} = \frac{\varepsilon_{2r}}{\mu_{2r}}. \qquad (55)$$

So that this condition is true, a specific sort of coating must be applied to medium (2). If medium (1) is air, then Eq. (55) means that between the dielectric permittivity and magnetic permeability of the second medium, there is a simple equation given by

$$\varepsilon_r = \mu_r. \qquad (55')$$

Comment: the antiecho condition for EM waves is just one part of the conditions under research in optics in order to find a material that is completely anti-reflective.

In such a case, the condition $R = 0$ means that $R = \left| \dfrac{n_1 - n_2}{n_1 + n_2} \right|^2 = 0$, so that $n_1 = n_2$. This condition has no sense in itself, as it would mean that the media (1) and (2) would be identical and that therefore there would be no interface. One method though is to insert between the two media a layer of a given thickness and index n such that $n_1 < n < n_2$ and then find the conditions for which a destructive interference is formed between the two reflected waves, one being at the interface of the materials with indices n_1 and n and the other being at the interface of the materials with interfaces n and n_2.

11.5.2. The Dallenbach layer: an anti-radar structure

11.5.2.1. The stealth concept

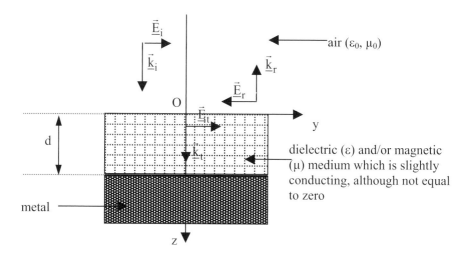

Figure 11.21. *Antiradar structure.*

Assuming that a wave with a frequency denoted by ω has a normal incidence to a metallic surface, the reflection is total. In order to make the metallic layer stealth-like, it is covered with a dielectric and/or magnetic material adapted so as to satisfy the antiecho condition for an air/coating interface. In addition, the wave transmitted in the layer also must be totally attenuated in order to suppress the total reflection that would be produced by the metallic layer. The attenuation is obtained with a slightly conducting layer of sufficient thickness (d) as indicated in Figure 11.21.

11.5.2.2. Condition (1): antiecho at the air/dielectric-magnetic interface
The classically used absorbent materials are generally composites based on carbon, iron carbonyls, or ferrites (see also Chapter 4, Section 4.2.5.2). These materials do, however, suffer from certain inconveniences such as their weight or their mechanical rigidity due to the high number of charges necessary to reach the required absorption. These composite materials are presently being replaced by conducting polymers, which can be acquired through doping a conductivity appropriate to the system.

11.5.2.2.1. Equation for the reflection coefficient
In the system under study, the conductivity (σ) is such that $\sigma \neq 0$ and in the layer $\underline{k} = \underline{k}_t$ so that from Eq. (5) of Chapter 8, Section 8.5.1, where "electrokinetic" notation was used to establish the equation we have:

$$\underline{k}_t = \omega\sqrt{\mu\varepsilon}\left(1 - i\frac{\sigma}{\omega\varepsilon}\right)^{1/2}. \qquad (56)$$

Under normal incidences, the preceding Eq. (52) can be written for the air/layer interface (where for air $k_i = k_r = k_0 = \dfrac{\omega}{c}$):

$\dfrac{k_0}{\mu_0}E_i^0 - \dfrac{k_0}{\mu_0}\underline{E}_r^0 = \dfrac{\underline{k}_t}{\mu}\underline{E}_t^0$. By dividing the two terms by E_i^0 it is determined that:

$\dfrac{k_0}{\mu_0}(1 - \underline{r}) = \dfrac{\underline{k}_t}{\mu}\underline{t}$, from which with the help of Eq. (51) we determine that:

$$\underline{r} = \frac{\dfrac{k_0}{\mu_0} - \dfrac{\underline{k}_t}{\mu}}{\dfrac{k_0}{\mu_0} + \dfrac{\underline{k}_t}{\mu}}, \text{ so that also } \underline{r} = \frac{\mu k_0 - \mu_0\underline{k}_t}{\mu k_0 + \mu_0\underline{k}_t} = \frac{\mu_r\dfrac{\omega}{c} - \underline{k}_t}{\mu_r\dfrac{\omega}{c} + \underline{k}_t}. \qquad (57)$$

11.5.2.2.2. Approximate calculation using $v = 5$ GHz, $\varepsilon_r = 15$,
$$\sigma = 5 \times 10^{-1} \ \Omega^{-1}m^{-1} = 5 \times 10^{-3} \ cm^{-1})$$

Figures 11.22a and b give a qualitative indication of the evolution of the conductivity and dielectric permittivity of polyaniline films with varying levels of doping as a function of frequency. On going from A to D the plots are for increasing levels of doping. It can be seen that for sufficiently high levels of doping, quite high values can be attained.

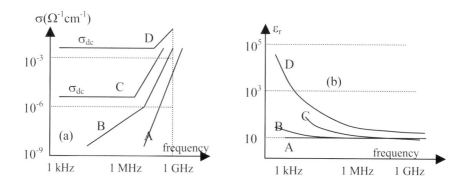

Figure 11.22. *Evolution of σ and ε_r' as a function of frequency for doped polyanilines.*

Using the values given in Eq. (56) for $\underline{k}_t$, the term

$$\frac{\sigma}{\omega\varepsilon} = \frac{5 \times 10^{-1}}{2\pi \times 5 \times 10^9 \times 15 \times 8.85 \times 10^{-12}} \approx 1.2 \times 10^{-1}$$ is very much less than 1. With a

limited development, from Eq. (56) for $\underline{k}_t$ we obtain $\underline{k}_t \simeq \omega\sqrt{\mu\varepsilon}\left(1 - i\frac{\sigma}{2\omega\varepsilon}\right)$.

By making $\beta = \frac{\sigma}{2\omega\varepsilon}$, we thus can write that:

$$\underline{k}_t \simeq \omega\sqrt{\mu\varepsilon}\left(1 - i\beta\right) = \frac{\omega}{c}\sqrt{\mu_r\varepsilon_r}\left(1 - i\beta\right). \qquad (56')$$

The reflection coefficient given by Eq. (57) then takes on the form:

$$\underline{r} = \frac{\mu_r - \sqrt{\mu_r\varepsilon_r}\,(1 - i\beta)}{\mu_r + \sqrt{\mu_r\varepsilon_r}\,(1 - i\beta)}, \text{ so that finally } \underline{r} = \frac{1 - \sqrt{\dfrac{\varepsilon_r}{\mu_r}}(1 - i\beta)}{1 + \sqrt{\dfrac{\varepsilon_r}{\mu_r}}(1 - i\beta)}.$$

Numerically, as $\beta \approx 10^{-1} \ll 1$, we have $(1 - i\beta) \approx 1$, and $\underline{r}$ can be reduced to:

$$\underline{r} \approx \frac{1 - \sqrt{\dfrac{\varepsilon_r}{\mu_r}}}{1 + \sqrt{\dfrac{\varepsilon_r}{\mu_r}}}$$
(58). The condition $R \approx \left[\dfrac{1 - \sqrt{\dfrac{\varepsilon_r}{\mu_r}}}{1 + \sqrt{\dfrac{\varepsilon_r}{\mu_r}}} \right]^2 = 0$ thus gives $\dfrac{\varepsilon_r}{\mu_r} = 1$,

so that $\varepsilon_r = \mu_r = 15$.

We once again find the condition of Eq. (55'), which is quite normal as the numerical calculation lead us to neglect the component due to conductivity, just as was assumed hypothetically in order to arrive at Eq. (55').

11.5.2.3. Condition (2) to predict total reflection at a metal: attenuation and depth of penetration of the wave into the coating

The transmitted wave is given by $\underline{\vec{E}}_t = \underline{t}\,\vec{E}_i\,\exp(i[\omega t - \underline{k}_t z])$ using electrokinetic notation. According to Eq. (56') $\underline{k}_t$ is of the form $\underline{k}_t = \alpha - i\gamma$, with $\alpha = \dfrac{\omega}{c}\sqrt{\mu_r \varepsilon_r}$

and $\gamma = \dfrac{\omega}{c}\beta\sqrt{\mu_r \varepsilon_r}$, so that:

$\gamma = \dfrac{\omega}{c}\sqrt{\mu_r \varepsilon_r}\,\dfrac{\sigma}{2\omega\varepsilon} = \dfrac{\sigma}{2c\varepsilon_0}\sqrt{\dfrac{\mu_r}{\varepsilon_r}}$. With the preceding condition (1), that is, $\varepsilon_r = \mu_r$,

we finally have $\gamma = \dfrac{\sigma}{2c\varepsilon_0}$. The wave is in the form:

$\underline{\vec{E}}_t = \underline{t}\,\vec{E}_i\,\exp(-\gamma z)\exp(i[\omega t - \alpha z])$, so that by making $\delta = \dfrac{1}{\gamma}$, we have

$$\underline{\vec{E}}_t = \underline{t}\,\vec{E}_i\,\exp\left(-\dfrac{z}{\delta}\right)\exp(i[\omega t - \alpha z]).$$

The wave is attenuated little by little along its propagation through the coating. When $z = \delta$, the wave amplitude is divided by $e = 2.7$.

This attenuation is obtained at a penetration depth (δ) such that:

$$\delta = \dfrac{1}{\gamma} = \dfrac{2c\varepsilon_0}{\sigma},$$

which here in numerical terms means that $\delta \approx 1.3$ cm.

11.5.2.4. Conclusion

It is notable that the thickness calculated is for σ measured at a given frequency, while the law for the evolution of current as a function is frequency is typically of the form $\sigma(\omega) = \sigma_{(0)} + A\omega^s$ (see also Figure 11.22 for a sufficiently doped material such as curve C or D). The absorption thus is limited to a certain band width. In order to absorb over a large enough band, a structure based on one single layer (called the Dallenbach layer) has to be improved upon by using multi-layer absorbing structures, in which each layer is optimized to obtain a minimum reflection coefficient for a given band width. A final restriction is imposed by the necessity of having thick enough layers to be practicable, to the extent that absorbing paints might be used.

11.6 Problems
11.6.1. Reflection and absorption at low and high frequencies by a conductor
The optical properties of a metal are treated through a series of questions (and answers!) as given below and follow on from the problem studied in Section 8.6.2, concluding with a look at the optical properties of a metal as derived from the relationships for dielectric permittivities as established in Section 8.6.2 from the equation for electronic polarization.

Recalling Section 8.6.2, the equations for the complex dielectric permittivity of a metal were:

- $\underline{\varepsilon_r} = 1 - \dfrac{i}{\omega\varepsilon_0[1 + i\omega\tau]}$ [question 3(c)] Eq. (1);

- $\underline{\varepsilon_r} = 1 - i\dfrac{\omega_p^2\tau}{\omega[1 + i\omega\tau]}$ [question 3(d)] Eq. (2).

The metal under consideration is copper, and in the equations for $\underline{\varepsilon_r}$:
- τ is the relaxation time, which is such that $\tau = 10^{-14}$s ;
- $\sigma(0)$ is the conductivity for a steady (DC) state, which will be taken equal to 6×10^7 $\Omega^{-1}m^{-1}$; and

- ω_p is the plasma angular frequency, that is defined by the equation $\omega_p^2 = \dfrac{Nq^2}{m\varepsilon_0}$,

and is typically $\omega_p = 10^{16}$ rad s^{-1} . It is worth noting that $\varepsilon_0 = 8.85 \times 10^{-12}$ MKS .

The incident wave at medium (1) is a MPPEM wave polarized in parallel with Ox in the form $\underline{\vec{E}_0} = \vec{E}_0 \exp(-ikz)$ in electrokinetic notation as the applied field was written in the form $\underline{\vec{E}} = \vec{E}_0 \exp(i\omega t)$ from the beginning of the problem in Chapter 8, Section 8.6.2. The propagation of the wave thus is much as usual, along positive values of z. The wave is considered to arrive at a normal incidence, so $\theta_1 = 0$ and the reflection is detailed at the air/metal interface where each, respectively, have indices n_1 such that $n_1 = 1$, or more simply $\underline{n} = n' - i n''$.

1. This part considers the optical properties of a metal at low frequencies.

(a) This domain is defined by $\omega << \dfrac{1}{\tau}$. From Eq. (1) for $\underline{\varepsilon}_r$, give the simplified

form which $\underline{\varepsilon}_r$ takes.

(b) By using orders of magnitude of different parameters, show that the expression for $\underline{\varepsilon}_r$ can be reduced to a single and explainable term.

(c) Calculate the complex index for $\underline{n} = n' - i\,n''$

(d) Determine the reflection coefficient for the amplitude $\underline{r}$ (which is thus in the

form $\underline{r} = \dfrac{n_1 - \underline{n}}{n_1 + \underline{n}}$ following the preceding Section 11.4.1). Equally, evaluate the

reflection factor in terms of energy.

(e) Give $\underline{k}$ as a function of $\underline{n}$ inside the metal. Study the form of penetration of the electric wave in the metal by writing the form of the wave in there as

$$\vec{\underline{E}}_t = \vec{\underline{E}}_{0t}\,\exp(i\omega t) = \vec{\underline{E}}_t^0\,\exp(i[\omega t - \underline{k}z]),$$ where the complex amplitude $\vec{\underline{E}}_{0t}$ is

thus of the form $\vec{\underline{E}}_{0t} = \vec{\underline{E}}_t^0\,\exp(-i\underline{k}z)$. In order to do this, determine the depth

(δ) of the penetration of the wave, which we will define here by using the term

for attenuation in the form $\exp(-\dfrac{z}{\delta})$.

(f) Establish the form of the energy transmission factor (T). Give this as a function of n' or n", and then as a function of δ (the so-called Hagen-Rubens equation). Give a numerical value for δ and then also for T given that $\nu = 1$ GHz. From this result, make a conclusion.

2. These questions now concern a zone of slightly higher frequencies which are

defined by $\omega << \dfrac{1}{\tau}$.

(a) Using Eq. (1) for $\underline{\varepsilon}_r$, give the simplified form of $\underline{\varepsilon}_r$ in this domain.

(b) When $\omega << \omega_p$, determine the form of $\underline{\varepsilon}_r$ as well as that of the MPPEM wave in the metal. Calculate the reflection coefficient for a normal incidence at an air/metal interface.

(c) When $\omega > \omega_p$ (high frequencies), give the range of variation possible for $\underline{\varepsilon}_r$. Give a value for $\underline{\varepsilon}_r$ when $\omega >> \omega_p$, as well as that of the reflection coefficient for an incidence normal to the air/metal interface.

3. Give a recapitulative scheme of the reflection and transmission properties of a metal in the EM spectrum.

Answers

1.

Low frequencies $\omega < \dfrac{1}{\tau}$

(a) We use $\underline{\varepsilon}_r = 1 - \dfrac{i\sigma(0)}{\omega\varepsilon_0[1 + i\omega\tau]}$,

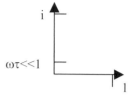

and with

$\omega\tau \ll 1$, $i\omega\tau \approx 0$ (negligeable with respect to 1), we have $\varepsilon_r = 1 - i\dfrac{\sigma(0)}{\omega\varepsilon_0}$.

(b)

$\sigma(0) = 6 \times 10^7 \ \Omega^{-1} \ m^{-1}$,

$\omega \ll \dfrac{1}{\tau} = 10^{14} \ s^{-1}$, $\Rightarrow \ \omega \, \varepsilon_0 \ll 10^3$ $\left. \begin{array}{c} \\ \\ \\ \end{array} \right\} \Rightarrow$

$\left[\dfrac{\sigma(0)}{\omega\varepsilon_0} \right]_{i\,min} = \dfrac{\sigma(0)}{\left[\omega\varepsilon_0 \right]_{max}} = \dfrac{\sigma(0)}{10^3} = 6 \times 10^4 \gg 1$.

With $\dfrac{\sigma(0)}{\omega\varepsilon_0}$ always being greater than 1, in the low-frequency range we have:

$$\varepsilon_r \approx - i\dfrac{\sigma(0)}{\omega\varepsilon_0} .$$

(c) Given $\underline{\varepsilon}_r = \underline{n}^2 \approx -i\dfrac{\sigma(0)}{\omega\varepsilon_0} = \dfrac{\sigma(0)}{\omega\varepsilon_0} e^{-i\frac{\pi}{2}}$, from which we deduce that

$\underline{n} = n' - i\,n'' = \sqrt{\dfrac{\sigma(0)}{\omega\varepsilon_0}} e^{-i\frac{\pi}{4}}$, so that with $e^{-i\frac{\pi}{4}} = \cos\dfrac{\pi}{4} - i \sin\dfrac{\pi}{4} = \dfrac{1}{\sqrt{2}}(1 - i)$, we

obtain:

$\underline{n} = n' - i\,n'' = \sqrt{\dfrac{\sigma(0)}{2\omega\varepsilon_0}}(1 - i)$, from which $n' = n'' = \sqrt{\dfrac{\sigma(0)}{2\omega\varepsilon_0}}$.

(d) We have $\left[\dfrac{\sigma(0)}{\omega\varepsilon_0} \right]_{i\,min} \gg 1 \ \Rightarrow \ \dfrac{\sigma(0)}{\omega\varepsilon_0} \gg 1$ and $n' \gg 1$ just as $n'' \gg 1$.

The result is that $\underline{r} = \dfrac{n_1 - \underline{n}}{n_1 + \underline{n}} = \dfrac{1 - n' + in''}{1 + n' - in''} \approx -\dfrac{n' - in''}{n' - in''} = -1 = e^{i\pi}$ and on reflection

there is a dephasing by π while the reflecting power is in the neighborhood of 1, i.e., $R = |\underline{r}^2| \approx 1$.

The ability to reflect energy is given by $R = |\underline{r}|^2 = \dfrac{\left(1 - n'\right)^2 + n''^2}{\left(1 + n'\right)^2 + n''^2} \simeq \dfrac{n'^2 + n''^2}{n'^2 + n''^2} = 1$

where $n' \gg 1$.

(e) With $\underline{k} = \dfrac{\omega}{c}\underline{n} = k' - ik''$, we have $k'' = \dfrac{\omega}{c}n'' = \dfrac{\omega}{c}\sqrt{\dfrac{\sigma(0)}{2\omega\varepsilon_0}} = \dfrac{1}{c}\sqrt{\dfrac{\omega\,\sigma(0)}{2\varepsilon_0}}$.

A wave that propagates in the form $\underline{E}_t = \underline{E}_{0t}e^{i\omega t} = \underline{E}_t^0 e^{-i\underline{k}z}e^{i\omega t}$ can be written for

this frequency range as $\underline{E} = \underline{E}_t^0\, e^{i(\omega t - k'z)}e^{-k''z}$. The equation carries a term for the

propagation as in $e^{i(\omega t - k'z)}$, and a term for attenuation as in $e^{-k''z}$ that also can be

rewritten as $e^{-\frac{z}{\delta}}$ where $\delta = \dfrac{1}{k''} = c\sqrt{\dfrac{2\varepsilon_0}{\omega\,\sigma(0)}}$. With δ representing the depth at

which the electric wave is attenuated by the ratio $\dfrac{1}{e}$, for distances greater than

several δ , the wave is practically zero.

As $\delta = c\sqrt{\dfrac{2\varepsilon_0}{\omega\,\sigma(0)}}$, we can see that as ω increases, δ decreases. For the

highest frequencies in this region ($\omega \le \dfrac{1}{\tau} \approx 10^{14}\,\mathrm{Hz}$) the EM waves are localized at

the surface of the conductors, in an effect called the skin effect.

(f) The equation for T is deduced from that of R, as in $T = 1 - R = \dfrac{4n_1 n'}{(n_1 + n')^2 + n''^2}$.

With n" = n' = n $\gg$ n$_1$ = 1, the result is that $T \simeq \dfrac{2}{n}$, so in addition $T \simeq 0$ (as n $\gg$

1). In more precise terms, when n' = n' = n , and with $\delta = \dfrac{1}{k''} = \dfrac{c}{\omega}\dfrac{1}{n''}$, we have

$T = \dfrac{2\,\omega\,\delta}{c} = 2\,k_0\,\delta$, so that with $\lambda_0 = c\dfrac{2\pi}{\omega} = \dfrac{2\pi}{k_0}$, we also have $T = \dfrac{4\pi\,\delta}{\lambda_0}$,

which is known as the Hagen-Rubens equation.

The energy received by the metal is transmitted by carriers to the lattice that dissipates the energy through the Joule effect. So, the factor T is often called the absorption power of the metal.

In numerical terms, for copper,

$\sigma(0) = 6 \times 10^7 \ \Omega^{-1}m^{-1}$, $\omega = 2\pi \ 10^9$ rad/s, $\varepsilon_0 = 8.85 \ 10^{-12}$ MKS so that

$$\delta = c\sqrt{\frac{2\varepsilon_0}{\omega\,\sigma(0)}} = 2.06 \ \mu m \ .$$

As $\lambda_0 = \dfrac{c}{\nu} = 0.3$ m (region of centimeter wavelengths), the Hagen-Rubens formula gives $T \approx 8 \times 10^{-5} \rightarrow 0$, to which $R \approx 0.99990 \rightarrow 1$. Hence the use of microwave frequency radars for police use in determining car speeds!

2.

(a) This part uses for $\underline{\varepsilon}_r$ the general formula (2) obtained from question 3(d) of Section 8.6.2, for metals, as in:

$$\underline{\varepsilon}_r = 1 - i\frac{\omega_p^2\tau}{\omega[1 + i\omega\tau]} \ .$$

When $\omega\tau \gg 1$, we have $\underline{\varepsilon}_r \approx 1 - i\frac{\omega_p^2\tau}{i\omega^2\tau^2} = 1 - \frac{\omega_p^2}{\omega^2}$.

(b) When $\omega < \omega_p$, so that $\dfrac{\omega_p^2}{\omega^2} \gg 1$, we have $\underline{\varepsilon}_r = \varepsilon_r'' - \dfrac{\omega_p^2}{\omega^2} < 0$. For the index then

$\underline{\varepsilon}_r = \underline{n}^2 = -\dfrac{\omega_p^2}{\omega^2} = i^2\dfrac{\omega_p^2}{\omega^2}$ must be true. The result is that $\underline{n} = n' - i\,n'' = \pm\,i\,\dfrac{\omega_p}{\omega}$ so

that $n' = 0$ and $n'' = \dfrac{\omega_p}{\omega}$ (only the positive solution is physically acceptable and it represented the only available absorption of the wave in the medium).

With $k' = 0$, the wave defined by $\underline{E}_t = \underline{E}_{0t}e^{i\omega t} = \underline{E}_t^0 e^{-i\underline{k}z}e^{i\omega t}$ takes on he form $\underline{E}_t = \underline{E}_t^0 e^{-k''z}e^{i\omega t}$, and as there is no longer a term for the propagation, in effect the signal no longer propagates through the metal. The wave oscillates in a standing position in the neighborhood of the interface, and in practical terms the wave is reflected, as shown in the following calculation.

In effect, $\underline{r} = \dfrac{1 + in''}{1 - in''}$ (≈ -1 with $n'' = \dfrac{\omega_p}{\omega} \gg 1$), which is such that $|\underline{r}| = 1$, so $R = |\underline{r}^2| = 1$.

(c) If $\omega > \omega_p$, so that $\dfrac{\omega_p^2}{\omega^2} < 1$, and with $\dfrac{\omega_p^2}{\omega^2} > 0$, we can state more succinctly

that $0 < \dfrac{\omega_p^2}{\omega^2} < 1$ and $0 < \varepsilon_r < 1$, and that ε_r is positive, real, and between 0 and 1.

When $\omega \gg \omega_p$, so that $\dfrac{\omega_p^2}{\omega^2} \ll 1$, we have $\underline{\varepsilon}_r = \varepsilon_r \approx 1$. Thus the metal behaves

as if it is a vacuum without charges simply because the latter cannot follow such a high frequency.

As $\underline{\varepsilon}_r = \underline{n}^2$, we have $\underline{n} = n' = 1$; $k' = \dfrac{\omega}{c} n'$ and $n'' = 0$ just   as $k'' = 0$ (no

absorption). So $r = \dfrac{n_1 - n}{n_1 + n} = \dfrac{1 - 1}{1 + 1} = 0$, and $R = 0$ and $T = 1$, and the transmission

is perfect.

3. To sum up the characteristics of a metal:

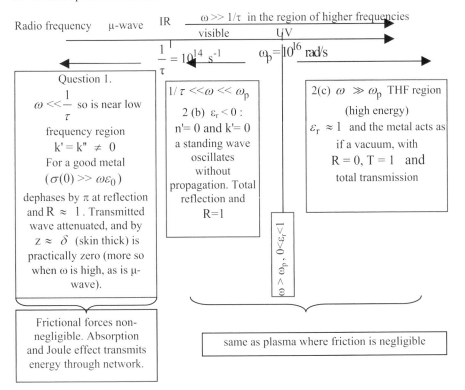

11.6.2. Limited penetration of Hertzian waves in sea water

Here the system is based on medium (1) which is air (which in electromagnetic terms is assumed to be the same as a vacuum) separated by a horizontal plane (Oxy) from medium (2) which is sea water. For Hertzian frequencies (the maximum of which is considered to be 100 MHz) we can take for the real parts of the dielectric permittivity (ε_r) and the conductivity (σ) values measured for a continuous stationary regime, as in $\varepsilon_r = 81$ and $\sigma = 4\,\Omega^{-1}\mathrm{m}^{-1}$. In addition, medium (2) has a permeability equal to μ_0 and is assumed to be non-magnetic.

1. The Maxwell equations
(a) Give the equation for the total current in medium (2).
(b) Write the four "Maxwell equations" for medium (2).
(c) Using a rapid calculation (resembling that used for a vacuum), establish from the Maxwell equations the equation for the propagation of an electric field (equation of partial derivatives followed by the electric field).
(d) Toward which forms does this equation tend on going from a nonconducting medium to a medium where conductivity predominates?

2. The question concerns the search for a solution to the electric field in the form of a MPPEM wave where $\vec{\underline{E}} = \vec{E}^0 \exp(i[\omega t - \underline{\vec{k}}\,\vec{r}])$ and $\underline{\vec{k}}$ is the vector of the complex wave.
(a) With the help of the equation for propagation, establish the equation that brings together $\underline{k}^2$, ε_r, and σ.
(b) From the general equation that exists for wavenumber and index, give the equation that ties $\underline{n}^2$, ε_r, and σ.

3. Numerically compare the value of two terms that intervene in $\underline{n}^2$. From this determine the approximate and literal form for $\underline{n}$ (the imaginary number which is such that $\underline{n} = n' - i\,n''$) that can be expressed as a function of $(\dfrac{\sigma}{2}\,\omega\varepsilon_0)^{\frac{1}{2}}$.

4. This question concerns, within the approximation of the preceding question, the propagation of a wave with an incidence normal to the air/sea interface.
(a) Give the theoretical form of the electric field that can be expressed as a function of the parameter: $\delta = (\dfrac{c}{\omega n'}) = (\dfrac{2}{\mu_0\omega\sigma})^{\frac{1}{2}}$.
(b) At which order of depth (when $\nu = 1$ MHz and $\nu = 100$ MHz) does the wave penetrate the water. Give a conclusion from the result.

(c) What is its phase velocity (literal expression given as a function of the frequency v and then a numerical value for when $v = 100$ MHz $v = 100$ MHz).

Answers

1.

(a) Taking into account the data given for the problem, for the medium with a real conductivity (σ) and a real dielectric permittivity (ε_r), we can state that the total current density in the medium is a sum of the conduction current and the displacement current, so $\vec{J}_T = \vec{J}_c + \vec{J}_D = \sigma\vec{E} + \dfrac{\partial\vec{D}}{\partial t} = \sigma\vec{E} + \varepsilon_0\varepsilon_r\dfrac{\partial\vec{E}}{\partial t}$.

(b) Here, with $\rho_\ell = 0$,

$$\text{div}\vec{E} = 0 \quad (1), \qquad \text{div}\vec{B} = 0 \qquad\qquad (3)$$

$$\overrightarrow{\text{rot}}\vec{E} = -\dfrac{\partial\vec{B}}{\partial t} \quad (2), \qquad \overrightarrow{\text{rot}}\,\vec{B} = \mu_0\vec{J}_T = \mu_0\sigma\vec{E} + \dfrac{\varepsilon_r}{c^2}\dfrac{\partial\vec{E}}{\partial t}. \qquad (4)$$

(c) The calculation of the rotational of Eq. (2) gives:

$$\overrightarrow{\text{rot}}(\overrightarrow{\text{rot}}\vec{E}) = -\dfrac{\partial}{\partial t}\overrightarrow{\text{rot}}\,\vec{B}, \text{ so that } \Delta\vec{E} = \mu_0\sigma\dfrac{\partial\vec{E}}{\partial t} + \dfrac{\varepsilon_r}{c^2}\dfrac{\partial^2\vec{E}}{\partial t^2}. \qquad (5)$$

(d) If the medium is a nonconductor, we immediately find (with $\sigma = 0$) the classic d'Alembert equation, as in $\Delta\vec{E} - \dfrac{\varepsilon_r}{c^2}\dfrac{\partial^2\vec{E}}{\partial t^2} = 0$, where $\varepsilon_r = n^2$ (see the relation obtained Section 7.2.1.5 for nonabsorbing materials)

If the term due to conductivity is dominant, i.e., the medium is well conducting, the propagation equation takes on the form $\Delta\vec{E} - \mu_0\sigma\dfrac{\partial\vec{E}}{\partial t} = 0$.

2.

(a) By looking for a solution to the general Eq. (5) of the form $\vec{E} = \vec{E}^0\exp(i[\underline{\vec{k}}\,\vec{r} - \omega t])$, we end up with the following equation to verify

(where $\Delta\vec{E} = i^2\underline{\vec{k}}^2\vec{E} = -\underline{k}^2\vec{E}$, $\dfrac{\partial\vec{E}}{\partial t} = i\omega\vec{E}$, and $\dfrac{\partial^2\vec{E}}{\partial t^2} = -\omega^2\vec{E}$):

$$\underline{k}^2 = \dfrac{\varepsilon_r}{c^2}\omega^2 - i\,\omega\,\mu_0\sigma.$$

Accordingly, afterward it can be verified that $\underline{k}$, just as $\underline{k}^2(\neq|k|^2 = \underline{k}\,\underline{k}^*)$ are complex.

(b) With the general equation introduced into that for absorbent media [Eq. (25) of Chapter 7] as in $\underline{k} = \dfrac{\omega}{c}\underline{n}$, we obtain

$$\underline{n}^2 = \frac{c^2}{\omega^2}\underline{k}^2 = \varepsilon_r - i\frac{\sigma}{\varepsilon_0\omega} .$$

3. Numerically, and for each of the two terms:
- for the first term, $\varepsilon_r = 81$;
- for the second term, the minimum value is obtained at a maximum value of ω given in the Hertzian domain, which is $\omega_{max} = 100\ \mathrm{MHz} = 10^8\ \mathrm{Hz}$. Thus,

$$\left[\frac{\sigma}{\varepsilon_0\omega}\right]_{min} \approx 719 .$$

So numerically, $\dfrac{\sigma}{\varepsilon_0\omega} \gg \varepsilon_r$ and the conducting component is largely dominant, so that in practical terms we can state that:

$\underline{n}^2 \approx - i\ \dfrac{\sigma}{\varepsilon_0\omega}$, so that with $-i = \exp\left(-i\dfrac{\pi}{2}\right)$, then $\underline{n}^2 \approx \dfrac{\sigma}{\varepsilon_0\omega}\exp\left(-i\dfrac{\pi}{2}\right)$ from which can be deduced that

$$\underline{n} \approx \sqrt{\frac{\sigma}{\varepsilon_0\omega}}\exp\left(-i\frac{\pi}{4}\right) = \sqrt{\frac{\sigma}{\varepsilon_0\omega}}\left(\cos\left[-\frac{\pi}{4}\right] + i\sin\left[-\frac{\pi}{4}\right]\right) = \sqrt{\frac{\sigma}{2\,\varepsilon_0\omega}}(1 - i).$$

With $\underline{n} = n' - i\,n"$ (using electrokinetic notations), we finally have:

$$n' = n" \approx \sqrt{\frac{\sigma}{2\,\varepsilon_0\omega}} .$$

Comment: The numerical condition, $\dfrac{\sigma}{\varepsilon_0\omega} \gg \varepsilon_r$, is the same as neglecting the displacement current with respect to the conduction current. The complex index ($\underline{n}$) which is tied to the complex relative permittivity by the equation $\underline{n}^2 = \underline{\varepsilon}_r$ is therefore such that $\underline{n}^2 = \underline{\varepsilon}_r = \varepsilon_r - i\ \dfrac{\sigma}{\varepsilon_0\omega} \approx - i\ \dfrac{\sigma}{\varepsilon_0\omega}$. The complex relative permittivity is practically purely imaginary as is the case for a conductor subject to low frequencies for which exactly $n' = n" = \sqrt{\dfrac{\sigma(0)}{2\,\varepsilon_0\omega}}$ is found [see also the preceding exercise, questions 1(b) and (c) with concerning the notation: $\sigma \equiv \sigma(0)$].

4.

(a) For a normal incidence, $(\theta_i = 0)$, with n_1 denoting the index of medium (1), we have for the transmitted wave:

$$k'_{ty} = k_0\, n_1\, \sin\theta_i = 0\,,\ \ k'_{tz} = k_0 R(\underline{n}^2 - n_1^2\sin^2\theta_i\,)^{\tfrac{1}{2}} = k_0 n'$$

$$k''_{tz} = k_0 I(\underline{n}^2 - n_1^2\sin^2\theta_i\,)^{\tfrac{1}{2}} = k_0 n''$$

By using electrokinetic notation, the wavevector for the wave transmitted in medium (2) is thus $\ \underline{k} = \underline{k}_t = \left(k'_t - ik''_t\right)\vec{e}_z = k_0\,(n' - i\,n'')\vec{e}_z = k_0\underline{n}\,\vec{e}_z = \underline{k}_t\,\vec{e}_z$.

By making $\qquad \delta = \dfrac{1}{k_0 n'} = \dfrac{1}{k_0 n''}\,,\qquad$ so that with $\qquad n' = n'' \approx \sqrt{\dfrac{\sigma}{2\,\varepsilon_0\omega}}\,,$

$\delta = \sqrt{\dfrac{2}{\mu_0\omega\sigma}} = c\sqrt{\dfrac{\varepsilon_0}{\pi\,\sigma\,v}}\,,$ and $\underline{k}_t = \dfrac{1}{\delta} - \dfrac{i}{\delta}$; the form of the electric field in medium (2) is therefore:

$$\underline{\vec{E}} = \vec{E}^0\,\exp(i[\omega t - \underline{k}\,\vec{r}\,]) = \vec{E}^0\,\exp(i[\omega t - \underline{k}_t z\,]) = \vec{E}^0\,\exp\left(-\dfrac{z}{\delta}\right)\exp(i[\omega t - \dfrac{z}{\delta}\,])\,.$$

The depth of the penetration is of the order of δ. The z noted above is   equal   to several δ and the transmitted wave is practically zero.

(b)   In   terms   of   actual   numbers,   when $v = 100\ \text{MHz} = 10^8\ \text{Hz}$ and

$$\lambda = \dfrac{c}{v} = \dfrac{3.10^8}{10^8} = 3\ \text{m}\,,\ \text{we find that}\ \delta = c\sqrt{\dfrac{\varepsilon_0}{\pi\,\sigma\,v}} = 2.52\ 10^{-2}\ \text{m} = 2.52\ \text{cm}\,.$$

When $v = 1\ \text{MHz}$, $(\lambda = 300\ \text{m})$, we find $\delta = 25.2\ \text{cm}$.

To conclude, we can see that for a depth z of the order of several δ, being here at most a meter or so, the wave signal is practically all absorbed, and it therefore is not possible to communicate using Hertzian waves with a submarine. In effect, underwater communications are established using sonar with acoustic waves.

(c) The speed of the phase is given by $v_\varphi = \dfrac{c}{n'}$, and with $n' = \dfrac{1}{k_0\delta}$, we have

$$v_\varphi = \omega\delta = \sqrt{\dfrac{4\pi\,v}{\mu_0\,\sigma}}\ .\ \text{Thus, if } v \text{ decreases, } v_\varphi \text{ decreases also, and we arrive at a}$$

dispersion of waves.

In numerical terms, when $v = 100\ \text{MHz}$, we have $v_\varphi = 1.6 \times 10^7\ \text{m/s}$.

Chapter 12

Total Reflection and Guided Propagation of Electromagnetic Waves in Materials of Finite Dimensions

12.1. Introduction

As described in Chapter 11, there are two forms of total reflection:

- a reflection at a vacuum/perfect conductor interface where $\dfrac{1}{\tau} < \omega < \omega_p$ and $R = 1$,

with $\omega < \omega_p$ and $\sigma(0) \gg \omega \varepsilon_0$, so that $R \approx 1$; and

- a reflection between two dielectrics such that $n_1 > n_2$ and $\theta_1 > \theta_\ell$.

The superposition of the incident wave with its reflected wave can lead only to propagation, on average, when parallel to the surface. The resulting wave is in effect guided. By having a second interface, as shown in Figure 12.1a, the propagation can be channeled between the two surfaces and the system constitutes a wave guide.

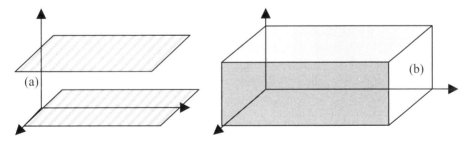

Figure 12.1. *Wave guides.*

Wave-guides can be made with different geometrical configurations:
- with a rectangular cross section as shown in Figure 12.1b formed of two parallel metallic planes and generally used for guiding waves with wavelengths of the order of centimeters;

- with a circular cross section so that the signal propagates in a central dielectric core which is bound within a metal surface to give a coaxial cable;
- for the optical domain, the phenomenon of total internal reflection provided by two dielectrics is used, so that the internal dielectric at the heart of the cable has an index higher than that of the external dielectric which makes up the gain. When the cross section is circular the result is an optical fiber, and when the cross section is rectangular, the result is an optical guide. In the latter case, in order to assure wave guiding, i.e., confinement of the optical wave, different geometries are used such as the buried guide shown in Figure 12.2 a or the strip guide in Figure 12.2 b.

(a)   (b)

Figure 12.2. *(a) a buried guide; and (b) a strip guide.*

If the extremities of the guides are closed, then we end up with resonance cavities that are used in oscillators and lasers.

This chapter will look first at the form of the electromagnetic (EM) wave between two conductors in a coaxial cable. In a second part it will then describe the metallic total reflection for a perfect conductor along with the generation of stationary waves. Following this there will be a study of the propagation of a wave between two plane conductors which will then yield a more general description of the properties of a wave guide, most notably those termed buried optical guides.

Generally, metallic wave guides are hollow and have constant cross-sectional widths and are used for the propagation of EM energy of relatively high frequency, such as microwaves with attenuations being less than those for wires. It is worth noting that attenuation in wave guides is due to small imperfections in the conducting walls or imperfect characteristics of the conductor and dielectric losses in the insulator in the case of coaxial cables. These losses will not be covered in any detail, although losses due to material characteristics in an optical wave guide will be discussed.

12.2. A Coaxial Line

12.2.1. Form of transverse EM waves in a coaxial cable

Here the coaxial cable is assumed to have infinite length and has a structure, shown in Figure 12.3 a, made up of two conducting surfaces (C_1) and (C_2) which are cylindrical and around the same axis Oz. The C_1 has a radius denoted by a, and C_2 is assumed to have walls thick enough so that its internal radius (b) and its external radius (e) are $a < b < e$. It also is assumed that the two conductors are separated by a dielectric that exhibits no losses and has an absolute permittivity denoted by ε.

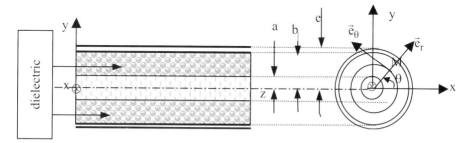

Figure 12.3(a) *Cross section of a coaxial cable.*

Following the application of a sinusoidal current—which circles in one sense with respect to Oz in the internal conductor (radius = a) and on passing to the external conductor circles in the opposite sense—there is between the two conductors a radial and symmetrical electric field that has a form given by

$$\vec{E}\,(r,z,t) = E_0(r)\exp\big(i[\omega t - kz]\big)\vec{e}_r\,.$$

It is supposed that the amplitude of the electric field at the surface of the internal core is given by E_a. Initially assuming that the two conductors are perfect and for a point (M) with cylindrical coordinates r, θ, z, we will look for forms of $\vec{E}$ and $\vec{B}$, along with the intensity.

12.2.1.1. Form of $E_0(r)$

For a point M in a dielectric, where there is no real charge, the Maxwell-Gauss equation is written as $\mathrm{div}\,\vec{E} = 0$. Changing this to cylindrical coordinates, in that

$$\mathrm{div}\,\vec{E} = \frac{1}{r}\frac{\partial(rE_r)}{\partial r} + \frac{1}{r}\frac{\partial(E_\theta)}{\partial \theta} + \frac{\partial(E_z)}{\partial z} \quad \text{with } \vec{E} \text{ having components } E_r\,, E_\theta = 0$$

and $E_z = 0$, this equation can be reduced to $\dfrac{1}{r}\dfrac{\partial\left(rE_r\right)}{\partial r} = 0$, so that

$r\, E_0(r) =$ constant $= a\, E_a$ (value of constant of r is $r = a$).

From this can be deduced that

$$E_0(r) = \frac{a\, E_a}{r}\, ,\ \text{and}\ \vec{E}\, (r,z,t) = \frac{a\, E_a}{r}\, \exp(i[\omega t - kz])\vec{e}_r\, . \tag{1}$$

The amplitude of the electric field varies by $\dfrac{1}{r}$ in the dielectric and passes,

between the core and the gain, from the value E_a to $\dfrac{a}{b}E_a$. From this can be deduced the graphical representation given in Figure 12.3 b for perfect conductors which exhibit internally $\vec{E} = 0$.

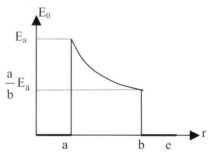

Figure 12.3(b). *Field within a coaxial cable.*

12.2.1.2. *Form of* $\vec{B}$

Here the Maxwell-Faraday equation, as in $\overrightarrow{\mathrm{rot}}\,\vec{E} = -\dfrac{\partial\vec{B}}{\partial t}$, is used but with the cylindrical coordinates:

$$\overrightarrow{\mathrm{rot}}\,\vec{E} = \left(\frac{1}{r}\frac{\partial E_z}{\partial\theta} - \frac{\partial E_\theta}{\partial z}\right)\vec{e}_r + \left(\frac{\partial E_r}{\partial z} - \frac{\partial E_z}{\partial r}\right)\vec{e}_\theta + \left(\frac{1}{r}\frac{\partial rE_\theta}{\partial r} - \frac{\partial E_r}{\partial\theta}\right)\vec{e}_z\, .$$

As $E_\theta = E_z = 0$, and $E = E_r$ is independent of θ, we have more simply:

$\dfrac{\partial E_r}{\partial z}\vec{e}_\theta = -\dfrac{\partial B_r}{\partial t}\vec{e}_r - \dfrac{\partial B_\theta}{\partial t}\vec{e}_\theta - \dfrac{\partial B_z}{\partial t}\vec{e}_z$. By identification according to the components

and following a simple integration, we obtain:

$$B_r = \text{constant} \, , B_z = \text{constant} \, .$$

Solutions that do not represent propagation are rejected (see Section 6.3.1 of this volume) and constants are assumed to be equal to zero, so that

$$B = B_\theta = -\int \frac{\partial E_r}{\partial z} dt = jk \int E_0(r) \, \text{expi}(\omega t - kz) \, dt = \frac{k}{\omega} E_0(r) \exp(i[\omega t - kz]) .$$

In terms of vectors, we thus obtain:

$$\boxed{\vec{B} = \frac{a}{r} \frac{k}{\omega} E_a \exp\left[i(\omega t - kz)\right] \vec{e}_\theta } . \qquad (2)$$

The $\vec{B}$ field thus is orthoradial. In addition the $\vec{E}$ and $\vec{B}$ fields are in phase and orthogonal. The ratio of their amplitude is such that $\dfrac{E}{B} = \dfrac{\omega}{k}$, but as a difference of plane waves, their amplitude varies by $\dfrac{1}{r}$.

12.2.2. Form of the potential, the intensity, and the characteristic impedance of the cable

12.2.2.1. Form of the vector potential

The vector potential, denoted by $\vec{A}(r,z,t)$ and directed along Oz, is such that $\vec{B} = \overrightarrow{\text{rot}} \, \vec{A}$, so that with respect to $\vec{e}_\theta$ it is given by $B_\theta = -\dfrac{\partial A}{\partial r}$. By integrating with respect to r, we have

$$A = - \int B_\theta dr = - a \frac{k}{\omega} E_a \exp\left[i(\omega t - kz)\right] \int \frac{dr}{r} , \text{ from which}$$

$$\vec{A}(r,z,t) = \left[- a \frac{k}{\omega} E_a \ \text{Ln} \, r + \text{Cte} \right] \exp\left[i(\omega t - kz)\right] \vec{e}_z .$$

By taking the origin of the vector potentials as $r = a$ (on the core of the central conductor), the constant can be fixed as: $\text{constant} = a \dfrac{k}{\omega} E_a \ \text{Ln} \, a$, so that finally,

$$\vec{A}(r,z,t) = \left[a \frac{k}{\omega} E_a \ \text{ln} \, \frac{a}{r} \right] \exp\left[i(\omega t - kz)\right] \vec{e}_z .$$

12.2.2.2. Form of the scalar potential

Along Oz the equation $\vec{E} = -\overrightarrow{grad}V - \dfrac{\partial \vec{A}}{\partial t}$ gives $0 = -\dfrac{\partial V}{\partial z} - \dfrac{\partial A}{\partial t}$, so that with the

equation for A, we have $\dfrac{\partial V}{\partial z} = -\dfrac{\partial A}{\partial t} = -j\omega\left[a\dfrac{k}{\omega}E_a \ln \dfrac{a}{r} \right]\exp\left[i(\omega t\text{-}kz)\right]$. By

integrating with respect to z and also by taking the origin of (scalar) potentials as r = a, then it is possible to state that:

$$V(r,z,t) = \left[a\,E_a\,Ln\,\dfrac{a}{r} \right]\exp\left[i(\omega t\text{-}kz)\right] \ .$$

12.2.2.3. Form of the intensity

Ampère's theorem applied to a circular cross section of the cable and passing through a point M is written as

$$\oint_{(C)} \vec{B}.d\vec{l} = \iint_{\Sigma}\left(\mu_0\vec{j} + \mu_0\varepsilon\dfrac{\partial \vec{E}}{\partial t} \right)d\vec{\Sigma} = \iint_{\Sigma}\left(\mu_0\vec{j} + i\omega\mu_0\varepsilon\vec{E} \right)d\vec{\Sigma}.$$ With $\vec{E}$ being radial

and $\vec{E} \perp \vec{\Sigma}$ (section perpendicular to the plane of the circle), Ampère's theorem can be written as

$$\oint_{(C)} \vec{B}.d\vec{l} = \iint_{\Sigma}\mu_0\vec{j}\,d\vec{\Sigma} = \mu_0\,I\ .$$

The result is that $I = \dfrac{2\pi\,r\,B}{\mu_0}$. Given the expression for B, and with

$v_\varphi = \dfrac{\omega}{k} = \dfrac{c}{n} = \dfrac{c}{\sqrt{\varepsilon_r}}$, we obtain

$$I = 2\,\pi\,a\,c\,\varepsilon_0\sqrt{\varepsilon_r}\,E_a\exp\left[i(\omega t - kz)\right]. \tag{3}$$

Equation (3) also can be rewritten as $I = I_0\,\exp(i[\omega t - kz])$, with

$$I_0 = 2\,\pi\,a\,c\,\varepsilon_0\sqrt{\varepsilon_r}\,E_a. \tag{3'}$$

12.2.2.4. Characteristic impedance (Z_c)

We have $Z_c = \dfrac{V_a\text{-}V_b}{I} = \dfrac{V_a\,(a,z,t)\,\text{-}V_b\,(b,z,t)}{I}$. By using the expressions for V and

for I we can directly obtain $Z_c = \dfrac{a\,E_a Ln\left(\dfrac{b}{a}\right)}{2\pi c\,a\,\varepsilon_0\sqrt{\varepsilon_r}\,E_a}$, so that $Z_c = \dfrac{Ln\left(\dfrac{b}{a}\right)}{2\pi c\,\varepsilon_0\sqrt{\varepsilon_r}}$. $\tag{4}$

With $\varepsilon_0 = \dfrac{1}{36\,\pi\,10^9}$ S.I. , and c = 3 x 10^8 m/s , we have $2\pi c\ \varepsilon_0 = \dfrac{1}{60}$, from which

$$Z_c = \frac{60}{\sqrt{\varepsilon_r}}\ Ln\left(\frac{b}{a}\right). \qquad (4')$$

12.2.3. Electrical power transported by an EM wave

From Ampère's theorem we obtained $B = \dfrac{\mu_0}{2\pi r}I$, so that from Eqs. (2) and (3'), it is

possible to state that $\vec{B} = \dfrac{\mu_0}{2\pi r}I_0\ \{exp[i(\omega t - kz)]\}\ \vec{e}_\theta$.

Using an equation developed in Section 12.2.1.2, i.e., $\dfrac{E}{B} = \dfrac{\omega}{k} = v_\varphi = \dfrac{c}{\sqrt{\varepsilon_r}}$,

we have $E = B\dfrac{c}{\sqrt{\varepsilon_r}} = \dfrac{\mu_0}{2\pi r}\dfrac{c}{\sqrt{\varepsilon_r}}I = \dfrac{1}{2\pi r\ \varepsilon_0\ c\ \sqrt{\varepsilon_r}}I$, so that in vectors:

$$\vec{E} = \frac{I_0}{2\pi r\ \varepsilon_0\ c\ \sqrt{\varepsilon_r}}\ .$$

The Poynting vector therefore is in the form $\vec{S} = \dfrac{\vec{E}\times\vec{B}}{\mu_0}$, so that by taking the real

solutions for $\vec{E}$ and $\vec{B}$ gives us $\vec{S} = \dfrac{1}{\left(2\pi r\right)^2\ \varepsilon_0 c\sqrt{\varepsilon_r}}I_0^2\ cos^2(\omega t\text{-}kz)\ \vec{e}_z$.

The average value of $cos^2(\omega t\text{-}kz)$ with respect to time is $\dfrac{1}{2}$, so that:

$$\langle\vec{S}\rangle = \frac{1}{8\pi^2 r^2\ \varepsilon_0 c\sqrt{\varepsilon_r}}I_0^2.\,\vec{e}_z\ .$$

The average power transported by the EM wave is given by the flux of the average Poynting vector through a ring composed of the circles with radii a and b.

$$P = \iint\langle\vec{S}\rangle.d\vec{\Sigma} = \frac{I_0^2}{8\pi^2\ \varepsilon_0 c\sqrt{\varepsilon_r}}\int_a^b\frac{2\pi\ rdr}{r^2} = \frac{I_0^2}{4\pi\ \varepsilon_0 c\sqrt{\varepsilon_r}}Ln\frac{b}{a}\ .$$

12.2.4.   Conductor with an imperfect core that exhibits a resistance and the attenuation length

Power dissipated along the length of an element (dz) of a cable with resistance given by $dR = \dfrac{1}{\sigma} \dfrac{dz}{\pi a^2}$ is $dp = I^2(z,t)dR$, so that with Eq. (3'), we have

$$dp = \frac{1}{\sigma} \frac{dz}{\pi a^2} I_0^2 \cos^2[\omega t - kz].$$ The average power loss with time for a length dz is

thus $dP = <dp> = \dfrac{1}{\sigma} \dfrac{1}{\pi a^2} \dfrac{I_0^2}{2} dz$.

The loss relative to the power is finally given by:

$$-\frac{dP}{P} = \frac{\dfrac{1}{\sigma} \dfrac{1}{\pi a^2} \dfrac{I_0^2}{2}}{\dfrac{I_0^2}{4\pi \varepsilon_0 c \sqrt{\varepsilon_r}} Ln \dfrac{b}{a}} dz = \frac{2\varepsilon_0 c \sqrt{\varepsilon_r}}{\sigma a^2 Ln \dfrac{b}{a}} dz.$$

By introducing Eq. (4) for the characteristic impedance we also have:

$-\dfrac{dP}{P} = \dfrac{1}{\pi a^2 \sigma Z_c} dz$. Integration with respect to z gives for the z abscissa

$$P(z) = P(0) \exp\left(-\frac{z}{\pi\sigma a^2 Z_c}\right).$$

This equation is in the form $P(z) = P(0) \exp\left(-\dfrac{z}{L}\right)$, with $L = \pi\sigma a^2 Z_c$ where L is the attenuation length of the cable, i.e., the distance over which the power is divided by $e \approx 2.7$.

In numerical terms, with $a = 10$ mm and $b = 40$ mm, $\sigma = 5 \times 10^7 \; \Omega^{-1}m^{-1}$, and $\varepsilon_r = 4$, we have

$$Z_c = \frac{60}{2} \ln 4 \approx 42 \; \Omega, \text{ and } L \approx 660 \text{ km}.$$

12.3. Preliminary Study of the Normal Reflection of a Rectilinearly Polarized MPPEM Wave on a Perfect Conductor, Stationary Waves, and Antennae

12.3.1. Properties of a perfect conductor and equations of continuity at the surface

As shown in Chapter 11, problem 1, for relatively low frequencies, such that $\omega << \dfrac{1}{\tau}$, and with the condition that the continuous conductivity is sufficiently large so that $\sigma(0) \gg \omega\varepsilon_0$, and that within a range of frequencies that does not go too high ($\omega \ll \omega_p$), the reflection at a conductor can be assumed to be total, as in $r \approx -1$, with as a consequence a dephasing equal to π.

A medium is assumed to be a perfect conductor if its conductivity is quasi-infinite. The result is that Ohm's law for a volume of the conductor, written $\vec{j} = \sigma\vec{E}$, shows how $\vec{j}$ stays infinite as $\sigma \to \infty$ (perfect conductor), and the field in the conductor, which is here denoted by $\vec{E}_C$, can be equal to zero.

With respect to the magnetic field, the Maxwell-Faraday relation, which is written $\overrightarrow{\text{rot}}\, \vec{E} = -\dfrac{\partial \vec{B}}{\partial t}$, indicates that $\vec{B} = \text{constant}$. A static magnetic field cannot exist without a distribution of permanent currents, and in the absence of their application (as is assumed here) the field in the conductor denoted $\vec{B}_C$ can be equal only to zero. In other words, there is no magnetic field in a perfect conductor under a varying regime. Finally, as $\vec{E}_C = 0$ and $\vec{B}_C = 0$, the EM wave cannot penetrate into a perfect conductor under a varying regime.

In addition, the volume of a perfect conductor can be assumed to be electrically neutral. The charges are sufficiently mobile so that an excess in one given charge, for example, positive, is immediately compensated for by the movement of opposing charges. Further details can be found in Section 1.3.5. In very simple, and perhaps too simple terms, the charges can be assumed to be infinitely mobile (as $\sigma = qn\mu$, so if $\sigma \to \infty$, then the mobility $\mu \to \infty$), so that there can be an instantaneous return to neutrality. Finally, any charges can only appear at a discontinuity such as a surface of the metal where from the point of view of a given charge, there cannot materially exist opposite charges in a vacuum (an excess of charges of a given type thus is possible at the surface). The contents of the envelope of the conductor are electrically equivalent to a vacuum and the laws of continuity at the vacuum/conductor interface only bring in the permittivity and permeability of a vacuum, along with the possible surface charge (σ_s) and surface current ($\vec{j}_s$) densities.

By denoting as $\vec{n}_{21}$ ($\equiv -\vec{e}_z$ as indicated in Figure 12.4) the normal that goes from medium (2) (the metal) to medium (1) (the vacuum or air), we can apply to the vacuum/metal interface the following classic conditions of continuity (generally cited in the first years of university courses):

$$\begin{cases} \vec{E}_{1t} = \vec{E}_{2t} \\ \vec{B}_{1n} = \vec{B}_{2n} \end{cases} \qquad \begin{cases} \vec{E}_{1n} - \vec{E}_{2n} = \dfrac{\sigma_s}{\varepsilon_0}\vec{n}_{21} \\ \vec{B}_{1t} - \vec{B}_{2t} = \mu_0 \vec{j}_s \times \vec{n}_{21} \end{cases} \begin{array}{l} \text{, so with } \vec{E}_C = \vec{E}_2 = 0 \\[1em] \text{, so with } \vec{B}_C = \vec{B}_2 = 0 \end{array} \qquad \begin{cases} \vec{E}_{1n} = \dfrac{\sigma_s}{\varepsilon_0}\vec{n}_{21} \\ \vec{B}_{1t} = \mu_0 \vec{j}_s \times \vec{n}_{21} \end{cases}$$

12.3.2. *Equation for the stationary wave following reflection*

The system under study is that of a vacuum/metal interface in the presence of a monochromatic (with angular frequency denoted by ω) EM wave undergoing a plane incidence (with respect to the normal at the interface so that $\theta_1 = 0\,°$) and progressing along Oz as shown in Figure 12.5 a with the incident wavevector $\vec{k}_i = k_0\,\vec{e}_z$. The wave also is polarized rectilinearly along Ox and is assumed to be of the form $\underline{\vec{E}}_i = E_i^0\left(\exp[i(k_0 z - \omega t)]\right)\vec{e}_x$.

   The wave reflected into the vacuum will have the same angular frequency as the incident wave (due to the properties of the media) and the wavevector of the reflected wave ($\underline{\vec{E}}_r$) will have the same modulus (k_0) as the incident wave. Following a reflection under a normal incidence, the wavevector of the reflected wave is according to the first Snell-Descartes law given by $\vec{k}_r = -\,\vec{k}_i = -\,k_0\,\vec{e}_z$. In addition, div $\underline{\vec{E}}_r = 0$, from which $\vec{k}_r\,\underline{\vec{E}}_r = 0$, showing that the reflected wave also is transversal (parallel to Oxy). In the medium (1) (vacuum) the resultant wave thus is $\underline{\vec{E}}_1 = \underline{\vec{E}}_i + \underline{\vec{E}}_r$ and is parallel to Oxy (in a tangential plane).

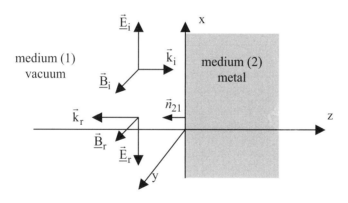

Figure 12.4. *Vacuum/metal interface.*

   As mentioned above, with the conductor being perfect, the fields that have been denoted $\vec{E}_C$ and $\vec{B}_C$ in the conductor are zero. At $z = 0$ (where $\vec{E}_{2t}(z = 0) = \vec{E}_{Ct} = 0$), we can state that $\vec{E}_{1t} = \vec{E}_{2t} = 0$, so that $0 = \vec{E}_{1t}(z = 0) = \underline{\vec{E}}_i(z = 0) + \underline{\vec{E}}_r(z = 0)$, from which :

$$\vec{E}_r(z = 0) = -\underline{\vec{E}}_i(z = 0).$$

The reflected wave is polarized parallel to the incident wave, that is to say that along $\vec{e}_x$ with an amplitude such that $E_r^0 = - E_i^0$. Thus, the reflected wave has the form:

$$\vec{E}_r = - E_i^0 \left(\exp[i(- k_0 z - \omega t)]\right) \vec{e}_x .$$

In medium (1) (vacuum), the resulting wave thus is given by:

$$\vec{E}_1 = \vec{E}_i + \vec{E}_r$$
$$= E_i^0 \left\{\left(\exp[i(k_0 z - \omega t)]\right) - \left(\exp[i(-k_0 z - \omega t)]\right)\right\} \vec{e}_x .$$
$$= 2i \, E_i^0 \, \sin k_0 z \, \left(\exp[-i\omega t]\right) \vec{e}_x$$

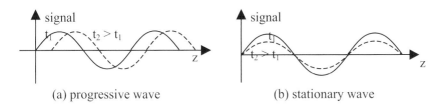

| (a) progressive wave | (b) stationary wave |

Figure 12.5. *(a) Progressive; and (b) stationary waves.*

The physical solution for the resultant wave in a vacuum thus is finally given by

$$\vec{E}_1 = R(\vec{E}_1) = 2 \, E_i^0 \, \sin k_0 z \, \sin \omega t \, \vec{e}_x$$

where the spatial and temporal dependences are now separated. The phase velocity thus is zero, and the wave is stationary, as shown in Figure 12.5 b (see also Section 6.2.5).

12.3.3. Study of the form of the surface charge densities and the current at the metal

12.3.3.1. The surface charge density of zero

The normal components of the electric field are zero in medium (1) ($\vec{E}_1 \, // \, \vec{e}_x$) as in medium (2) (where $\vec{E}_C = 0$), and the normal component of the electric field does not give rise to any discontinuity, so $\vec{E}_{2n} = \vec{E}_{1n} (= 0)$, where $\sigma_s = 0$.

12.3.3.2. Form of the surface current density

12.3.3.2.1. Equation for $\vec{j}_s$

As can be seen in Figure 12.4, which schematizes the results obtained for a reflected wave, following reflection $\underline{\vec{B}}_i$ and $\underline{\vec{B}}_r$ remain in phase. In medium (1) the resultant magnetic field thus is given by:

$$\underline{\vec{B}}_1 = \underline{\vec{B}}_i + \underline{\vec{B}}_r = \frac{1}{\omega}\left(\left[\vec{k}_i \times \underline{\vec{E}}_i\right] + \left[\vec{k}_r \times \underline{\vec{E}}_r\right]\right),$$

so that with $-\vec{k}_r = \vec{k}_i = k_0\,\vec{e}_z$ and $-E_r^0\,\vec{e}_x = E_i^0\,\vec{e}_x$ we have

$$\underline{\vec{B}}_1 = \frac{\vec{e}_z \times \vec{e}_x}{\omega}\, k_0 E_i^0\left(\exp\left[ik_0 z\right] + \exp\left[-ik_0 z\right]\right)\exp\left[-i\omega t\right].$$

With $k_0 = \dfrac{\omega}{c}$, we obtain for the physical solution for a vacuum

$$\vec{B}_1 = R(\underline{\vec{B}}_1) = \frac{2E_i^0}{c}\,\vec{e}_y\,\cos(k_0 z)\,\cos(\omega t)\ .$$

The magnetic field thus is directed parallel to the interface.

As in a metal $\vec{B}_C = 0$ (and $\vec{B}_{2t} = 0$), the tangential component of the magnetic field undergoes a discontinuity at the interface. Therefore, there must appear a surface current density ($\vec{j}_s$) at the interface so that the equation for continuity (end of preceding Section 11.3.1)

$$\left(\vec{B}_{1t}\right)_{z=0} = \mu_0 \vec{j}_s \times \vec{n}_{21} = -\mu_0 \vec{j}_s \times \vec{e}_z,\ \text{so that}\ \frac{\left(\vec{B}_{1t}\right)_{z=0}}{\mu_0} = \vec{e}_z \times \vec{j}_s\ .$$

By multiplying by vectors the right-hand side by $\vec{e}_z$ and on noting that $\vec{e}_z \perp \vec{j}_s$, we obtain:

$$\vec{j}_s = \frac{\left(\vec{B}_{1t}\right)_{z=0}}{\mu_0} \times \vec{e}_z = \frac{2E_i^0}{\mu_0 c}\,\vec{e}_x\,\cos(\omega t).$$

Of note then is that $\vec{j}_s\ // \vec{E}_i$ ($//\vec{e}_x$) and the current density is collinear to the incident electric field. Physically, the current density is the result of the polarization of the metal by the electric field.

12.3.3.2.2. Comment

As Figure 12.4 illustrates, the magnetic field vectors of the incident and reflected waves are parallel to the interface. Their resultant normal component in medium (1) thus is zero, thus assuring the continuity of the normal component of $\vec{B}$, as in medium (2) the magnetic field is zero ($\vec{B}_C = 0$ implies that $\vec{B}_{2n} = 0$).

12.3.3.3. Applications

This property is the basis of the use of metallic antennae for detecting EM waves. With the antenna lines parallel to the direction of polarization of the incident wave, the current density detected is at a maximum and directly proportional to the intensity of the incident electric field. If the lines of the antenna are not parallel to the incident wave, then the detected current density is no longer proportional to the intensity of the component of the electric field at a parallel incidence to the antenna lines, as shown in Figure 12.6.

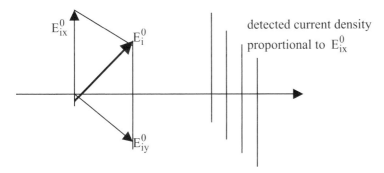

Figure 12.6. *Relation between the disposition of the antenna lines and the detected current.*

With a metal never being a perfect conductor, in effect the current circulates over a thickness several times greater than the "skin" thickness (δ). If the thickness (d) of the metal is such that $d < \delta$, a part of the wave in fact can be transmitted. However, if $d \gg \delta$, the metal becomes a real obstacle to the propagation of the EM wave and will create an electromagnetic shield.

12.4. Study of Propagation Guided between Two-Plane Conductors: Extension to Propagation through a Guide with a Rectangular Cross Section

12.4.1. Wave form and equation for propagation between two conducting planes

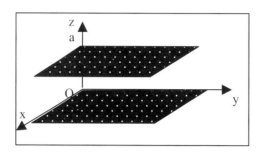

Figure 12.7. *Propagation between two-plane conductors.*

This section considers a wave propagated along Oy in a vacuum without attenuation between two-plane and perfect conductors separated by a given distance (a) and assumed to be of great dimensions (with respect to a) in directions Oy and Ox.

With the system being invariant with respect to Ox, the field ($\vec{\underline{E}}$) does not depend on x and it can be assumed that $\vec{\underline{E}}$ is a monochromatic wave in the form

$$\vec{\underline{E}} = \vec{\underline{E}}_m(z)\, \exp[i(k_g y - \omega t)], \quad (1)$$

where k_g is a positive constant.

Equation (1) is valid uniquely between the two-plane conductors. Above and below these planes, the electric field is zero (metallic planes inside which $\vec{\underline{E}} = 0$). Additionally, in a plane normal to the direction of propagation, the wave is not uniform and cannot be assumed to be planar.

Between the two plane conductors, dominated by a vacuum, the equation for propagation is $\overrightarrow{\Delta \vec{\underline{E}}} - \dfrac{1}{c^2}\dfrac{\partial^2 \vec{\underline{E}}}{\partial t^2} = 0$ [from Eq. (5) of Chapter 6]. Taking Eq. (1) into account and by making a calculation resembling that in Section 11.2.6, we have

$$\frac{d^2\vec{\underline{E}}_m(z)}{dz^2} + \left(\frac{\omega^2}{c^2} - k_g{}^2\right)\vec{\underline{E}}_m(z) = 0. \quad (2)$$

The tangential component of the electric field, situated in a plane parallel to Oxy, exhibits two components denoted E_x and E_y which are such that the limiting conditions are [with $\underline{\vec{E}}_m(0) = 0$ and $\underline{\vec{E}}_m(a) = 0$]

$$\underline{E}_{mx}(0) = \underline{E}_{my}(0) = 0, \text{ and } \underline{E}_{mx}(a) = \underline{E}_{my}(a) = 0. \qquad (3)$$

For its part, the continuity of the normal component of $\underline{\vec{B}}$ is

$$\underline{B}_{mz}(0) = \underline{B}_{mz}(a) = 0. \qquad (4)$$

12.4.2. Study of transverse EM waves

Here the possibility of the transverse wave propagating at a velocity (c) along Oy with, on the one side $v_\varphi = c$ (nondispersion solution as v_φ = constant), and on the

other $v_\varphi = \dfrac{\omega}{k_g}$ [equation for v_φ in agreement with Eq. (1)], we have $k_g = \dfrac{\omega}{c}$.

We shall now determine the corresponding exact form of the electric and magnetic fields.

12.4.2.1. Form of the electric field (given by $\underline{\vec{E}}_m(z)$)

Equation (2) for propagation thus is reduced to $\dfrac{d^2\underline{\vec{E}}_m(z)}{dz^2} = 0$. Two successive

integrations with respect to z give $\underline{\vec{E}}_m(z) = \vec{\underline{C}}_1 z + \vec{\underline{C}}_2$, $\qquad (5)$

where $\vec{\underline{C}}_1$ and $\vec{\underline{C}}_2$ are   two   constant   vectors.   The   conditions $\underline{E}_{mx}(0) = 0$ and $\underline{E}_{my}(0) = 0$, respectively, give:

$$C_{2x} = 0 \text{ and } C_{2y} = 0$$

and similarly, from $\underline{E}_{mx}(a) = \underline{E}_{my}(a) = 0$ we can deduce that

$$C_{1x} = 0 \text{ and } C_{1y} = 0.$$

Finally the vectors $\vec{\underline{C}}_1$ and $\vec{\underline{C}}_2$ only have components in the z direction, and are of the form $\vec{\underline{C}}_1 = \underline{C}_{1z} \vec{e}_z$ and $\vec{\underline{C}}_2 = \underline{C}_{2z} \vec{e}_z$. Eq. (5) can therefore be rewritten as

$$\underline{\vec{E}}_m(z) = \left(\underline{C}_{1z} z + \underline{C}_{2z}\right) \vec{e}_z = \underline{E}_m(z) \vec{e}_z. \qquad (6)$$

Gauss's equation gives then for its part $\operatorname{div} \underline{\vec{E}} = 0$, where $\dfrac{d\underline{E}_m(z)}{dz} = 0$,

while Eq. (6) gives $\dfrac{d\underline{E}_m(z)}{dz} = \underline{C}_{1z}$. From this can be deduced that $\underline{C}_{1z} = 0$, such that

$$\underline{\vec{E}}_m(z) = \underline{C}_{2z} \vec{e}_z = \underline{E}_m \vec{e}_z$$

where $\underline{E}_m$ is a constant. For this type of wave, $\underline{\vec{E}}_m(z)$ is a constant vector parallel to $\vec{e}_z$. The electric field, also parallel to $\vec{e}_z$, thus is in the form

$$\underline{\vec{E}} = \underline{E}_m \{\exp[i(k_g y - \omega t)]\}\, \vec{e}_z. \qquad (7)$$

12.4.2.2. Form of the magnetic field

With the wave assumed to be monochromatic, the Maxwell-Faraday equation makes it possible to state that $\operatorname{rot}\underline{\vec{E}} = -\dfrac{\partial \underline{\vec{B}}}{\partial t} = -(-i\omega\underline{\vec{B}})$, from which can be deduced directly

that $\underline{\vec{B}} = -\dfrac{i}{\omega}\operatorname{rot}\underline{\vec{E}}$.

From Eq. (7) for $\underline{\vec{E}}$, we can deduce that:

$\operatorname{rot}\underline{\vec{E}} = \dfrac{\partial}{\partial y}\underline{E}_m\exp[i(k_g y-\omega t)]\,\vec{e}_x = ik_g\underline{E}_m\exp[i(k_g y-\omega t)]\,\vec{e}_x$, with the result that

$\underline{\vec{B}} = -\dfrac{i^2 k_g}{\omega}\underline{E}_m\exp[i(k_g y-\omega t)]\,\vec{e}_x$, so with $k_g = \dfrac{\omega}{c}$ we have

$$\underline{\vec{B}} = \dfrac{\underline{E}_m}{c}\left\{\exp[i(k_g y-\omega t)]\right\}\vec{e}_x. \qquad (8)$$

We can note that this result is identical to that which would be obtained had we used the equation for plane waves as in $\underline{\vec{B}} = \dfrac{\vec{k}_g \times \underline{\vec{E}}}{\omega}$, which *a priori* cannot be used here directly as the wave cannot be assumed to be completely planar. Nevertheless, Eqs. (7) and (8) demonstrate that here again $\underline{\vec{E}}$ and $\underline{\vec{B}}$ are transverse, and hence the term "transverse electromagnetic wave" or TEM wave.

12.4.2.3. Comment: The rectangular cross-sectional wave guide

If we envisage that in a rectangular guide, as shown in Figure 12.1 b with an axis as in Figure 12.7, there is a wave that propagates at a given speed (c), then the limiting conditions imposed on that wave are:
• in one part by the parallel metallic planes with sides $z = 0$ and $z = a$, which result in the electric field taking on a form given by the equation (see also results from Section 12.4.2.1):

$$\underline{\vec{E}} = \underline{E}_m \{\exp[i(k_g y - \omega t)]\}\, \vec{e}_z$$

where $\underline{E}_m$ is a constant and $\underline{\vec{E}}$ also is parallel to $\vec{e}_z$ (at whatever value of x);

• and in another part, fixed by the second system of metallic planes parallel at the sides to $x = 0$ and $x = b$ with respect to Ox (0 being on the side of the plane Oyz and b on the side of the plane to which it is parallel), impose that for $x = 0$ and $x = b$, $\vec{E} = 0$ so that $\underline{E}_m = 0$. This trivial solution is not a physical one, as the possibility of propagation at the speed c along Oy is not available to a guide with a rectangular cross section.

12.4.3. Study of transverse electric waves (TE waves)

Here we study the possibility of a wave propagating in the guide plane Oxy. The wave is thus polarized in accord with the breakdown given by:

$$\vec{\underline{E}} = \underline{E}_x \vec{e}_x + \underline{E}_y \vec{e}_y, \quad (9)$$

and has a form given by Eq. (1), as in $\vec{\underline{E}} = \underline{\vec{E}}_m(z) \exp[i(k_g y - \omega t)] = \vec{\underline{E}}(y,z)$.

12.4.3.1. Polarization direction of the electric field

Under these conditions, Gauss's equation gives:

div $\vec{\underline{E}} = 0$

$$\left. \begin{array}{l} = \dfrac{d\underline{E}_x}{dx} + \dfrac{d\underline{E}_y}{dy} + \dfrac{d\underline{E}_z}{dz} = ik_g\, \underline{E}_y \\[6pt] \underbrace{}_{} \qquad \underbrace{}_{} \\[2pt] = 0 \qquad\quad = 0 \\ (\text{as } \underline{E}_x = \underline{E}_x(y,z)) \quad (\text{as } \underline{E}_z = 0) \end{array} \right\} \Rightarrow \underline{E}_y = 0$$

Equation (9) thus is reduced to $\vec{\underline{E}} = \underline{E}_x \vec{e}_x$ and the electric field is polarized along Ox perpendicularly to the direction of propagation Oy. The wave is termed "transverse electric" and denoted TE. The electric field can be written:

$$\vec{\underline{E}} = \underline{E}_{mx}(z)\, \vec{e}_x \, \exp[i(k_g y - \omega t)] = \vec{\underline{E}}_{mx}(z) \exp[i(k_g y - \omega t)]. \quad (10)$$

12.4.3.2. Solutions for the propagation equation

Equation (2) for propagation becomes:

$$\frac{d^2\underline{E}_{mx}(z)}{dz^2} + \left(\frac{\omega^2}{c^2} - k_g^2 \right) \underline{E}_{mx}(z) = 0. \quad (11)$$

12.4.3.2.1. First case

If $\left(\dfrac{\omega^2}{c^2} - k_g^{\ 2} \right) = -\alpha^2 < 0$ (where α is real), so that $k_g > \dfrac{\omega}{c}$, the solution to Eq. (11) is:

$$\underline{E}_{mx}(z) = \underline{C}_1 \exp(\alpha z) + \underline{C}_2 \exp(-\alpha z).$$

With $\underline{E}_{mx}(0) = 0$ and $\underline{E}_{mx}(a) = 0$, we can deduce that, respectively,

$$\left. \begin{array}{l} \underline{C}_1 = -\underline{C}_2 \\ \underline{C}_1 \exp(\alpha a) + \underline{C}_2 \exp(-\alpha a) = 0 \end{array} \right\} \Rightarrow 2\underline{C}_1 \, sh\alpha a = 0$$

With $sh\alpha a \neq 0$, we deduce that $\underline{C}_1 = -\underline{C}_2 = 0$, so that $\underline{E}_{mx}(z) = 0$, and there is no possible physical solution in this case.

12.4.3.2.2. Second case

If $\left(\dfrac{\omega^2}{c^2} - k_g^{\ 2} \right) = K^2 > 0$ (with K being real), then $k_g < \dfrac{\omega}{c}$ and the solution to Eq. (11) is given by

$$\underline{E}_{mx}(z) = \underline{C}_1 \exp(iKz) + \underline{C}_2 \exp(-iKz). \qquad (12)$$

The limiting conditions, $\underline{E}_{mx}(0) = 0$ and $\underline{E}_{mx}(a) = 0$, respectively, give:

$$\left. \begin{array}{l} \underline{C}_1 = -\underline{C}_2 \\ \underline{C}_1 \exp(ika) + \underline{C}_2 \exp(-ika) = 0 \end{array} \right\} \Rightarrow 2i\underline{C}_1 \, \sin ka = 0,$$

so that $K = \dfrac{m\pi}{a}$, with m being a whole nonzero number.

Substituting these results into Eq. (12), we gain:

$$\underline{E}_{mx}(z) = 2 \, i \, \underline{C}_1 \sin \dfrac{m\pi}{a} z. \qquad (13)$$

The electric field, according to Eq. (10), is of the form $\vec{E} = E_{mx} \vec{e}_x \exp[i(k_g y - \omega t)]$,

so that with Eq. (13), we find that $\vec{E} = 2 \, i \, \underline{C}_1 \sin \dfrac{m\pi}{a} z \, \vec{e}_y \exp[i(k_g y - \omega t)]$.

By choosing the origin of the phases as being at $y = 0$ at a time $t = 0$, then it is imposed that $2 \, i \, \underline{C}_1$ is real, as in $2 \, i \, \underline{C}_1 = E°$, and we have

$$\underline{\vec{E}} = E° \left(\sin \frac{m\pi}{a} z \right) \vec{e}_x \exp[i(k_g y - \omega t)] . \qquad (14)$$

12.4.3.3. Proper vibrational modes

With $K = \dfrac{m\pi}{a}$, we thus have:

$$\frac{\omega^2}{c^2} = k_g^2 + \frac{m^2\pi^2}{a^2} \quad , \qquad (15)$$

from which $\omega = c\sqrt{k_g^2 + m^2 \dfrac{\pi^2}{a^2}}$. The equation $\omega = f(k_g)$ is not a linear relation and the system undergoes dispersion.

When $m = 1$ and $k_g \to 0$, $\omega \to \omega_c = \dfrac{\pi c}{a}$ (Figure 12.8).

For any value of m and and $k_g \to 0$, $\omega \to \omega_m = m\omega_c = m\dfrac{\pi c}{a}$.

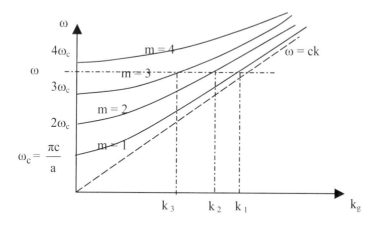

Figure 12.8. *Dispersion plots for $\omega = f(k)$.*

For a given value of ω, there are many possible values of k_g : k_1, k_2, k_3 . They are associated with whole values of m which are of a limited number as in the second case, $\dfrac{\omega^2}{c^2} - \dfrac{m^2\pi^2}{a^2} = k_g^2 > 0$, so that $\dfrac{m^2\pi^2}{a^2} < \dfrac{\omega^2}{c^2}$, which is:

$$m < \frac{\omega a}{\pi c} = \frac{\omega}{\omega_c} .$$

For each value of m, or of k_g, there is associated a solution of the electric field that assures a propagation of the field without modification of its intensity along the guide (for given values of m and z, the intensity is given by $E° \sin \dfrac{m\pi}{a} z = Cte$, $\forall y$). These are the guide modes, and the different values of k_g ($k_1, k_2, k_3...$) are called the proper modes of spatial vibration of the guide.

Only signals with angular frequencies higher than $\omega_c = \dfrac{\pi c}{a}$ (termed the cutoff frequency) can propagate in the guide.

This angular frequency corresponds to a wavelength in the vacuum given by $\lambda_{0c} = \dfrac{2\pi c}{\omega_c} = 2a$. The condition $\omega \geq \omega_c$ also can be written as $\lambda_0 \leq \lambda_{0c} = 2a$.

12.4.3.4. Form of $\vec{B}$

The magnetic field can be taken from the electric field given in Eq. (14) with the help of the Maxwell-Faraday relation, $\overrightarrow{rot}\vec{E} = -\dfrac{\partial\vec{B}}{\partial t} = i\,\omega\vec{B}$, so that:

$$\vec{B} = -\frac{i}{\omega}\overrightarrow{rot}\vec{E} \begin{vmatrix} 0 \\[2mm] (\dfrac{-i}{\omega})(\dfrac{\partial E_x}{\partial z}) = -i(\dfrac{m\pi}{a\omega})\,E^0\,[\cos(\dfrac{m\pi z}{a})]\,\exp[i(k_g y - \omega t)] \\[4mm] (\dfrac{-i}{\omega})(\dfrac{-\partial E_x}{\partial y}) = -(k_g/\omega)\,E^0\,[\sin(\dfrac{m\pi z}{a})]\,\exp[i(k_g y - \omega t)] \end{vmatrix} . \qquad (16)$$

The $E = E_x$ is in phase with B_z and in the squared phase, or quadrature, of the longitudinal component $B_y \neq 0$; $\vec{B}$ therefore is not transverse and the wave is simply a transverse electric wave and not a TEM.

12.4.3.5. Propagation of energy

With $\underline{\vec{E}} = \vec{E}_m \exp(i[k_g y - \omega t])$ and $\underline{\vec{B}} = \vec{B}_m \exp(i[k_g y - \omega t])$, the average over time for the Poynting vector is given by (see also Section 7.3.3) :

$$\langle \vec{S} \rangle = \frac{1}{2\mu_0} R(\underline{\vec{E}}_m \times \underline{\vec{B}}_m^*) .$$

Taking the form of the electric and magnetic fields into account, the direct calculation of $\underline{\vec{E}}_m \times \underline{\vec{B}}_m^*$ shows that the vector has a purely imaginary component with respect to Oz. The associated average value, which corresponds to the real part, is therefore zero. With respect to Oy, the component of $\underline{\vec{E}}_m \times \underline{\vec{B}}_m^*$, however is real so that the propagation of the energy is through Oy, the direction of wave propagation.

12.4.3.6. Phase and group velocities

The phase of the wave is given by $(k_g y - \omega t)$ and its phase velocity is given by $v_\varphi = \frac{\omega}{k_g}$. With $k_g^2 = \frac{\omega^2}{c^2} - \frac{m^2 \pi^2}{a^2}$, we have $v_\varphi = \frac{\omega}{\left(\frac{\omega^2}{c^2} - \frac{m^2 \pi^2}{a^2} \right)^{1/2}}$, and as $\omega_c = \frac{\pi c}{a}$,

therefore

$$\boxed{v_\varphi = \frac{c}{\sqrt{\left(1 - \frac{m^2 \omega_c^2}{\omega^2} \right)}} > c} .$$

For its part, the group velocity is $v_g = \frac{d\omega}{dk}$. By differentiating Eq. (15), we obtain $\frac{2\omega}{c^2} d\omega = 2k_g dk_g$, from which:

$$v_g = \frac{d\omega}{dk_g} = c^2 \frac{k_g}{\omega} = \frac{c^2}{v_\varphi} .$$

With $v_\varphi > c$, then $v_g < c$ while noting that $v_g v_\varphi = c^2$ also is verified.

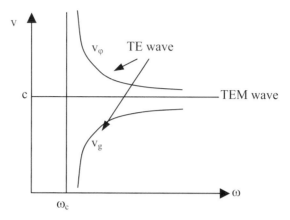

Figure 12.9. *Plots of $v_g = f(\omega)$ and $v_\varphi = g(\omega)$.*

Variations in v_g and in v_φ as a function of ω are indicated in Figure 12.9 for $m = 1$. When $\omega \gg \omega_c$, so that also $\lambda_0 \ll 2a$, we have $\omega \to ck$, and v_g just as v_φ tends toward c. The wave guide is no longer dispersing. This is found in metallic guides with "a" being of the order of a millimeter, the optical waves which are such that $\lambda_0 \ll 2a$ is true propagate with the same speed as if in a vacuum (TEM wave).

12.4.4. Generalization of the study of TE wave propagation to a guide with a rectangular cross section, and the physical origin of the form of solutions for the electric wave

12.4.4.1. Generalization of the study of TE wave propagation to a guide with a rectangular cross section

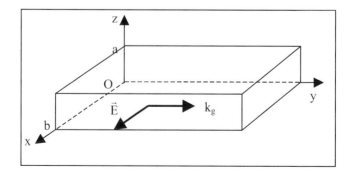

Figure 12.10. *Propagation in a guide with rectangular cross section.*

In place of studying the propagation of a TE wave, polarized with respect to Ox, between only the metallic planes parallel to Oxy as noted above, here the propagation is through a wave guide with a rectangular cross section so that it has two supplementary metallic sides at 0 and b with respect to Ox and parallel to Oyz (Figure 12.10).

The form of the preceding solutions (Section 12.4.3) is still, *a priori*, valid, as the conditions required for the continuity of the tangential component of the electric field in the planes parallel to Oxy and on the sides z = 0 and z = a still should be true, so the solution given in Eq. (14), i.e.,

$$\vec{\underline{E}} = E° \sin(\frac{m\pi}{a} z) \, \vec{e}_x \, \exp[i(k_g y - \omega t)]$$

must still be a solution, and the procedure therefore is to simply find out the consequences of the supplementary limiting conditions due to the presence of a second system of metallic planes, as in:

$$E_t(0) = E_t(b) = 0,$$

where E_t represents the tangential composition of the electric field in the planes Oyz of the sides $x = 0$ and $x = b$.

In effect, as the solution for the TE wave for $\vec{\underline{E}}$ is parallel to $\vec{e}_x$, the tangential component of the electric field in the planes parallel to Oyz is zero, which accords to the new limiting conditions for whatever value of b. This latter dimension of the guide therefore plays no role in the solutions for the electric field, which thus remain unchanged.

Note: However, to state that the solutions for a system based on two parallel planes also are solutions for a rectangular guide does not mean that in the latter system there are not other solutions for transverse waves. To the contrary, we can show that the general solutions for the transverse electric waves (TE waves and as a consequence with $[\underline{E}]_y = 0$ and $B_y \neq 0$) give rise to components along Ox and Oz. For this, it would be correct to look for solutions for $\vec{\underline{E}}$ in the form:

$$\vec{\underline{E}} \begin{cases} \underline{E}_x = \underline{E}_{mx}(x,z) \, \exp[i(k_g y - \omega t)] \\ \underline{E}_y = 0 \\ \underline{E}_z = \underline{E}_{mz}(x,z) \, \exp[i(k_g y - \omega t)] \end{cases}$$

(and thus with $\vec{\underline{E}} \, /\!/ \, Oxz$, the solutions would again correspond to those for a TE wave).

For the equation of propagation, we therefore look for solutions to a variable separated such that, for example $\underline{E}_{mx}(x,z) = f(x).g(z)$. This can be done by proceeding as was the case for the parallelepiped potential box. The rather long calculations show that in general terms $\vec{\underline{E}}$ has the components:

$$\underline{E}_x = \underline{A}_x \sin\left(m\pi\frac{z}{a}\right) \cos\left(n\pi\frac{x}{b}\right) \exp[i(k_g y - \omega t)], \text{ and}$$

$$\underline{E}_z = \underline{A}_z \cos\left(m\pi\frac{z}{a}\right) \sin\left(n\pi\frac{x}{b}\right) \exp[i(k_g y - \omega t)],$$

where $\underline{A}_x$ and $\underline{A}_z$ are complex constants.

The corresponding wave is denoted TE_{mn} and Eq. (15) for the dispersion is replaced by :

$$\frac{\omega^2}{c^2} = k_g^2 + \frac{m^2\pi^2}{a^2} + \frac{n^2\pi^2}{b^2}. \qquad (15')$$

In even more general terms, if the vacuum in the guide is replaced by a dielectric without loss in permittivity (ε), then Eq. (15') would become $\omega^2\varepsilon\mu_0 = k_g^2 + \frac{m^2\pi^2}{a^2} + \frac{n^2\pi^2}{b^2}$.

Thus Eq. (15') shows that when $b > a$, the mode associated with the lowest angular frequency is the wave TE_{01} (such that $E_z \neq 0$ while $\underline{E}_x = \underline{E}_y = 0$). When $a > b$, the low frequency mode, also called the dominant mode, corresponds to the TE_{10} wave ($m = 1$, $n = 0$), which is thus such that $\underline{E}_z = \underline{E}_y = 0$, and:

$$\underline{E}_x = \underline{A}_x \sin\left(\frac{\pi}{a}z\right) \exp[i(k_g y - \omega t)] .$$

The latter equation is the form of the TE wave studied thus far, which we will continue to use in its general form (TE_{m0} wave polarized with respect to Ox) as an example solution in order to interpret the physical origin.

12.4.4.2. Physical origin of the solution form of a TE wave: TE_{m0} wave given to support argument for a wave polarized with respect to Ox

$$\vec{\underline{E}} = \underline{E}_{mx}(z) \vec{e}_x \exp[i(k_g y - \omega t)] = E^\circ \sin\frac{m\pi}{a}z \vec{e}_x \exp[i(k_g y - \omega t)] .$$

Here we will study a monochromatic plane EM wave rectilinearly polarized with respect to Ox in a guide that is alternately reflected against the planes sides z = 0 and z = a parallel to the plane 0xy and as shown in Figure 12.11.

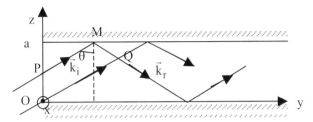

Figure 12.11. *Propagation of a monochromatic plane EM wave with successive reflections on plane sides z = a and z = 0.*

At any point (Q), the wave is the result of incident and reflected waves that have fields in the form:

$$\vec{E}_i = E_i^0 \left[\sin(\vec{k}_i \vec{r} - \omega t) \right] \vec{e}_x \text{ and } \vec{E}_r = E_r^0 \left[\sin(\vec{k}_r \vec{r} - \omega t) \right] \vec{e}_x .$$

In a vacuum, we have:

$\left|\vec{k}_i\right| = \left|\vec{k}_r\right| = k_0 = \dfrac{\omega}{c}$, and Descartes first law gives $\theta_i = \theta_r = \theta$, for which:

$$\vec{k}_i \begin{cases} 0 \\ k_0 \sin\theta \\ k_0 \cos\theta \end{cases} \qquad \vec{k}_r \begin{cases} 0 \\ k_0 \sin\theta \\ -k_0 \cos\theta \end{cases}$$

The wavevector $\vec{k}_i$ of the incident wave depends on the direction of the incidence and belongs to a plane parallel to Oyz, and is by consequence normal to the direction Ox of wave polarization. It is worth noting that in fact $\vec{k}_i = \vec{k}_0$, and that by denoting the wavevector of the oblique wave in the guide as $\vec{k}_0$, we can state that:

$$\vec{k}_0 = k_0 \sin\theta \, \vec{e}_y + k_0 \cos\theta \, \vec{e}_z).$$

The dephasing by π of the reflection at the perfect conductor (due to the limiting condition at the metallic plane, at z = 0 or z = a, where the resultant tangential field must be equal to zero) gives:

$$E_r^0 = E_i^0 \, e^{j\pi} = -E_i^0 .$$

At the point Q, the resulting wave is thus:

$$\vec{E} = \vec{E}_i + \vec{E}_r$$

$$= E_i^0[\sin(y\, k_0\, \sin\theta + z\, k_0\, \cos\theta - \omega t)]\vec{e}_x - E_i^0\, [\sin(y\, k_0\, \sin\theta - z\, k_0\, \cos\theta - \omega t)]\vec{e}_x$$

With $\sin p - \sin q = 2 \cos\dfrac{p+q}{2} \sin\dfrac{p-q}{2}$, we obtain:

$$\vec{E} = 2\, E_i^0\, \{\sin([k_0\, \cos\theta]\, z)\}\, \{\cos([k_0\, \sin\theta]y - \omega t)\}\vec{e}_x . \qquad (17)$$

This form of the solution of the wave contains two parts:
- one part due to propagation and associated with the term $\cos([k_0\, \sin\theta]y - \omega t)$, which shows that the propagation is carried out with a wavevector given by $\vec{k}_g = k_0\, \sin\theta\, \vec{e}_y$ (such that $\vec{k}_g.\vec{r} = y\, k_0\, \sin\theta$) where $\vec{k}_g$ thus represents the definitive wavevector which "pilots" the wave guided along $\vec{e}_y$; and
- a part due to a stationary term given by $\sin([k_0\, \cos\theta]z$). This term corresponds to a wavevector $\vec{k}_s = k_0\, \cos\theta\, \vec{e}_z$ and as such must verify $\vec{k}_s.\vec{r} = [k_0\, \cos\theta]\, z$.

The wavevector $\vec{k}_0 = k_0\, \sin\theta\, \vec{e}_y + k_0\, \cos\theta\, \vec{e}_z$ therefore is such that $\vec{k}_0 = \vec{k}_g + \vec{k}_s$.

With respect to Oz, the presence of a node (E = 0) at z = a demands that $\sin([k_0\cos\theta]a) = 0$ so that $([k_0\cos\theta]a) = m\pi$, from which with $k_0\cos\theta = k_s$ we have $k_s = \dfrac{m\pi}{a}$. From this can be deduced that $k_s = k_0\, \cos\theta = \dfrac{\omega}{c}\cos\theta = \dfrac{m\pi}{a}$, so that in addition $\cos\theta = m\dfrac{\pi c}{\omega a} = m\dfrac{\omega_c}{\omega}$. This occurs for values of θ_m of θ such that:

$$\theta_m = \text{Arc cos } m\dfrac{\omega_c}{\omega} . \qquad (18)$$

We also can state that following Pythagoras's law as shown in Figure 12.12, that $k_0^2 = k_g^2 + k_s^2$, which means that $\dfrac{\omega^2}{c^2} = k_g^2 + \dfrac{m^2\pi^2}{a^2}$, which is an equation identical to that of Eq. (15).

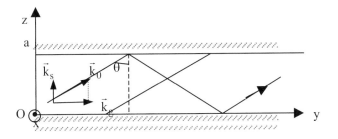

Figure 12.12. *Breakdown of the vector* $\vec{k}_0$ *such that* $k_0^2 = k_g^2 + k_s^2$.

With $k_g = k_0 \sin\theta$, we can deduce that for varying values of m there are corresponding different values of θ which can be denoted by θ_m, and are such that:

$$\frac{\omega^2}{c^2} = k_0^2 \sin^2\theta_m + \frac{m^2\pi^2}{a^2},$$ so that in addition, $1 = \sin^2\theta_m + m^2\frac{\omega_c^2}{\omega^2}$, hence the equation equivalent to Eq. (18), as in:

$$\theta_m = \text{Arc sin} \sqrt{1 - m^2\frac{\omega_c^2}{\omega^2}}. \qquad (19)$$

In numerical terms, for a value of a of 3 cm, we can determine that $v_c = \frac{\omega_c}{2\pi} = \frac{c}{2a} = 5$ GHz . When $v = 12$ GHz (which verifies $\omega > \omega_c$), we obtain for when $m = 1$ that $\theta_1 \approx 66°$ and when $m = 2$, we have $\theta_2 \approx 37°$.

Finally, it is the angle of incidence (i') at the entrance of the guide that determines the mode of propagation (detailed further on Section 12.5.2 and Figure 12.17).

12.4.4.3. Physical representations

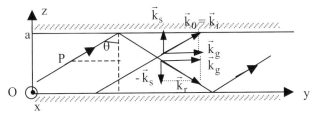

Figure 12.13. *Geometrical representation of the* $\vec{k}$ *vectors.*

12.4.4.3.1. Wave breakdown

The superposition zones of the incident wave with the reflected wave on the metallic planes give (see Figure 12.13):

- with respect to the Oy axis, a progressive wave with a doubled amplitude due to the superposition of two progressive waves with the same amplitudes and both with the same wavevectors given by $\vec{k}_g$, such that for the incident wave $\vec{k}_g = |k_i| \sin\theta \ \vec{e}_y$ and for the reflected wave $\vec{k}_g = |k_r| \sin\theta \ \vec{e}_y$; and

- with respect to the Oz axis, a stationary wave that by superposition of the two progressive waves with the same amplitude but propagating in opposing senses such that the wavevector of the incident wave is $\vec{k}_s$ and for the reflected wave is $-\vec{k}_s$.

12.4.4.3.2. Different speeds

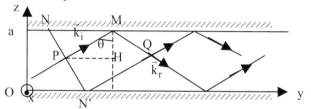

Figure 12.14. *Definition of the points P, M, N, and H used to illustrate the various speeds.*

Figure 12.14 shows an incident ray along PM which is a progressive wave such that $PM = c\,\tau$. We thus have:

- $\dfrac{PM}{NM} = \cos\left(\dfrac{\pi}{2} - \theta\right) = \sin\theta = \dfrac{k_g}{k_0}$, so that with (see Section 12.4.3.6)

  $k_g = \dfrac{\omega}{v_\varphi}$, and $k_0 = \dfrac{\omega}{c}$ then:

  $\dfrac{PM}{NM} = \sin\theta = \dfrac{c}{v_\varphi}$. From this can be deduced that $NM = \dfrac{PM}{\sin\theta} = \dfrac{c\tau}{c/v_\varphi}$,

  which in turn means that $NM = v_\varphi\,\tau$.

- $\dfrac{PH}{PM} = \sin\theta$, from which $PH = PM\sin\theta = NM\sin^2\theta$, so that

  $$PH = (v_\varphi\,\tau)\left(\dfrac{c}{v_\varphi}\right)^2 = \dfrac{c^2}{v_\varphi}\tau = v_g\,\tau .$$

The three speeds, c, v_φ, and v_g, respectively, correspond to the following:

- c is the speed of point P on the zigzag trajectory of incident and reflected waves
  $(k_i = k_r = k_0 = \dfrac{\omega}{c})$;

- v_g is the displacement speed from P to H, in other words the displacement of P along a trajectory globally parallel to Oy, such that $v_g < c$ (the resultant speed given by v_g of the point P on the resultant direction thus is inferior to the speed along the instantaneous zigzag trajectory); and

- v_φ is the speed of a point N with respect to the direction Oy. The N has no material reality and as it physically represents none of the characteristics of a wave it is not subject to relativistic physical laws. To give an idea of what it represents, it can be thought of as a "running shadow" on the line z = a of the point P following its illumination from N'. This means that $v_\varphi > c$ simply states that the immaterial shadow is faster than the material point that moves at the speed c.

12.4.4.3.3. Representation of the resultant amplitude for the resultant progressive wave

According to Eq. (14), the resultant wave is given by:

$$\underline{\vec{E}} = E^\circ \sin(\frac{m\pi}{a} z)\, \vec{e}_x \, \exp[i(k_g y - \omega t)] = E_x^0 \, \vec{e}_x \, \exp[i(k_g y - \omega t)],$$

where $E_x^0 = E^\circ \sin(\frac{m\pi}{a} z)$.

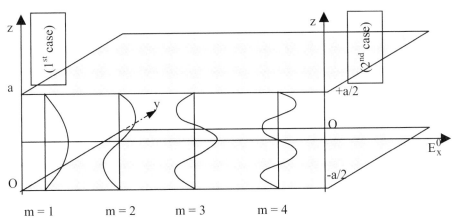

Figure 12.15. *Representation of the amplitudes* $E_x^0 = f(z)$ *for two axial systems where the origin O can be taken on the lower metallic plane or as between the two metallic planes.*

Whatever the y point of the trajectory, the amplitude (E_x^0) of the wave only depends on z and the excited m mode (which is fixed by the choice of the incidence angle).

So, Figure 12.15 represents the form of the electric field amplitudes associated with the various possible modes, and $E_x^0 = f(z)$ is represented in the plane Oxz on the y = 0 side. It is notable that the wave is polarized with respect to Ox, so for the representation that has to be in the 0xz plane, the Oy axis is directed toward the back of the figure.

Comment: Role of the choice of the origins for the guide planes with respect to stationary solutions

In general terms, the propagation is given by Eq. (11) so that with $\frac{\omega^2}{c^2} - k_g^2 = K^2$ and $\underline{E}_{mx} = \underline{E}_m$ (as E // Ox), then $\frac{d^2\underline{E}_m(z)}{dz^2} + K^2 \underline{E}_m(z) = 0$. The general solutions, which also can be written as $\underline{E}_m(z) = A \cos(Kz) + B \sin(Kz)$, can be placed into one of the two following forms, depending on the how the guide planes parallel to Oxy are assumed to be placed:

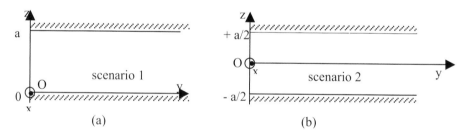

Figure 12.15. bis. *The origin (O) placed (a) on a plane; or (b) midway between planes.*

1. At z = 0 and z = a as shown in Figure 12.15bis, scenario 1, which is the same as the representation used up till now, so that the limiting conditions are thus:

- $\underline{E}_m(0) = 0$, for which A = 0, so that $\underline{E}_m(z) = B \sin(Kz) = E^0 \sin(Kz)$, with $B = E^0$ (constant); and

- $\underline{E}_m(a) = 0$, for which $E^0 \sin(Ka) = 0$, so that $Ka = m\pi$ with m = 1, 2, 3...., so $K = \frac{m\pi}{a}$, and finally $E_m = E^0 \sin\frac{m\pi}{a}z$ (m = 0 means that $E_m = 0$, which is a nonphysical solution).

The form of the wave is thus:

$$E = E_m \exp[i(k_g y - \omega t)] = E^0 \sin\left(\frac{m\pi}{a} z\right) \exp[i(k_g y - \omega t)]),$$

which is what has been used until now.

2. At $z = \dfrac{-a}{2}$ and $z = \dfrac{+a}{2}$ (scenario 2 in Figure 12.15bis). The limiting conditions thus are now:

$$\underline{E}_m(\frac{a}{2}) = A \cos(K\frac{a}{2}) + B \sin(K\frac{a}{2}) = 0 = \underline{E}_m(-\frac{a}{2}) = A \cos(K\frac{a}{2}) - B \sin(K\frac{a)}{2});$$

Two solutions are possible for A and B, either:

- $B = 0$ and $A \neq 0$ ($A =$ constant denoted E^0) with $\cos(K\frac{a}{2}) = 0$, so that

  $$K\frac{a}{2} = \frac{\pi}{2} + n\pi, \quad \text{for which} \quad K = \frac{\pi}{a}[1 + 2n] = \frac{M\pi}{a} \quad \text{with odd M, and}$$

  $$E_m = E^0 \cos\frac{M\pi}{a} z \text{ with odd whole M; or}$$

- $A = 0$ and $B \neq 0$ ($B =$ constant denoted E^0) with $\cos(K\frac{a}{2}) = 0$, so that

  $$K\frac{a}{2} = n\pi, \quad \text{for which } K = 2n \frac{\pi}{a} = \frac{M\pi}{a} \quad \text{with an even value of M.}$$

The form of the wave $E = E_m \exp[i(k_g y - \omega t)]$ thus is given by either:

- $E = E^0 \cos\left(\frac{M\pi}{a} z\right) \exp[i(k_g y - \omega t)]$ with a whole odd value for M; or

- $E = E^0 \sin\left(\frac{M\pi}{a} z\right) \exp[i(k_g y - \omega t)]$ with M being whole and even.

We can verify that the plots of $E_m(y)$ in scenario (1) where $m = 1, 2, 3, 4...$ or for scenario (2) with the two forms of solutions for even or odd values of M, where $M = 1, 2, 3, 4... (\equiv m)$, are identical, which is reassuring as the problem is the same! (This remark also can be made in terms of the quantum mechanics of wells with symmetrical potentials, where the origin can be taken either as at an extremity of the well where there is a potential wall, or at the center of the well.)

12.4.4.4. Multimodal fields

It should be noted that it is not only the field corresponding to a given mode that can be guided, meaning in other words those fields that have a transversal intensity (with respect to Oz) independent of the axial position (with respect to Oy) as shown in Figures 12.16 a and b corresponding, respectively, to the given modes (using notation of scenario 1):

- complex amplitude, with $m = 1$

$$\underline{E}_1^0 = 2\,E_i^0\, \sin([k_0 \cos\theta_1]z)\,\exp(i[k_0 \sin\theta_1]y)\,,\ \text{so that its intensity is independent of } y;\ \text{and}$$

- complex amplitude, with $m = 2$

$$\underline{E}_2^0 = 2\,E_i^0\, \sin([k_0 \cos\theta_2]z)\,\exp(i[k_0 \sin\theta_2]y)\,,\ \text{so that its intensity is independent of } y.$$

The fields simply have to verify the limiting conditions ($E = 0$ on metallic planes) and thus can correspond to the superposition of several modes. Thus in the case given in Figure 12.16 c, taking into account the analytical forms of the fields, it is possible to see that the transversal intensity (with respect to Oz) of the wave is dependent on the axial position (with respect to Oy) in which it is placed.

In effect, the complex amplitude of the wave $\vec{E}_{1,2}$, as a superposition of the modes $m = 1$ and $m = 2$, is in the form

$$2\,E_i^0\{\ \sin([k_0 \cos\theta_1]z)\,\exp(i[k_0 \sin\theta_1]y) + \sin([k_0 \cos\theta_2]z)\,\exp(i[k_0 \sin\theta_2]y)\}\,,$$

and its intensity gives rise to a variation in its distribution with respect to Oy as $k_0 \sin\theta_1 \neq k_0 \sin\theta_2$.

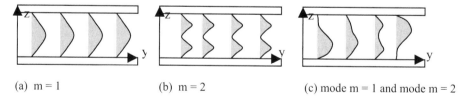

(a) $m = 1$ (b) $m = 2$ (c) mode $m = 1$ and mode $m = 2$

Figure 12.16. *(a and b) Guided waves corresponding to a mode; and (c) without a mode.*

12.4.4.5. Comment: TM modes

TM modes occur when it is the magnetic field that is applied with respect to Ox. They can be studied in the same way as TE modes and as such can be guided. The electric thus has components with respect to Oy and Ox.

12.5. Optical Guiding: General Principles and How Fibers Work

12.5.1. Principle

Optical fibers are used, by way of a support, to guide an incident luminous wave from its point of injection at an entrance face to its exit point at the other. Typically, there are two main types of fiber: fibers with a jump in indices and fibers with a gradient in indices.

This text will limit itself to detailing the first type, which are constituted of a core made from a material with a circular cross section and a given index (n) surrounded by a gain with an index denoted n_1 such that $n > n_1$. Generally, the whole structure sits in air, which has an index given by $n' \approx 1$.

The luminous wave injected at the entrance face of the fiber is under an angle i', which is such that $n' \sin i' = n \sin i$. This wave meets the core/gain interface at an angle of incidence given by $\theta_i = \theta$, such that $i + \theta = \dfrac{\pi}{2}$. When $\theta > \theta_\ell$ where θ_ℓ is a limiting angle above which the phenomenon of total reflection occurs when $n > n_1$, the wave makes no penetration of the gain material and is guided along the fiber in a zigzag trajectory associated with successive total reflections at the core/gain interface, and as illustrated in Figure 12.17.

The calculations carried out below in the plane of the longitudinal section of the fiber are equally valid for the longitudinal plane section of a symmetrical rectangular guide where the Or axis is simply replaced by a Oz axis.

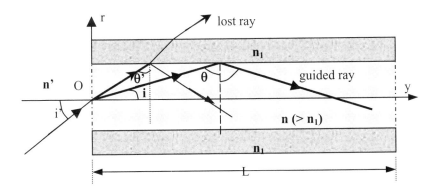

Figure 12.17. *Trajectories when $\theta > \theta_\ell$ and $\theta' < \theta_\ell$.*

12.5.2. Guiding conditions

12.5.2.1. Condition on i

The limiting angle θ_ℓ is defined by $n \sin\theta_\ell = n_1$, wherein at the entry face an angle

i has a value given by $i_\ell = \frac{\pi}{2} - \theta_\ell$. A condition for the total reflection $\theta \geq \theta_\ell$ is

therefore $i \leq i_\ell$. In addition, by moving $\theta_\ell = \frac{\pi}{2} - i_\ell$ into $n \sin\theta_\ell = n_1$, we directly

obtain $\cos i_\ell = \frac{n_1}{n}$.

This condition for wave guiding thus brings us to a condition for the incidence at the

entry face, which must be such that $i \leq i_\ell$, so that $\cos i \geq \cos i_\ell = \frac{n_1}{n}$, which

means that $i \leq i_\ell = \text{Arc} \cos \frac{n_1}{n}$.

All rays with $i > i_\ell$ have at the core/gain interface an incidence such that $\theta' < \theta_\ell$. A fraction of the wave is transmitted toward the gain, and following several successive reflections, the remaining fraction of the wave in the core becomes smaller and smaller so finally it is extinguished along its journey.

In numerical terms, with $n = 1.50$ and $n_1 = 1.49$, we obtain

$i_\ell = \text{Arc} \cos \frac{n_1}{n} = 6.62°$.

When $n = 1.60$ and $n_1 = 1.50$, we have $i_\ell = \text{Arc} \cos \frac{n_1}{n} = 20.36°$.

12.5.2.2. Condition on i' (acceptance angle): numerical aperture

The condition $\theta \geq \theta_\ell$ makes it possible to state that:

$$\cos\theta \leq \cos\theta_\ell = \sqrt{1 - \sin^2\theta_\ell} = \sqrt{1 - \frac{n_1^2}{n^2}}.$$

We also have $i = \frac{\pi}{2} - \theta$, from which with $n' \sin i' = n \sin i$

$n' \sin i' = n \sin(\frac{\pi}{2} - \theta) = n \cos\theta$, so that $\sin i' = \frac{n}{n'}\cos\theta$ and the preceding equation

therefore leads to

$$\sin i' = \frac{n}{n'}\cos\theta \le \frac{n}{n'}\cos\theta_\ell = \frac{n}{n'}\sqrt{1-\frac{n_1^2}{n^2}} = \frac{n}{n'}\left(\frac{n^2-n_1^2}{n^2}\right)^{1/2} = \frac{\left(n^2-n_1^2\right)^{1/2}}{n'} , \text{ so that:}$$

$$\sin i' \le \frac{\left(n^2-n_1^2\right)^{1/2}}{n'} .$$

To conclude, all rays introduced under an angle of incidence i' verify the preceding inequality and therefore can propagate.

The quantity given by

$$NA = \left(n^2-n_1^2\right)^{1/2}$$

is the numerical aperture of the system, which although here is an optical fiber, also can be applied to a guide structure.

The acceptance angle is by definition:

$$i'_\ell = \text{Arc sin} \frac{\left(n^2-n_1^2\right)^{1/2}}{n'} .$$

In numerical terms

$n = 1.50$ - $n_1 = 1.49$ - $n' = 1$, so the NA = 0.173, so that $i'_\ell = 9.96\,°$.

When $n = 1.6$ - $n_1 = 1.5$ - $n' = 1$, and the NA = 0.557, so that $i'_\ell = 33.83\,°$.

12.5.3. Increasing the signals

Once a luminous impulsion is injected into a fiber, various pathways can be followed, as in:

• the shortest trajectory which is along the pathway Oy and corresponds exactly to the length (L) of the fiber. With the speed of propagation being given by $v_\varphi = \frac{c}{n}$, the pathway time (t_y) is $t_y = \frac{nL}{c}$;

• zigzag trajectories which each have a given value for the angle i and have a pathway length (L_i) given by $L_i = \frac{L}{\cos i} > L$. With the pathway speed still being $v_\varphi = \frac{c}{n}$, the pathway times (t_i) are now given by $t_i = \frac{L_i}{v_\varphi} = \frac{n\,L}{c\cos i} > t_y$ (t_i is great when cos i is small, so that i is great).

The difference in propagation times thus is given by:

$$\Delta t_i = t_i - t_y = \frac{n\,L}{c}\left(\frac{1}{\cos i} - 1\right) \approx \frac{n\,L}{c}\left(\frac{1 - 1 + \dfrac{i^2}{2}}{1 - \dfrac{i^2}{2}}\right) \approx \frac{n\,L}{2\,c}i^2 .$$

In numerical terms, when $L = 1$ km, $n = 1.5$ and $i = 6\,° < i_\ell$, so that $\Delta t_i = 28$ ns.

Therefore, an impulsion emitted at $y = 0$ at a time $t_0 = 0$ over an interval of time assumed to be $\delta t \approx 0$ will be emitted at $y = L$ starting from the moment $t_0 + t_y = t_y$ (for the wave that takes the shortest pathway) up to the moment $t_0 + (t_i)_{max} = (t_i)_{max}$ where $(t_i)_{max}$ is the pathway time for the longest possible trajectory. This latter value comes from the highest possible value of i, i.e., $i = i_\ell$. The corresponding interval in time therefore is given by $\Delta t_{i\ell} = \dfrac{n\,L}{2\,c}i_\ell^2$ where $\Delta t_{i\ell}$ is the duration time of the emission at the exit.

The interval in time (T) between two impulsions at the entrance which is required so that they do not mix at the exit therefore must be such that $T \geq \Delta t_{i\ell}$, which is the same as stating that the frequency of the impulsions must be lower than $\nu_{i\ell} = \dfrac{1}{\Delta t_{i\ell}}$ (we can state that we should have $T = \dfrac{1}{\nu} \geq \Delta t_{i\ell}$ so that $\nu < \dfrac{1}{\Delta t_{i\ell}} = \nu_{i\ell}$. and in terms of actual numbers, this means that for the aforementioned values, $\nu_{i\ell} \approx 30$ MHz).

12.6. Electromagnetic Characteristics of a Symmetrical Monomodal Guide

Here we look at a buried guide, which for technological reasons is simplified to the symmetrical structure shown in Figure 12.18.

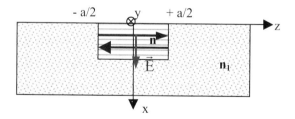

Figure 12.18. *Buried guide.*

   The propagation of an electromagnetic wave with an angular frequency given by ω is along Oy and is polarized with respect to Ox. The field is reflected at the interface, between a medium with an index given by n and another medium with an index denoted by n_1, along planes parallel to Oxy with the EM wave alternating its reflection at the planes defined by $z = \dfrac{-a}{2}$ and $z = \dfrac{+a}{2}$. This configuration has the same geometry as that given in Section 12.4, and as that presented in Figure 12.19a, making direct comparisons possible between systems that only differ by virtue of their planes of interface. The metallic planes, which are imposed as a limiting condition where $\vec{E} = 0$, are now replaced by dielectric diopters with the aforementioned index n_1 where $n_1 < n$ (and as a consequence, with different limiting conditions). In addition, the two dielectrics are assumed to be nonabsorbent. In the zones (1), (2), and (3) the electric field can be written by notation as $\vec{E}^{(1)}, \vec{E}^{(2)}$, and $\vec{E}^{(3)}$, respectively, and in a manner analogous to that of Section 12.4, taking into account the symmetry of the problem, the solutions for the TE wave polarized with respect to Ox are looked for in each of the three zones in the form:

$$\vec{E}^{(p)} = E^{(p)}_{mx}(z)\, \vec{e}_x \exp[j(k_{yp}y - \omega t)]$$

where $p = 1, 2, 3$ characterizes the zone under consideration.

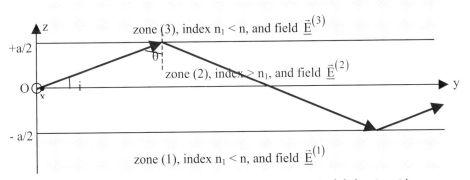

Figure 12.19(a). *Propagation in a symmetrical dielectric guide.*

12.6.1. General form of the solutions

The equations for continuity of the tangential component for the different field $\vec{E}^{(1)}, \vec{E}^{(2)}$, and $\vec{E}^{(3)}$ directed exactly along Ox, true for all moments t and whatever values of y, lead to $k_{y1} = k_{y2} = k_{y3} = k_y$ [see also Section 11.2.3, concerning Eq. (7) or (7') with respect to Oy].

With the media being nonabsorbent, $\varepsilon_{rp} = n_p^2$, we then have:

$\overrightarrow{rot}\ \underline{\vec{B}}^{(p)} = \mu_0\varepsilon_0\varepsilon_{rp}\dfrac{\partial\underline{\vec{E}}^{(p)}}{\partial t} = \dfrac{n_p^2}{c^2}\dfrac{\partial\underline{\vec{E}}^{(p)}}{\partial t}$. By taking the rotational of the two terms

of the equation $\overrightarrow{rot}\ \underline{\vec{E}}^{(p)} = -\dfrac{\partial\underline{\vec{B}}^{(p)}}{\partial t}$, and by using the preceding equation, we obtain

an equation for propagation in a medium with an index denoted n_p [see Chapter 7, Eq. (5) this volume, and with $\mu_r = 1$]

$$\Delta\underline{\vec{E}}^{(p)} - \dfrac{n_p^2}{c^2}\dfrac{\partial^2\underline{\vec{E}}^{(p)}}{\partial t^2} = 0 .$$

With the wave being monochromatic, we also can state, using Eq. (7") again from Chapter 7,

$$\Delta\underline{\vec{E}}^{(p)} + \dfrac{\omega^2}{c^2}n_p^2\dfrac{\partial^2\underline{\vec{E}}^{(p)}}{\partial t^2} = 0 .$$

Taking into account the form of the required solution, this equation gives [using a calculation analogous to that yielding Eq. (11) of Section 12.4.3.2]:

$$\dfrac{d^2\underline{E}_{mx}^{(p)}(z)}{dz^2} + \left(\dfrac{\omega^2}{c^2}n_p^2 - k_y^2\right)\underline{E}_{mx}^{(p)}(z) = 0 .$$

Making $\alpha_p^2 = \dfrac{\omega^2}{c^2}n_p^2 - k_y^2$, and as $\left(\dfrac{\omega^2}{c^2}n_p^2 - k_y^2\right)$ can be, *a priori*, positive or negative, and α_p can be real or purely imaginary, under these conditions the solutions are in the general form:

$$\underline{E}_{mx}^{(p)}(z) = A\ \exp(j\alpha_p\ z) + B\ \exp(-j\alpha_p\ z) .$$

12.6.2. Solutions for zone (2) with an index denoted by n

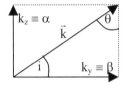

Figure 12.19 (b) *Geometrical component of* $\vec{k}$.

In this zone, the wavenumber is given by $k = \dfrac{n\omega}{c}$ and its projection along the axis of propagation Oy is $k_y = k \sin\theta = k \cos i \equiv \beta$ (by notation) which is a constant bound to the propagation phase.

The result is that $\alpha_2{}^2 = \dfrac{\omega^2}{c^2} n^2 - k_y{}^2 = k^2 - k^2 \cos^2 i = k^2(1 - \cos^2 i) = k^2 \sin^2 i$, so that $\alpha_2 = \pm k \sin i$. The sign is not important with respect to the general solution given by $\underline{E}_{mx}^{(2)}(z)$, which contains both signs as exponentials, so by notation we use $\alpha_2 = k \sin i = k \cos\theta = k_z \equiv \alpha$. This magnitude is finally tied to the amplitude of a stationary wave due to reflections at the interfaces (see the metallic guide in Sections 12.4.2 and 12.4.3).

Therefore, in the zone (2) we have, with $\alpha = k \sin i$,

$$\underline{E}_{mx}^{(2)}(z) = D_1 \exp(j\alpha z) + D_2 \exp(-j\alpha z)$$

where D_1 and D_2 are the constants to be determined for the limiting conditions.

In fact, given the symmetry of the problem, we should have $D_1 = \pm D_2$, so that $D_1 = D$ and $D_2 = \varepsilon D$ with $\varepsilon = \pm 1$. In effect, given this symmetry of the guide, with respect to the z axis and for two points symmetrical placed about O located by $z = z_2$ and $z = -z_2$, the intensity of the wave should be the same (in quantum mechanics, the probabilities of presence would be stated to be equal).

Thus,

$$| D_1 \exp(j\alpha z_2) + D_2 \exp(-j\alpha z_2)|^2 = | D_1 \exp(-j\alpha z_2) + D_2 \exp(j\alpha z_2)|^2,$$

so that

$$| D_1 + D_2 \exp(-2j\alpha z_2)|^2 = | D_1 \exp(-2j\alpha z_2) + D_2 |^2,$$

for which $|D_1|^2 = |D_2|^2$, and $D_1 = \pm D_2$. Finally,

$$\vec{\underline{E}}^{(2)} = \left[D \exp(j\,\alpha\,z) + \varepsilon\, D \exp(-j\,\alpha\,z) \right] \vec{e}_x \exp[j(k_y\,y - \omega t)].$$

12.6.3. Solutions for the zones (1) and (3)

In the titled zones [zones (1) and (3)], we have $\alpha_1{}^2 = \alpha_3{}^2 = \dfrac{n_1^2 \omega^2}{c^2} - k_y{}^2$, where

$\dfrac{n_1^2 \omega^2}{c^2} = \dfrac{n_1^2 n^2 \omega^2}{n^2 c^2} = \dfrac{n_1^2}{n^2} k^2$, so that $k_y = k \cos i$ is always true. The result is that

$$\alpha_1{}^2 = \alpha_3{}^2 = k^2 \left(\frac{n_1^2}{n^2} - \cos^2 i \right),$$

and as:

- on the one hand, $n > n_1$, $\dfrac{n_1^2}{n^2} < 1$,

- and on the other, in order to guide the incident ray (i) to the entry face shown in Figure 12.19a, we should have $i \leq i_\ell$ as a result of Section 12.5.2.1 which uses the same geometry, so that

$$\cos i \geq \cos i_\ell = \frac{n_1}{n}$$

We have $\left(\dfrac{n_1^2}{n^2} - \cos^2 i \right) < 0$, so that we can make:

$$\alpha_1{}^2 = \alpha_3{}^2 = k^2 \left(\frac{n_1^2}{n^2} - \cos^2 i \right) = j^2 \gamma^2 < 0, \text{ for which } \alpha_1 = \alpha_3 = j\gamma, \text{ where:}$$

$$\gamma = k \sqrt{\cos^2 i - \frac{n_1^2}{n^2}} > 0.$$

In zone (1) we thus find that $\underline{E}_{mx}^{(1)}(z) = C_1 \exp(-j \alpha_1 z) + C_3 \exp(j \alpha_1 z)$, and in this zone where $z < 0$, so that the wave is evanescent (corresponding to the guiding condition as in $\theta > \theta_\ell$ or $i < i_\ell$), we should find $C_3 = 0$ [If not, the component $C_3 \exp(j \alpha_1 z)$ yields for $\vec{\underline{E}}^{(1)}$ a contribution of the form

$$C_3 \exp(j^2 \gamma z) = C_3 \exp(-\gamma z),$$

which will diverge when $z \to -\infty$ to give the amplification term rather than the term for attenuation as required].

In zone (1), the solution thus is in the form $\underline{E}_{mx}^1(z) = C_1 \exp(-j \alpha_1 z)$, so that:

$$\underline{E}_{mx}^{(1)}(z) = C_1 \exp(\gamma z).$$

In the zone (1), the final expression for the wave is:

$$\vec{\underline{E}}^{(1)} = C_1 \exp(\gamma z) \, \vec{e}_x \exp[j(k_y y - \omega t)].$$

This wave is evanescent with respect to Oz and progressive with respect to Oy. It is in effect an inhomogeneous MPP wave that has an amplitude that is dissimilar at different points along the plane of the wave defined by y = constant.

Similarly, in zone (3), we have:

$$E_{mx}^{(3)}(z) = C_1 \exp(-j\,\alpha_3\,z) + C_3 \exp(j\,\alpha_3\,z)$$

and in this zone where z > 0, the physically acceptable solution (yielding an attenuation and not an amplification) is $E_{mx}^3(z) = C_3 \exp(j\,\alpha_3\,z)$, so that:

$$E_{mx}^{(3)}(z) = C_3 \exp(-\gamma\,z) \ .$$

In zone (3) then, the final equation for the wave is given as

$$\underline{\vec{E}}^{(3)} = C_3 \exp(-\gamma\,z)\,\vec{e}_x \exp[j(k_y y - \omega t)]$$

which also is an evanescent wave with respect to Oz and a progressive wave with respect to Oy.

It is worth now looking at the relationship between C_1 and C_3 due to the symmetry of the guide. With the problem being one of symmetry around O with respect to the z axis, we should have the same wave intensity at two symmetrical points, one located with $z = z_1$ and in zone (1), and the other located with $z = z_3 = -z_1$ and in zone (3). So,

$$\left. \begin{array}{l} |C_1 \exp(\gamma\,z_1)|^2 \\ = \{|C_3 \exp(-\gamma\,z_3)|^2\}_{z_3 = -z_1} = |C_3 \exp(\gamma\,z_1)|^2 \end{array} \right\} \Rightarrow C_1 = \pm C_3$$

By making $C_1 = C$ and $C_3 = \varepsilon C$ with $\varepsilon = \pm 1$, then we will have as solutions for zone (1):

$$\boxed{E_{mx}^{(1)}(z) = C \exp(\gamma\,z), \text{ and } \underline{\vec{E}}^{(1)} = C \exp(\gamma\,z)\,\vec{e}_x \exp[j(k_y y - \omega t)]} \ .$$

On zone (3), the solutions will be the same:

$$\boxed{E_{mx}^{(3)}(z) = \varepsilon\,C \exp(-\gamma\,z), \text{ and } \underline{\vec{E}}^{(3)} = \varepsilon\,C \exp(-\gamma\,z)\,\vec{e}_x \exp[j(k_y y - \omega t)]} \ .$$

12.6.4. Equations for the magnetic field

Zone (1)

From Maxwell-Faraday law, which states $\overrightarrow{rot}\,\vec{E}^{(1)} = -\dfrac{\partial \vec{B}^{(1)}}{\partial t}$, we have :

$$\overrightarrow{rot}\,\vec{E}^{(1)} = \begin{cases} 0 \\ \partial Ex^{(1)}/\partial z = \gamma\,\underline{E}^{(1)} \\ -\,\partial Ex^{(1)}/\partial y = -jk_y\underline{E}^{(1)} \end{cases} \quad \text{and} \quad -\dfrac{\partial \vec{B}^{(1)}}{\partial t} = j\omega\vec{B}^{(1)} \quad \Rightarrow \vec{B}^{(1)} = \dfrac{1}{\omega}\begin{cases} 0 \\ -j\,\gamma\,\underline{E}^{(1)} \\ -k_y\underline{E}^{(1)} \end{cases}$$

Zone (3)

$$\overrightarrow{rot}\,\vec{E}^{(3)} = \begin{cases} 0 \\ \partial Ex^{(3)}/\partial z = -\gamma\,\underline{E}^{(1)} \\ -\,\partial Ex^{(3)}/\partial y = -jk_y\underline{E}^{(1)} \end{cases} \quad \text{and} \quad -\dfrac{\partial \vec{B}^{(3)}}{\partial t} = j\omega\vec{B}^{(3)} \quad \Rightarrow \vec{B}^{(3)} = \dfrac{1}{\omega}\begin{cases} 0 \\ j\,\gamma\,\underline{E}^{(3)} \\ -k_y\underline{E}^{(3)} \end{cases}$$

Zone (2)

$$\overrightarrow{rot}\,\vec{E}^{(2)} = \begin{cases} 0 \\ \partial Ex^{(2)}/\partial z = -j\alpha\left[D\exp(j\,\alpha\,z) - \varepsilon\,D\exp(-j\,\alpha\,z)\right]\exp[j(k_y y-\omega t)] \\ -\,\partial Ex^{(2)}/\partial y = -jk_y\underline{E}^{(2)} \end{cases} \text{and}$$

$$-\dfrac{\partial \vec{B}^{(2)}}{\partial t} = j\omega\vec{B}^{(2)} \quad \Rightarrow \vec{B}^{(2)} = \dfrac{1}{\omega}\begin{cases} 0 \\ \alpha\left[D\exp(j\,\alpha\,z) - \varepsilon\,D\exp(-j\,\alpha\,z)\right]\exp[j(k_y y-\omega t)] \\ -k_y\underline{E}^{(2)} \end{cases}$$

12.6.5. *Use of the limiting conditions: determination of constants*

Here we make $\delta = \exp\left(\dfrac{\gamma\,a}{2}\right)$ and $\mu = \exp\left(j\dfrac{\alpha\,a}{2}\right)$.

The continuity of the tangential components of E at $z = \dfrac{a}{2}$ are written:

$$E^{(3)}(\frac{a}{2}) = E^{(2)}(\frac{a}{2}) ,$$

from which

$$\varepsilon\, C \exp(-\gamma\, \frac{a}{2}) = D \exp(j\alpha\, \frac{a}{2}) + \varepsilon\, D \exp(-j\alpha\, \frac{a}{2}) ,$$

so that

$$\frac{\varepsilon C}{\delta} = D(\mu + \frac{\varepsilon}{\mu}) . \qquad (1)$$

With the media being nonmagnetic, we can equally write that the continuity of the tangential components for B at $z = \frac{a}{2}$ are such that we have

$$B_y^{(3)}(\frac{a}{2}) = B_y^{(2)}(\frac{a}{2}) , \text{ from which}$$

$$j\gamma\varepsilon C \exp(-\gamma\, a/2) = \alpha\big[D \exp(j\,\alpha\, a/2) - \varepsilon\, D \exp(-j\,\alpha\, a/2)\big] , \text{ so that:}$$

$$j\, \gamma\, \varepsilon\, C/\delta = \alpha\, [\mu - \varepsilon/\mu]\, D . \qquad (2)$$

We thus obtain two linearly independent equations that allow a determination of the unknowns C and D. The other equations of continuity give for their part equations leading to the same relations as Eqs. (1) or (2). If the reasoning permitted by the effect of symmetry had not been possible, then there would have been four constants, namely C_1, C_2, D_1, and D_2, which these additional equations would have permitted to determine.

For example, at $z = -\frac{a}{2}$, we have:

$$E^{(1)}(-\frac{a}{2}) = E^{(2)}(-\frac{a}{2}) ,$$

for which

$$C \exp(-\gamma\, \frac{a}{2}) = D \exp(-j\alpha\, \frac{a}{2}) + \varepsilon\, D \exp(j\alpha\, \frac{a}{2}) ,$$

so that

$$\frac{C}{\delta} = \frac{D}{\mu} + \varepsilon\mu D .$$

By multiplying by ε we find Eq. (1).

12.6.6. Modal equation

12.6.6.1. Placing into equations

First, we write the Eqs. (1) and (2) for $\varepsilon = +1$, as in:

$$C/\delta = (\mu + 1/\mu) D \qquad (1')$$
$$j\,\gamma\, C/\delta = \alpha\,[\mu - 1/\mu]\, D \qquad (2')$$
$$\left.\right\} \quad \Rightarrow \text{ the ratio (2')/(1') gives}$$

$$j\,\gamma = \alpha\,\frac{\mu^2 - 1}{\mu^2 + 1}, \text{ so that } \frac{\gamma}{\alpha} = \frac{1}{j}\frac{e^{\frac{j\alpha a}{2}} - e^{-\frac{j\alpha a}{2}}}{e^{\frac{j\alpha a}{2}} + e^{-\frac{j\alpha a}{2}}} = \tan\frac{\alpha a}{2}, \text{ and therefore}$$

$$\tan\frac{\alpha a}{2} = \frac{\gamma}{\alpha}. \qquad (3)$$

Similarly, when $\varepsilon = -1$, we obtain:

$$- C/\delta = (\mu - 1/\mu) D \qquad (1'')$$
$$- j\,\gamma\, C/\delta = \alpha\,[\mu + 1/\mu]\, D \qquad (2'')$$
$$\left.\right\} \quad \Rightarrow \text{ the ratio (2'')/(1'') gives}$$

$$j\,\gamma = \alpha\,\frac{\mu^2 + 1}{\mu^2 - 1}, \text{ for which } \frac{\gamma}{\alpha} = -j\,\frac{e^{\frac{j\alpha a}{2}} + e^{-\frac{j\alpha a}{2}}}{e^{\frac{j\alpha a}{2}} - e^{-\frac{j\alpha a}{2}}} = \frac{-1}{\tan\dfrac{\alpha a}{2}}, \text{ so that}$$

$$\tan\frac{\alpha a}{2} = -\frac{\alpha}{\gamma}. \qquad (4)$$

Equation (3) gives $\dfrac{\alpha a}{2} = \text{Arctan}\left(\dfrac{\gamma}{\alpha}\right) + r\pi$, where r is a whole number ≥ 0 as

$\dfrac{\alpha a}{2} > 0$. So, we have $\dfrac{\alpha a}{2} = \dfrac{\pi}{2} - \text{Arctan}\left(\dfrac{\alpha}{\gamma}\right) + r\pi$, by using the equation

$\text{Arctan } x + \text{Arctan}\dfrac{1}{x} = \dfrac{\pi}{2}$, $\left(\tan(\dfrac{\pi}{2} - y) = \dfrac{1}{\tan y}\right.$ gives $\dfrac{\pi}{2} - y = \text{Arctan}\left(\dfrac{1}{\tan y}\right)$ by

making $x = \tan y$). We deduce that:

$$\frac{\alpha a}{2} + \text{Arctan}\left(\frac{\alpha}{\gamma}\right) = (2r + 1)\frac{\pi}{2} \qquad [\text{r is an integer } \geq 0 \;]. \qquad (5)$$

From Eq. (4) we pull out:

$$\frac{\alpha\ a}{2} = -\,\text{Arctan}\left(\frac{\alpha}{\gamma}\right) + s\,\pi, \quad \text{and with s integer} > 0 \text{ as } \frac{\alpha\ a}{2} > 0 \text{ , we thus have}$$

$$\frac{\alpha\ a}{2} + \text{Arctan}\left(\frac{\alpha}{\gamma}\right) = \frac{(2\,s)\pi}{2}. \qquad (6)$$

The solutions from Eqs. (5) and (6) can be regrouped into a single expression, as in:

$$\frac{\alpha\ a}{2} + \text{Arctan}\left(\frac{\alpha}{\gamma}\right) = \frac{q\,\pi}{2}, \qquad (7)$$

where q is an integer and $q \geq 1$ as $r \geq 0$ and $s > 0$.

12.6.6.2. General equation and solutions

In addition, we made $\alpha = k\sin i$, and $\gamma = k\sqrt{\cos^2 i - \frac{n_1^2}{n^2}}$, with $\frac{n_1}{n} = \cos i_\ell$. The left-hand side of Eq. (7) is by consequence a function of the angle i, and Eq. (7) can be rewritten:

$$f(i) = k\frac{a}{2}\sin i + \text{Arctan}\ \frac{\sin i}{\sqrt{\cos^2 i - \cos^2 i_\ell}} = q\frac{\pi}{2}. \qquad (8)$$

Equation (8) is a modal equation that permits a determination of q when $i \in [0, i_\ell]$.

For this interval, $\sin i$ and $\dfrac{1}{\sqrt{\cos^2 i - \cos^2 i_\ell}}$ are strictly increasing functions of i, just like the arc tangent function. Indeed, f(i) is also a strictly incremental function, so that for given values of a and q, there is a single value of i as a solution to f(i). The maximum value of the function f(i) is obtained for the maximum acceptable value of i, that is $i = i_\ell$. For this value, we therefore have:

$$[f(i)]_{max} = k\frac{a}{2}\sin i_\ell + \text{arc tan}\infty = k\frac{a}{2}\sin i_\ell + \frac{\pi}{2} = q\frac{\pi}{2} \ .$$

The solution $q = 1$ therefore is always obtained, even when $a \to 0$, as it suffices to take $i \to i_\ell$ as the arc tangent function tends toward $\pi/2$.

12.6.6.3. Monomodal solution

To remain in a monomodal regime where $q = 1$, the solution $q = 2$ must remain unattainable. For this to be true $[f(i)]_{max} < \pi$ must be imposed, so that

$$k\frac{a}{2}\sin i_\ell < \frac{\pi}{2}, \text{ which means that } a < a_1 = \frac{\pi}{k \sin i_\ell}. \text{ With } k = \frac{n\omega}{c} = n k_0 = n\frac{2\pi}{\lambda_0}$$

and $\sin i_\ell = \sqrt{1 - \cos^2 i_\ell} = \sqrt{1 - \frac{n_1^2}{n^2}}$, we obtain:

$$\boxed{a_1 = \frac{\lambda_0}{2\sqrt{n^2 - n_1^2}}}.$$

In numerical terms, with $\lambda_0 = 1.3$ μm (a typical telecommunications wavelength), $n = 1.50$ and $n_1 = 1.49$. For the monomodal condition, where $q = 1$ uniquely, the guide must have a width less than $a_1 = 3.76$ μm.

When $\lambda_0 = 1.55$ μm, $n = 1.50$ and $n_1 = 1$, we obtain $a_1 = 0.7$ μm.

12.6.6.4. Multimodal solutions

When $a > a_1$, the modal Eq. (8) in effect permits several different values of q as solutions. In order to find the general solution to Eq. (8), we can write that with $k = \frac{2\pi}{\lambda}$ and $\sqrt{\cos^2 i - \cos^2 i_\ell} = \sqrt{\sin^2 i_\ell - \sin^2 i}$, we have

$$\pi\frac{a}{\lambda}\sin i + \text{Arctan} \frac{\sin i}{\sqrt{\sin^2 i_\ell - \sin^2 i}} = q\frac{\pi}{2},$$

so that on applying $\text{Arctan } x + \text{Arctan}\frac{1}{x} = \frac{\pi}{2}$, we find that

$$\pi\frac{a}{\lambda}\sin i - (q-1)\frac{\pi}{2} = \text{Arctan} \frac{\sqrt{\sin^2 i_\ell - \sin^2 i}}{\sin i}.$$

By setting $m = (q - 1)$ and as $q \geq 1$, m therefore is an integer such that $m = 0, 1, 2, 3....$, and so on using the fact that $x = \text{Arctan } y$, we have $y = \tan x$, and

$$\tan\left(\frac{\pi a}{\lambda}\sin i - m\frac{\pi}{2}\right) = \left(\frac{\sin^2 i_\ell}{\sin^2 i} - 1\right)^{1/2}. \qquad (9)$$

The two terms in this equation are functions of the variable $\sin i$, and the solutions can be found at the intersection of the plots representing:

- the function $g(\sin i) = \tan\left(\dfrac{\pi a}{\lambda}\sin i - m\dfrac{\pi}{2}\right)$; and

- the function $h(\sin i) = \left(\dfrac{\sin^2 i_\ell}{\sin^2 i} - 1\right)^{1/2}$.

In reality, the function $g(\sin i)$ can be more simply seen as the function given by:

1. $g_1(\sin i) = -\cotan\left(\dfrac{\pi a}{\lambda}\sin i\right)$ when m is odd (as

$$\tan\left[x - \dfrac{\pi}{2}\right] = -\tan\left[\dfrac{\pi}{2} - x\right] = -\cotan x\ );\ \text{and}$$

2. $g_2(\sin i) = \tan\left(\dfrac{\pi a}{\lambda}\sin i\right)$ when m is even $\left(\text{as } \tan\left[x \pm \pi\right] = \tan x\right)$.

The resolution of Eq. (9) thus leads to plotting, as a function of $\sin i$, the two systems given by:

- the plots for **2** defined for when $0 < \dfrac{\pi a}{\lambda}\sin i < \dfrac{\pi}{2}$ when m = 0,

$$\pi < \dfrac{\pi a}{\lambda}\sin i < \dfrac{3\pi}{2}\ \text{ when m = 2 , etc..., so that}$$

$$0 < \sin i < \dfrac{\lambda}{2a}\ (\text{m} = 0),\ \dfrac{\lambda}{a} < \sin i < \dfrac{3\lambda}{2a}\ (\text{m} = 2),\ \text{and so on; and}$$

- the plots for **1** defined for when $\dfrac{\pi}{2} < \dfrac{\pi a}{\lambda}\sin i < \pi\ \ (\text{m} = 1)$,

$$\dfrac{3\pi}{2} < \dfrac{\pi a}{\lambda}\sin i < 2\pi\ \ (\text{m} = 3),\ \text{etc..., so that}$$

$$\dfrac{\lambda}{2a} < \sin i < \dfrac{\lambda}{a}\ (\text{m} = 1),\ \dfrac{3\lambda}{2a} < \sin i < \dfrac{2\lambda}{a}\ (\text{m} = 3),\ \text{and so on.}$$

The solutions for $\sin i$ in each interval precisely defined by the value of m (hence the notation $\sin i_m = m\dfrac{\lambda}{2a}$ for the solution), thus can be found as shown in Figure 12.20 at the intersection with the plot of $h(\sin i) = \left(\dfrac{\sin^2 i_\ell}{\sin^2 i} - 1\right)^{1/2}$, which is a function that decreases monotonically with $\sin i$. When $i = i_\ell$, this function cancels out so that $h(\sin i_\ell) = 0$.

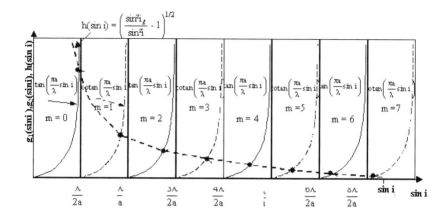

Figure 12.20. *Graphical solutions for the modal equations obtained by the intersection of the plots h(sin i) and g(sin i).*

For given values of a, λ, n, and n_1, for each value of m there is a corresponding value of I denoted I_m and between 0 and i_ℓ. Associated with I_m is the wavevector with a component $nk_0 \cos I_m$ with respect to the axis of propagation along Oy. $\beta_m = k_y = nk_0 \cos i_m$ is the propagation constant. With respect to the axis Oz, the component of the wavevector associated with I_m is for its part given by $k_z = nk_0 \sin i_m \equiv \alpha_m$.

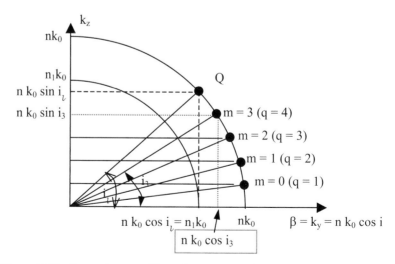

Figure 12.21. *Determination of I_m angles corresponding to m modes.*

As $\cos i_m$ varies between 1 and $\cos i_\ell = \dfrac{n_1}{n}$, β_m varies between nk_0 and n_1k_0.

The i_m angles corresponding to the m modes, as well as the corresponding components given by $k_y \equiv \beta_m$ and $k_z \equiv \alpha_m$, are shown in Figure 12.21.

12.6.6.5. Number of nodes

From Figure 12.20, each interval associated with a mode has a width given by $\lambda/2a$, which means that for a width defined by L as a function of sin i such that $L = \sin i_\ell$, the number of possible modes (M) is equal to the first integer immediately above the number given by the ratio $\dfrac{\sin i_\ell}{\lambda/2a}$. By convention, we can write that $M \doteq \dfrac{\sin i_\ell}{\lambda/2a}$.

12.6.6.6. Cutoff frequency

Returning to Eq. (8) and the reasoning used in Sections 12.6.6.2 and 12.6.6.3, for a mode given by $q\,(=m+1)$ to not be affected, then $[f(i)]_{max} < q\,\dfrac{\pi}{2} = (m+1)\,\dfrac{\pi}{2}$

must be true, so that $k\dfrac{a}{2}\sin i_\ell < (m+1)\,\dfrac{\pi}{2}$, and from which with $k = \dfrac{2\pi}{\lambda_0}n$ and

$\dfrac{n_1}{n} = \cos i_\ell = \sqrt{1 - \sin^2 i_\ell}$ included we can deduce that $\sin i_\ell = \dfrac{\sqrt{n^2 - n_1^2}}{n}$ and

$\lambda_{0m} > \lambda_{0mc} = \dfrac{c\,(m+1)}{2a\sqrt{n^2 - n_1^2}}$.

12.6.7. Comments: alternative methodologies

12.6.7.1. Comment 1: by analogy to solutions for potential wells

The type of mode, whether even or odd, can be apprehended directly through Eqs. (3) and (4) which are identical to those found in a resolution of potential wells with the given width as a.

Considering Eq. (3), and by making $X = \dfrac{\alpha\,a}{2}$ and $Y = \dfrac{\gamma\,a}{2}$;

Eq. (3) can be written as $Y = X \tan X$

Similarly, Eq. (4) gives $Y = -X \cot an\,X$. This then yields:

$$X^2+Y^2 = \frac{a^2}{4}\left[\alpha^2 + \gamma^2\right] = \frac{a^2}{4}\left[k^2 \sin^2 i + k^2 \cos^2 i - k^2 \frac{n_1^2}{n^2}\right]$$

$$= k^2\left(1 - \frac{n_1^2}{n^2}\right)\frac{a^2}{4} = k_0^2 (n^2 - n_1^2)\frac{a^2}{4}.$$

For a given guide, with indices n and n_1 and of dimension a, it is possible to state that $X^2+Y^2 = R^2$ is a constant characteristic of the guide. The solutions for X and Y can be found at the intersection of the plots of the preceding Y = f(X) with a circle of radius R. From these solutions α can be determined, thus i, and then, as will be detailed in Section 12.6.8, the distribution of the field that varies with cos αz (for even modes corresponding to the solutions to the equation Y = X.tan X) or with sin αz (for odd modes corresponding to solutions to the equation Y = - X cotan X).

12.6.7.2. Comment 2: methodology from principles of optics

We have assumed up till now that for a wave to be guided that (1) it has to undergo a reflection at an incidence such that $\theta > \theta_\ell$ at an interface between two dielectrics that make up the guide; and (2) the equation for the propagation should be verified using the limiting conditions of the problem. It is by this route that we have selected solutions corresponding to the modes. For a given mode, the solutions are such that the transversal intensity of the wave is independent of the position in Oy of the resultant direction of propagation.

The second condition can be directly replaced due to the fact that for the wave to propagate it must be part of the same system of plane waves, meaning that after several reflections the planes of the waves orthogonal to the direction of propagation are conserved. This is the same as saying that the dephasing between the wave that propagates along AB (subject to dephasing of φ_A and φ_B on reflection at A and B) and the wave that propagates directly from A to C is a multiple of 2π, as shown in Figure 12.22. Thus:

$$\left(k.\overline{AB} - k.\overline{AC}\right) + \varphi_A + \varphi_B = \frac{2\pi}{\lambda}[\overline{AB} - \overline{AC}] + 2\varphi_\perp = 2\pi\,m . \qquad (10)$$

In effect, the two interfaces are identical (the same dielectrics with indices n and n_1) and the wave is polarized perpendicularly to the plane of the figure so that $\varphi_A = \varphi_B = \varphi_\perp$, and therefore we either have $\varphi_A + \varphi_B = 2\varphi_\perp$, or $\varphi_\perp$ as determined in Section 11.3.2.2 with Eq. (37).

This with $(i + \theta) = \frac{\pi}{2}$, gives:

$$\tan\left(\frac{\varphi_\perp}{2}\right) = -\frac{\left(\sin^2\theta - \sin^2\theta_\ell\right)^{1/2}}{\cos\theta} = -\left(\frac{\sin^2 i_\ell}{\sin^2 i} - 1\right)^{1/2} . \qquad (11)$$

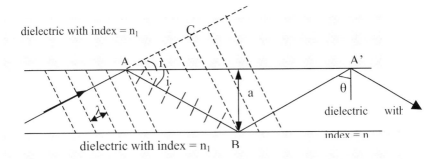

Figure 12.22. *Conservation of wave planes in guided propagation.*

With $[\overline{AB} - \overline{AC}] = 2\,a \sin i$ (classic dephasing calculated, for example, in establishing Braggs law, so that here $AB = \dfrac{a}{\sin i}$ and $AC = AB \cos 2i$, to give

$$AB - AC = \frac{a}{\sin i}[1 - \cos 2i] = 2a \sin i\,),\ \text{then Eq. (10) gives:}$$

$\dfrac{2\pi}{\lambda} 2\,a \sin i + 2\,\varphi_\perp = 2\pi\,m$, which in turn yields:

$$\left(\frac{\pi a}{\lambda}\sin i - m\frac{\pi}{2}\right) = -\frac{\varphi_\perp}{2}\ .$$

By taking the tangent of this equation into which is also plugged in Eq. (11), we obtain the preceding Eq. (9), as in:

$$\tan\left(\frac{\pi a}{\lambda}\sin i - m\frac{\pi}{2}\right) = \left(\frac{\sin^2 i_\ell}{\sin^2 i} - 1\right)^{1/2}\ .$$

12.6.8. Field distribution and solution parity

It is worth looking at the distribution of the electric field in each of the zones (1), (2), and (3) for the possible values of q.

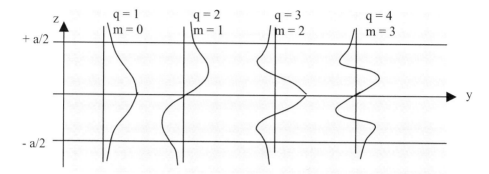

Figure 12.23. *Distribution of the electric field for various m modes.*

12.6.8.1. In zone (2)

The form of the field established from Section 12.6.2 is such that

$$\underline{E}_{mx}^2(z) = \left[D \exp(j\alpha_q z) + \varepsilon D \exp(-j\alpha_q z) \right] .$$

12.6.8.1.1. Even solutions (as cosines)
When $\varepsilon = +1$, we have:

$$\underline{E}_{mx}^{(2)}(z) = \left[D \exp(j\alpha_q z) + D \exp(-j\alpha_q z) \right] \propto \cos\left(\alpha_q z\right) = \cos\left[(k \sin i_q)z\right] .$$

In addition, $\varepsilon = +1$ corresponds to Eq. (3) which leads to q = 2r+1 [Eq. (5)], with r an integer ≥ 0. The r = 0 gives q = 1 so that m = 0; r = 1 gives q = 3 so that m = 2, while r = 2 gives q = 5 so that m = 4, and so on (see also Figure 12.21 and 12.23).

12.6.8.1.2. Odd solutions (as sins)
When $\varepsilon = -1$, we have:

$$\underline{E}_{mx}^{(2)}(z) = \left[D \exp(j\alpha_q z) - D \exp(-j\alpha_q z) \right] \propto \sin\left(\alpha_q z\right) = \sin\left[(k \sin i_q)z\right] .$$

Here, $\varepsilon = -1$ corresponds to Eq. (4) which leads to q = 2s (Eq. (6)), with s being an integer > 0, so that for s = 1, we have q = 2 and m = 1, and when s = 2 we have q = 4 and m = 3, and so on (see also Figures 12.21 and 12.23).

12.6.8.2. In zone (1)

The form of the field established in Section 12.6.3 is such that when $z < -a/2$,

$$\underline{E}_{mx}^{(1)}(z) = C \exp(\gamma\, z),$$

and with γ as the extinction coefficient, we can state that:

$$\gamma = k\sqrt{\cos^2 i - \frac{n_1^2}{n^2}} = k_0 n\sqrt{\cos^2 i - \cos^2 i_\ell}$$

$$= k_0 n \cos i_\ell \sqrt{\frac{\cos^2 i}{\cos^2 i_\ell} - 1} = k_0\, n_1 \sqrt{\frac{\cos^2 i}{\cos^2 i_\ell} - 1}$$

12.6.8.3. In zone (3)

The form of the field established in Section 12.6.3 is such that when $z > a/2$,

$$\underline{E}_{mx}^{(3)}(z) = \varepsilon\, C \exp(-\gamma\, z)$$

and γ takes on the same form as above.

12.6.8.4. The Goos–Hänchen effect

We can note that in the presence of a dielectric (with an index $= n_1$) in place of the plane conductors, the electric field no longer gives rise to nodes ($E = 0$) at $z = \pm a/2$ (see Figure 12.23) and the penetration to a depth of the order of $\delta \approx 1/\gamma$ of the ray undergoing a total reflection is called the Goos–Hänchen effect.

12.6.9. Guide characteristics

12.6.9.1. Form of the signal leaving the guide

If $a \gg \lambda$, numerous values are possible for q, which results in discreet but close values for the injection angle (i_q). The group of modes allows the guide to function in multimode.

To each value of i_q there is a different path length in the guide along with an associated time (t_q), and under multimodal regimes, there is a temporal increase of an impulsion leaving a wave with respect to that at the entrance, in what is an effect termed "intermodal dispersion" (Figure 12.24).

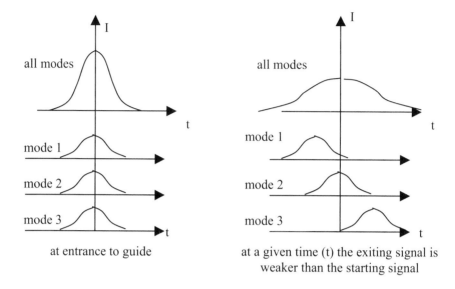

Figure 12.24. *Comparing guide entrance and exit signals.*

In effect, for a given value of q, i is dependent on ω [as $k = \dfrac{n\omega}{c}$ intervenes in the modal Eq. (8)], and $\beta = k \cos i$ thus is dependent on ω just as is $v_g = \dfrac{d\omega}{d\beta}$.

The pathway time also is dependent on ω (and hence also the wavelength), and hence the slight increase in the impulsions with each mode. This is termed the "intramodal effect", which is superimposed on that of dispersion due to a variation of the index with wavelength. The result is a chromatic dispersion.

12.6.9.2. Nature of losses in a guide

Losses in a guide have various origins and can be organized into two main classes. They are, first, characterized by a attenuation coefficient, or second, an attenuation factor.

12.6.9.2.1. Attenuation coefficient and the linear attenuation factor
The transmitted power (P) for a penetrated length (L) varies in accordance with the equation

$$P(L) = P(0) \exp(-2\gamma L).$$

P(0) is the power introduced at the entrance point, and γ here represents the linear attenuation coefficient for the amplitude.

For its part, the linear attenuation factor (α) is defined with the help of the expression:

$$P(L) = P(0)10^{-\alpha L/10} \text{ , so that } \alpha = \frac{10}{L}\log_{10}\frac{P_0}{P}$$

where α is expressed in decibels per meter (dB m^{-1}).

As an example, a fiber with 0.22 dB km^{-1} transmits around 95 % of the energy over 1 km.

12.6.9.2.2. Losses due to the physical configuration of the guide

These losses can include:

- losses due to a defective guiding (in principle, these losses do not count for guided modes);
- losses due to curves (unnatural losses due to deformations of the fiber) for example, when the radius of curvature is less than a centimeter, the higher modes can be refracted within the gain;
- losses at joints, which in turn can be divided into two groups:
    1. losses due to poor alignment of adjacent components;
    2. Fresnel losses, associated with the reflection of the injection signal at the entrance face (as in Figure 12.25). Under normal incidence, using classic experimental conditions, we have $R = \left[\dfrac{n - n_1}{n + n_1}\right]^2$, so that with $n = 1.5$ and $n_1 = 1$, we have $R \approx 0.04 = 4\%$. Thus the losses by Fresnel reflections are given by $\eta_{Fresnel} = 10\log_{10}(1-R) = -0.18$ dB .

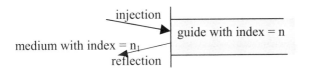

Figure 12.25. *Fresnel losses.*

12.6.9.2.3. Losses due to the material making up the guide

Losses due to the properties of the guide material can be divided into two physical phenomena, absorption and diffusion.

First, the losses due to absorption (which also are further detailed in Chapter 10 of this volume, or Chapter 3 of the second volume entitled *Applied Electromagnetism and Materials*) which can come about:

• in the visible domain (zone 3) due to electronic absorption;

• in the infrared domain (zone 4) due to network vibrations (vibrations of atoms or ions depending on the nature of the guide and/or fiber). In polymer based materials, the wavelength for the fundamental absorption by the C-H group is approximately 3.3 μm and over distances of any consequence, the window of transparency is in fact limited to 0.8 μm due to absorptions associated with harmonics. This phenomenon resembles absorptions due to hydroxyl bonds (-O-H) that exhibit a transparency window up to 2 μm, which includes in particular two transmission windows at 1.3 and 1.55 μm used in optical telecommunications. The use of fluorinated polymers permits an extension of this region of transparency;

• because of a wide number of impurities that present absorption characteristics, and hence the necessity of using highly purified materials. For silicon-based systems, absorptions due to impurities including hydroxyl groups are situated around the above-mentioned windows at 1.3 and 1.55 μm. The improvement of manufacturing techniques has meant that the presence of water has been reduced to less than 1 part per 10^7 and accordingly the performances of silicon-based fibers have increased considerably to less than 0.2 dB km^{-1} at 1.55 μm. The use of fluorinated glasses can extend the transparency window to 5 μm, and although the absorption bands are always present, they are well shifted to longer wavelengths.

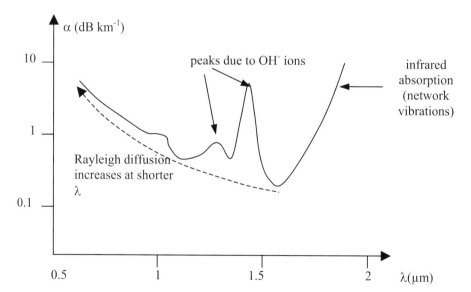

Figure 12.26. *Plot of* $\alpha = f(\lambda)$ *showing the origin of losses in a silicon based fiber.*

Second, losses due to diffusion that can be further separated into two groups:

- extrinsic diffusion which itself can be due to two possible causes, the first being tied to the possible inclusion of dust (of a size greater than $\dfrac{\lambda}{20}$) and the second being associated with the presence of microcrystallites. It is evident therefore that there is an interest in using extremely clean and amorphous materials.

- intrinsic diffusion due to Rayleigh diffusion (see also Chapter 10) which is caused by interactions of light with materials. It varies with respect to $\dfrac{1}{\lambda^4}$ so that the effect becomes more important at shorter wavelengths. Thus, at very short wavelengths, there is a deviation in the plot of $\alpha = f(\lambda)$ due to an attenuation of the system (see Figure 12.26 for silicon fiber optics).

12.7. Problems
Monomodal conditions
By way of recall, the total reflection at the interface between two dielectric media brings into consideration two different and real indices denoted here as n and n_1 and such that $n > n_1$. We thus have $n_1 = n - \Delta n$.

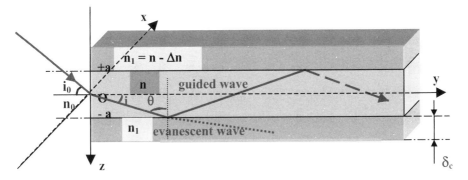

The figure shows the configuration of injection and propagation of light in a medium with an index denoted as n. When $\theta > \theta_\ell$, where θ_ℓ represents the limiting angle (and i_ℓ is the corresponding angle on the inside of the injection surface), electromagnetic theory tells us that the transmitted wave ($\vec{E}_t$) in the medium with index n_1 and for the zone defined by $z > a/2$, is in the form:

$\vec{E}_t = \overline{E_t^0} \exp(-\gamma z) \exp(i[k_y \ y - \omega t])$, avec $\gamma = k_0 \ n_1 \ (\cos^2 i/\cos^2 i_\ell) - 1]^{1/2}$.

1. Show that the intensity of the wave in the medium with index $= n_1$ can be written

in the form $I_t(z) = I_t(0) \exp\left(-\dfrac{2z}{\delta}\right)$. Give an equation for δ, which then can be used

to give an approximate value, assuming that θ is sufficiently close to $\pi/2$ to state that $\sin^2\theta \approx 1$.

2. Determine the condition for Δn that permits a weak penetration of a wave (evanescent) into the medium with an index $= n_1$. The condition for weak penetration is fixed using the condition $\delta \le \delta_c \approx 3 \ \lambda_1$ (wavelength in a medium with an index $= n_1$).

In numerical terms, determine the condition on Δn when $n \approx n_1 \approx 1.5$.

3. Monomodal condition. In this chapter it is shown that: $a < a_1 = \dfrac{\lambda_0}{2\sqrt{n^2 - n_1^2}}$. With

Δn being very small, which can be verified later on, and with the same numerical characteristics as given before in that $\lambda_0 = 1.3 \ \mu m$ and $a = 5 \ \mu m$, give the numerical condition that Δn must verify so that the guide is monomodal. Conclude.

Answers

1. We have $I_t(z) = \left|\vec{E}_t\right|^2 = \left|\overline{E_t^0}\right|^2 \exp(-2\gamma z) = I_t(0) \exp\left(-\dfrac{2z}{\delta}\right)$ with

$$\delta = \frac{1}{\gamma} = \frac{1}{k_0 \ n_1} \left(\frac{\cos^2 i}{\cos^2 i_\ell} - 1\right)^{-1/2}.$$

Expressing $\dfrac{\cos^2 i}{\cos^2 i_\ell}$ as a function of θ, we have in one

part, $\cos i = \cos(\pi/2 - \theta) = \sin \theta$, and in the other

$\cos i_\ell = \cos(\pi/2 - \theta_\ell) = \sin \theta_\ell = \dfrac{n_1}{n}$, with the result that

$$\delta = \frac{1}{k_0 \ n_1} \left(\frac{n^2}{n_1^2} \sin^2\theta - 1\right)^{-1/2} \approx \frac{1}{k_0 \ n_1} \left(\frac{n^2}{n_1^2} - 1\right)^{-1/2}.$$

2. With $k_0\, n_1 = \dfrac{\omega}{c} n_1$, $\delta < 3\,\lambda_1$ gives rise to the condition:

$$\delta \approx \frac{\lambda_1}{2\pi}\left(\frac{n^2}{n_1^2}-1\right)^{-1/2} < 3\,\lambda_1, \quad \text{and the condition for weak penetration therefore is}$$

$$\frac{1}{6\pi} < \left(\frac{n^2}{n_1^2}-1\right)^{1/2}, \quad \text{so that} \quad \left(\frac{1}{6\pi}\right)^2 < \frac{n^2}{n_1^2}-1 = \frac{\left(n_1 + \Delta n\right)^2}{n_1^2}-1 \approx \frac{2\,\Delta n}{n_1}.$$

We therefore have as a condition, finally,

$$\Delta n > \frac{n_1}{72\,\pi^2}.$$

In numerical terms with $n_1 = 1.5$, we have $\Delta n > 0.0021$.

3. The monomodal condition can be written as

$$a < a_1 = \frac{\lambda_0}{2\sqrt{n^2 - n_1^2}} = \frac{\lambda_0}{2\sqrt{\left(n_1 + \Delta n\right)^2 - n_1^2}} \approx \frac{\lambda_0}{2\sqrt{2 n_1\, \Delta n}}, \quad \text{so that:}$$

$$\Delta n < \frac{\lambda_0^2}{8 n_1}\frac{1}{a^2}.$$

In numerical terms, $a = 5$ μm, $n_1 = 1.5$, and $\lambda_0 = 1.3$ μm, we have $\Delta n < 0.0056$.

To conclude, the two conditions brought together at Δn have a common domain, namely $0.0021 < \Delta n < 0.0056$. In this region, the conditions of weak penetration and monomodal guiding are simultaneously true.

Index

Printed in the United States of America